Qualitative Analysis of
Large Scale Dynamical Systems

This is Volume 134 in
MATHEMATICS IN SCIENCE AND ENGINEERING
A Series of Monographs and Textbooks
Edited by RICHARD BELLMAN, *University of Southern California*

The complete listing of books in this series is available from the Publisher upon request.

QUALITATIVE ANALYSIS OF LARGE SCALE DYNAMICAL SYSTEMS

Anthony N. Michel

DEPARTMENT OF ELECTRICAL ENGINEERING
IOWA STATE UNIVERSITY
AMES, IOWA

Richard K. Miller

DEPARTMENT OF MATHEMATICS
IOWA STATE UNIVERSITY
AMES, IOWA

ACADEMIC PRESS   New York   San Francisco   London   1977

A Subsidiary of Harcourt Brace Jovanovich, Publishers

ACADEMIC PRESS, INC.
111 Fifth Avenue, New York, New York 10003

United Kingdom Edition published by
ACADEMIC PRESS, INC. (LONDON) LTD.
24/28 Oval Road, London NW1

Library of Congress Cataloging in Publication Data

Michel, Anthony N
     Qualitative analysis of large scale dynamical
systems.

     (Mathematics in science and engineering series ;
vol. 000)
     Bibliography: p.
     Includes index.
     1.   System analysis.   I.   Miller, Richard K.,
joint author.   II.   Title.   III.   Series.
QA402.M48   515'.7   76-50399
ISBN 0–12–493850–7

15?5277

To Wolfgang Hahn and Irwin W. Sandberg

Contents

Preface

The demands of today's technology have resulted in the planning, design, and realization of sophisticated systems that have become increasingly large in scope and complex in structure. It is therefore not surprising that over the past decade or more, many researchers have directed their attention to various problems that arise in connection with systems of this type, which are called large scale systems. Although it is reasonable to assume that in the near future there will evolve a well-defined body of knowledge on large systems, the directions of such a discipline have not been entirely resolved at this time. However, there are several well-established areas that have reached a reasonable degree of maturity. One is concerned with the qualitative analysis of large scale dynamical systems, the topic of this monograph.

There are numerous examples of large dynamical systems that provide great challenges to engineers of all disciplines, physical scientists, life scientists, economists, social scientists, and of course, applied mathematicians. Obvious examples of large scale dynamical systems include electric power systems, nuclear reactors, aerospace systems, large electric networks, economic systems, process control systems in the chemical and petroleum industries, different types of societal systems, and ecological systems. Most systems of this type have several general properties in common. They may often be viewed as an interconnection of several subsystems. (For this reason, such systems are often also called interconnected systems or composite systems.) In addition, such systems are usually endowed with a complex interconnecting structure and are frequently of high dimension.

In order that this monograph be applicable to many diverse areas and disciplines, we have endeavored to consider several important classes of equations that can be used in the modeling of a great variety of large scale dynamical systems. Specifically, we consider systems that may be represented by ordinary differential equations, ordinary difference equations, stochastic differential equations, functional differential equations, Volterra integrodifferential equations, and certain classes of partial differential equations. In addition, we consider hybrid dynamical systems, which are appropriately modeled by a mixture of different types of equations. Qualitative aspects of large scale dynamical systems that we consider include Lyapunov stability (stability, asymptotic stability, exponential stability, instability, and complete instability), Lagrange stability (boundedness and ultimate boundedness of solutions), estimates of trajectory behavior and trajectory bounds, input–output properties of dynamical systems (input–output stability, i.e., boundedness and continuity of the input–output relations that characterize dynamical systems), and questions concerning the well-posedness of large scale dynamical systems.

The qualitative analysis of large scale systems can be accomplished in a variety of ways. We present a unified approach of analyzing such systems at different hierarchical levels, namely, at the subsystem structure and interconnecting structure levels. This method of analysis offers several advantages. As will be shown, the method of analyzing complex systems in terms of lower order and simpler subsystems and in terms of system interconnecting structure often makes it possible to circumvent difficulties that usually arise in the analysis of high-dimensional systems with intricate structure. We shall also see that this method of analysis is somewhat universal in the sense that it may be applied to all the types of equations enumerated above. It will be seen that this approach is especially well suited for the qualitative analysis of hybrid dynamical systems (i.e., systems described by a mixture of different types of equations). In addition, analysis by this procedure yields trade-off information between qualitative effects of subsystems and interconnection components. This method of analysis also makes it possible to compensate and stabilize large systems at different hierarchical levels, making use of local feedback techniques. Furthermore, this method can be used as a guide in the planning of decentralized systems endowed with built-in reliability (i.e., safety) features. Because of these advantages, this method of analysis should be considered as being more important than the individual results presented. Indeed, all the subsequent results should be viewed as models; in particular applications, one should tailor the present method of analysis to specific problems.

This book consists of seven chapters. In the first chapter we provide an overview of the subject. Chapters II–V are concerned with Lyapunov stability, Lagrange stability, estimates of trajectory behavior and trajectory bounds, and questions of well-posedness of large scale systems. In Chapter II we consider systems described by ordinary differential equations, in Chapter III systems that can be represented by ordinary difference equations and also sampled data systems, and in Chapter IV systems that can be modeled by stochastic differential equations. In Chapter V we address ourselves to infinite-dimensional systems that can be represented by differential equations defined on Banach and Hilbert spaces. Such systems include those that can appropriately be described by functional differential equations, Volterra integro-differential equations, certain classes of partial differential equations, and infinite-dimensional hybrid systems described by a mixture of equations. Chapters VI and VII are devoted to input–output stability properties of large scale dynamical systems. The results in Chapter VI are rather general, while Chapter VII is confined to systems described by integrodifferential equations.

To demonstrate the usefulness of the method of analysis advanced and to point to various advantages and disadvantages, we have included several specific examples from diverse areas, such as problems from control theory, circuit theory, nuclear reactor dynamics, and economics. Because of their importance in applications, we have emphasized frequency domain techniques in several examples.

In order to make this book reasonably self-contained, we have included necessary background material on the following topics: the principal results from the Lyapunov stability theory (for finite-dimensional systems, infinite-dimensional systems, and systems described by stochastic differential equations), the main results for boundedness and ultimate boundedness of solutions, the principal comparison theorems (the comparison principle), results from the theory of M-matrices, selected results from semigroup theory, and pertinent results from systems theory (relating to input–output stability). In addition, we have provided numerous references for this background material.

Acknowledgments

We would like to thank Professor Richard Bellman for encouraging us to undertake this project, and we feel privileged to have this book published in his distinguished series. Likewise, thanks are due to the staff of Academic Press for advice and assistance.

A great part of this monograph is based on research conducted at Iowa State University by the authors and by former students of the first author, Drs. Eric L. Lasley, David W. Porter, and Robert D. Rasmussen. The work of both authors was supported in part by the National Science Foundation, and the first author was also supported by the Engineering Research Institute, Iowa State University. During 1972–1973, the first author's research was accomplished on a sabbatical leave at the Technical University of Graz, Austria, where he had the privilege of being associated with Professor Wolfgang Hahn.

We are particularly appreciative of the efforts of Mrs. Betty A. Carter in typing the manuscript. Above all, we would like to thank our wives, Leone and Pat, for their patience and understanding.

CHAPTER I

Introduction

In this chapter we first discuss somewhat informally the motivation for the method of analysis advanced in this monograph. We then briefly indicate the type of qualitative analysis with which we concern ourselves. This is followed by an overview of qualitative results for large scale systems which will be of interest to us. Finally, we give an indication of the contents of the subsequent chapters.

1.1 Introduction

In recent years many researchers have addressed themselves to various problems concerned with large systems. This is evidenced by an increasing number of publications in scientific journals and conference proceedings. At this time there also have appeared a number of monographs dealing with various aspects of large scale systems. For example, there is the fundamental work by Kron [1] on diakoptics, Tewarson [1] considers the theory of sparse matrices arising naturally in large systems, Lasdon [1] addresses himself to optimization theory of large systems, Mesarović, Macko, and Takahara [1] and Mesarović and Takahara [1] develop a general systems theory for hierarchical multilevel systems, and so forth. Although the state of the art in the area of qualitative analysis of large systems has reached a reasonable degree of maturity, no text summarizing the important results of this topic has appeared. We address ourselves in this book to this problem. As was pointed out in an editorial by Bellman [4], problems associated with large systems offer new and interesting challenges to researchers. We hope that the present monograph will in a small way further stimulate work in this new and exciting field.

It must be stated at the outset that no precise definition of large scale system can be given since this term has a different meaning to different workers. In this book we consider a dynamical system to be large if it possesses a certain degree of complexity in terms of structure and dimensionality. More specifically, we will be interested in dynamical systems which may be viewed as an interconnection of several lower order subsystems. This point of view motivates also the terms "composite system," "interconnected system," "multiloop system," and the like. In certain applications, the term "decentralized system" has also been used.

Roughly speaking, problems concerned with large scale systems may be divided into two broad areas: static problems (e.g., graph theoretic problems, routing problems) and dynamical problems. The latter may in turn be separated into quantitative problems (e.g., numerical solution of equations describing large systems) and into qualitative problems. All topics of this book are concerned with qualitative analysis of large scale dynamical systems.

The traditional approach in systems theory is to represent systems in certain "standard" or canonical forms. For example, the usual approach in classical as well as in modern control theory is to transform the equations describing a given system in such a fashion that the system in question may be represented, for example, in the familiar block diagram form of Fig. 1.1. Once this is accomplished, long-established and well-tested methods are

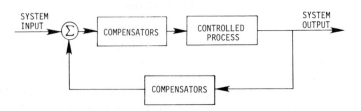

Figure 1.1 Typical feedback control system.

employed to treat the problem on hand. Frequently, however, certain complications arise with this approach. For example, often difficulties are encountered in analysis and synthesis procedures which can be attributed to the dimensionality and to structural complexity of a system. Another disadvantage of the traditional approach is that after the system in question has been cast into a standard mold (such as Fig. 1.1), the effects of individual components, subloops, etc., are often no longer explicitly apparent because several transformations had to be performed to put the system into a desired canonical form (e.g., Fig. 1.1). In the subsequent chapters we will develop a unified qualitative theory for large systems, in which the objective will always be the same: to analyze (and synthesize) large scale systems in terms of their lower order (and hopefully simpler) subsystems and in terms of their interconnecting structure. In this way, complications which usually arise in the qualitative analysis of high order systems with complex interconnecting structure may often be circumvented. Furthermore, such an approach provides insight into system structure in its original form, yielding information on effects of individual system components, subsystems or subloops, trade-off information between various subsystems and interconnecting structure, and the like. This type of information is usually of great value to the designer and to the analyst. In addition, the viewpoint advanced herein makes it often possible to compensate large systems (i.e., improve their performance) at several hierarchical levels (i.e., at the subsystem level and interconnecting structure level), using local feedback methods. A typical large scale system, in its original form, may conceptually be represented as shown in the block diagram of Fig. 1.2.

The method of analysis advanced herein is of course not without disadvantages. Thus, if a system is decomposed into too many subsystems, one may obtain overly conservative results. However, we will demonstrate by means of several specific examples that this need not be the case, provided that the method is applied properly.

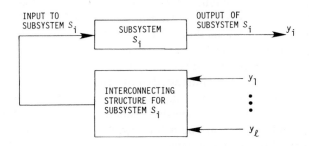

Figure 1.2 Typical large scale dynamical system where $i = 1, \ldots, l$.

1.2   Qualitative Analysis of Dynamical Systems:
General Remarks

Qualitative aspects of dynamical systems which we will primarily address ourselves to include the following: stability and instability in the sense of Lyapunov; boundedness of solutions (i.e., Lagrange stability); estimates of trajectory behavior and trajectory bounds; and input–output properties of dynamical systems.

The direct method of Lyapunov has found a wide range of applications in engineering and in the physical sciences. Although most of these applications were originally concerned with the stability analysis of systems described by ordinary differential equations (see, e.g., Hahn [1, 2], LaSalle and Lefschetz [1], Kalman and Bertram [1]), the direct method of Lyapunov has also been extended to systems described by difference equations (see, e.g., Hahn [1, 2], Kalman and Bertram [2]), and more recently to systems represented by partial differential equations (see, e.g., Zubov [1], Wang [1, 5], Sirazetdinov [1], Chaffee [1], Hahn [2]), differential difference equations (see, e.g., Bellman and Cooke [1], Yoshizawa [1], Krasovskii [1], Hahn [2]), functional differential equations (see, e.g., Hale [2] Yoshizawa [1], Krasovskii [1], Hahn [2], Halanay [1]), integrodifferential equations (see, e.g., Driver [1], Levin [1], Miller [1], Suhadloc [1], Bronikovski, Hall and Nohel [1]), systems of countably infinite many ordinary differential equations (see, e.g., Bellman [1], Shaw [1]), stochastic differential equations (see, e.g., Kushner [1], Arnold [1], Kozin [1], Kats and Krasovskii [1], Bertram and Sarachik [1]), stochastic difference equations (see, e.g., Kushner [3]), and the like. The stability theory of general dynamical systems is still of current interest (see, e.g., Bhatia and Szegö [1], Hale [1], Krein [1], Ladas and Lakshmikantham [1], Walter [1], Slemrod [1]).

In the case of many systems (e.g., systems exhibiting nonlinear oscillations) it is not the Lyapunov stability or instability that is of interest. Yoshizawa and others (see, e.g., Yoshizawa [1], Hahn [2], LaSalle and Lefschetz [1]) have extended the direct method of Lyapunov to establish conditions for boundedness (Lagrange stability), ultimate boundedness, unboundedness, and the like, of solutions of dynamical systems.

In practice one is not only interested in the qualitative type of information obtainable from the Lyapunov stability and the Lagrange stability of a dynamical system, but also in specific estimates of trajectory behavior and trajectory bounds. A system could for example be stable or bounded and still be completely useless because it may exhibit undesirable transient characteristics (e.g., its solutions may exceed certain limits or specifications imposed by the designer on the trajectory bounds). Estimates of trajectory behavior and trajectory bounds have been obtained using the Lyapunov-type approach by defining stability with respect to time-varying subsets (of the state space) which are *prespecified* in a given problem (this is not the case in Lyapunov and Lagrange stability). The boundaries of these sets yield estimates of system trajectory behavior and trajectory bounds (see, e.g., Matrosov [1, 2], Michel [1–3], Michel and Heinen [1–4]). This concept includes the notions of practical stability and finite time stability (see, e.g., LaSalle and Lefschetz [1], Weiss and Infante [1], Michel and Porter [4]).

In a radical departure from the classical approach described thus far, it is possible to view dynamical systems in a "black box" sense as relations mapping system inputs into system outputs on an extended function space. In the context of this formulation, the qualitative analysis of such systems is accomplished in terms of system input–output properties (see, e.g., Sandberg [2, 8], Zames [3, 4], Willems [1], Desoer and Vidyasagar [1], Holtzman [1]). Many important results have been obtained along these lines and for many important problems a connection between input–output stability and Lyapunov stability has been established.

1.3   Qualitative Analysis of Large Scale Systems: General Remarks

Despite its elegance and generality, the usefulness of the Lyapunov approach is severely limited when applied to problems of high dimension and complex interconnecting structure. For this reason it is frequently advantageous to view high order systems as being composed of several lower order subsystems which when interconnected in an appropriate fashion, yield the original *composite* or *interconnected system*. The stability analysis of such systems can often then be accomplished in terms of the simpler subsystems and in terms of the interconnecting structure of such composite systems. In this way, complications which usually arise when the direct method is applied to high order systems may often be avoided. This is precisely the approach which we will employ. Indeed, since statements similar to the above apply equally as well to problems involving Lagrange stability, estimates of trajectory behavior and trajectory bounds, input–output stability, and the like, throughout this book we will pursue a method of qualitative analysis of large systems at different hierarchical levels. As will be pointed out repeatedly, in a certain sense, the method of analysis advanced in this book should be viewed as being more important than the individual results presented.

There is a sizable body of literature concerned with various aspects of qualitative analysis of large dynamical systems. Since this subject is rather new, and since the literature is still growing, it is difficult and quite pointless to make an attempt at citing every reference. Indeed, if we were to list all sources concerned with this topic, such a list would be too long and not justified. On the other hand, if we were to confine the listings only to immediate references used, such a list would be too short and would hardly reflect the importance of this subject. For this reason we shall compromise and mention only some selected results which are in the spirit of the method of analysis proposed herein.

1. *Systems Described by Ordinary Differential Equations* (Lyapunov Stability). Using *vector Lyapunov functions*, originally introduced by Bellman [2], Bailey [1, 2] invoked the comparison principle (see, e.g., Walter [1]) to establish sufficient conditions for exponential stability of interconnected systems described by nonlinear nonautonomous ordinary differential equations with exponentially stable subsystems and with linear time–invariant interconnecting structure. Subsequently, results for exponential stability, uniform asymptotic stability, instability, and complete instability of composite systems represented by nonlinear nonautonomous ordinary differential equations with nonlinear and time varying interconnections and with subsystems that may be exponentially stable, uniformly asymptotically stable (and sometimes unstable) were obtained by Piontkovskii and Rutkovskaya [1], Thompson [1, 2], Porter and Michel [1, 2], Michel and Porter [3], Thompson and Koenig [1], Matrosov [1], Araki [1, 4], Araki and Kondo [1], Michel [5–7, 9], Matzer [1], Grujić and Siljak [2], Weissenberger [1], Bose [1], Bose and Michel [1, 2], and others. In some of these results, vector Lyapunov functions as well as scalar Lyapunov functions consisting of a weighted sum of Lyapunov functions for the free or isolated subsystems are employed. In addition, absolute stability results for interconnected systems endowed with several nonlinearities were established by several authors (see, e.g., McClamroch and Ianculescu [1], Bose [1], Bose and Michel [1, 2], and Blight and McClamroch [1]). Additional related Lyapunov stability results for large systems are contained in the paper by Tokumaru, Adachi, and Amemiya [2] and in the survey paper by Athans, Sandell, and Varaiya [1].

2. *Systems Described by Difference Equations and Sampled Data Systems* (Lyapunov Stability). Sufficient conditions for uniform asymptotic stability, exponential stability, and instability of composite systems described by nonautonomous nonlinear difference equations were established by Araki, Ando, and Kondo [1], Michel [5, 7], and Grujić and Siljak [1, 3]. Results for uniform asymptotic stability of sampled data composite systems (i.e., hybrid finite-dimensional systems) were obtained by Michel [5, 7].

3. *Systems Described by Stochastic Differential Equations and Stochastic Difference Equations* (Lyapunov Stability). The Lyapunov stability of large scale systems described by stochastic differential equations (Ito equations as well as other types of stochastic differential equations) and stochastic difference equations are treated in papers by Michel [8, 10, 11], Michel and Rasmussen [1–3], Rasmussen and Michel [1, 3], and Rasmussen [1]. Scalar and vector Lyapunov functions are used in the analysis. In addition to Wiener processes, disturbances modeled by Poisson step processes, jump Markov processes, and the like, are considered. The approach in these references is general enough to allow disturbances to enter into the subsystem structure and into the interconnecting structure. Composite systems with stable as well as unstable subsystems are treated. The obtained results yield sufficient conditions for asymptotic stability with probability one and in probability and exponential stability with probability one, in probability and in the quadratic mean.

4. *Infinite-Dimensional Systems* (Lyapunov Stability). Matrosov [2] uses vector Lyapunov functions while Michel [5, 7], Rasmussen and Michel [2, 4], and Rasmussen [1] use scalar Lyapunov functions (consisting of a weighted sum of Lyapunov functions for the free or isolated subsystems) to obtain conditions for uniform asymptotic stability and exponential stability of composite systems described by differential equations which are defined on Hilbert and Banach spaces. These results are general enough to allow analysis of large systems which can be represented by ordinary differential equations, differential difference equations, functional differential equations, integrodifferential equations, certain classes of partial differential equations, as well as other types of evolutionary systems. In addition, these results provide a systematic procedure for the stability analysis of hybrid dynamical systems, i.e., dynamical systems described by a mixture of different types of equations.

5. *Lagrange Stability and Estimates of Trajectory Behavior and Trajectory Bounds.* Sufficient conditions for uniform boundedness and uniform ultimate boundedness of solutions of interconnected systems described by ordinary differential equations were obtained by Bose and Michel [1, 2] and Bose [1]. Estimates of trajectory behavior and trajectory bounds of composite systems described by ordinary differential equations were established in references by Michel [3, 4, 9], Matrosov [1, 2], and Michel and Porter [1, 2]. These results include earlier ones for finite time stability of systems described on product spaces (see Weiss and Infante [1]).

6. *Input–Output Stability.* Conditions for the input–output stability (input–output boundedness and continuity) for large classes of inter-connected dynamical systems are established in Porter [1], Porter and Michel [3–5], Tokumaru, Adachi, and Amemiya [1], Lasley and Michel [1–6],

Lasley [1], Cook [2], Miller and Michel [1, 2], Vidyasagar [1], and Araki [3]. Input–output instability results are obtained by Vidyasagar [1] and Miller and Michel [1]. In the majority of these references large scale systems are analyzed in terms of their lower order subsystems and interconnecting structure. Overall stability conditions for the systems considered are phrased in terms of figures of merit which characterize the degree of stability of each subsystem and in terms of parameters expressing the degree of coupling of the interconnecting structure. Although the majority of these results are phrased in terms of general input–output relations, graphical frequency domain interpretations are emphasized when appropriate (e.g., circle criteria, Popov conditions). In this connection, stability on L_p- and l_p-spaces, $1 \leq p \leq \infty$, are of special interest.

Closely related to the above, are recent results by McClamroch [1] and Callier, Chan, and Desoer [1, 2].

7.   *Stabilization.* Most, but not all, of the results cited above are in a form which make stabilization and compensation procedures of dynamical systems at different hierarchical levels feasible. Such procedures can in principle be implemented by the use of local compensating feedback at the subsystem and interconnecting structure levels.

Significant case studies worthy of mention, where the method of analysis of the above references has been applied, have apparently not been conducted at this time. Nevertheless, great potential for applications of results of this type to significant problems in many diversified disciplines is apparent. Such disciplines include most branches of engineering (electrical, mechanical, aerospace, chemical, nuclear, biomedical, etc.), the physical sciences (physics, etc.), the life sciences (biology, etc.), economics, numerical analysis (e.g., iteration theory), and others. Typical examples of large scale systems with great promise for applications include electric power systems, nuclear power systems, aerospace systems, processes in the chemical and petroleum industry, circuits, economic systems, transportation systems, ecological systems, societal systems, and many others. Solutions to problems of this type will have to be effected by specialists with expertise in these areas, for in the case of truly large scale systems, great insight and understanding of modeling is crucial.

1.4   An Overview of the Subsequent Chapters

In the subsequent chapters we develop a unified approach to qualitative analysis of large scale systems described by many diversified types of equations. While our presentation constitutes an expansion and adaptation of most of the references cited in the previous section, we emphasize that many of the results presented here are new and have not appeared elsewhere. A brief overview of the contents of the remaining chapters follows.

In Chapter II we concern ourselves with the analysis of finite-dimensional large scale systems described by ordinary differential equations. Items of interest are uniform stability, uniform asymptotic stability, exponential stability, instability, and complete instability (all in the sense of Lyapunov), as well as uniform boundedness and uniform ultimate boundedness of solutions, and estimates of trajectory behavior and trajectory bounds. Both scalar Lyapunov functions and vector Lyapunov functions (involving suitable comparison theorems) are employed in our approach. In addition, extensive use of the properties of Minkowski matrices is made, when applicable, to simplify some of the results.

Our exposition in Chapter II is intentionally somewhat extensive and detailed. One of the reasons for this is to make many extensions and results which are not explicitly stated in later chapters apparent to the reader.

In Chapter III we modify some of the results of Chapter II to treat finite-dimensional discrete time systems described by difference equations. In this chapter we also consider a class of (continuous time) sampled data systems (i.e., finite-dimensional hybrid systems). It will become apparent that the results of this chapter do not always involve obvious modifications of corresponding results of Chapter II.

In all of the subsequent chapters we concern ourselves with deterministic dynamical systems, except in Chapter IV, where large scale systems described by stochastic differential equations are considered. Because of the presence of second order effects which need to be taken into account, such systems must be treated separately. In Chapter IV we primarily consider Ito differential equations and we use scalar as well as vector Lyapunov functions to obtain sufficient conditions for asymptotic stability with probability one, exponential stability with probability one, and exponential stability in the quadratic mean.

In Chapter V we generalize most of the results of Chapter II to infinite-dimensional dynamical systems. Using the theory of C_0-semigroups and nonlinear semigroups as a basis to characterize dynamical systems, we consider systems which may be described by differential equations defined on Hilbert and Banach spaces. In addition to Lyapunov stability and boundedness of solutions of large scale systems, we also address ourselves to questions of well posedness of such systems. The results of this chapter are general enough to be applicable to large systems described by ordinary differential equations, functional differential equations, Volterra integro-differential equations, certain types of partial differential equations, and the like. Furthermore, these results make it possible to analyze hybrid dynamical systems (i.e., systems described by a mixture of equations) in a systematic manner. We use scalar as well as vector Lyapunov functions in our approach.

Chapter VI is concerned with input–output properties of large dynamical systems. Of primary interest is the input–output stability and instability of such systems. By input–output stability we mean boundedness and sometimes also continuity of relations (and operators) which connect system inputs to system outputs in the system description. Continuous time systems as well as discrete time systems are considered. Of particular interest are results in the setting of L_2- and l_2-spaces and L_∞- and l_∞-spaces. Whenever appropriate, we emphasize results involving frequency domain methods (e.g., Nyquist plots, gain sector conditions, circle criterion, Popov-type results, and the like).

Because of their special structure and properties, large scale systems described by integrodifferential equations are treated in Chapter VII, where we further apply the stability and instability results of Chapter VI.

To establish the required notation and to make our presentation reasonably self contained, we shall introduce extensive background material from several diverse areas of applied mathematics. Since most of these results can be found in several excellent books, we will omit all proofs involving background material. On the other hand, we will prove in detail all of the results presented which are concerned with the qualitative analysis of large scale systems. Background material will include the following: Lyapunov results and Lyapunov related results for the types of systems considered in Chapters II–V; results from the theory of Minkowski matrices, which are used throughout; selected results from the theory of semigroups and dynamical systems; and appropriate material from system theory involving input–output properties of dynamical systems.

To demonstrate the usefulness of the method of analysis advanced, we consider in each chapter several specific examples and applications. The choice of these examples was in part dictated by the following objectives: to demonstrate the applicability of the method presented to problems in a variety of diverse disciplines; to illustrate the ease with which the results can be applied to complex problems; and to show that when used properly and applied to the right kinds of problems, the present method of analysis yields results which are not overly conservative and in fact, are not necessarily more conservative than corresponding results obtained by other methods. Specific examples which we consider include problems which arise in automatic control theory (e.g., direct control problem, indirect control problem), in circuit theory (e.g., networks with linear resistors and time varying capacitors, networks with linear resistors and nonlinear transistors), in the theory of nuclear reactors (point kinetics model of a coupled multicore nuclear reactor), in economics (Hicks conditions for the model of a multimarket economic system), and so forth. In several of the examples considered (e.g., systems endowed with several troublesome elements, such as nonlinear gains, time varying gains, hysteresis elements, time delays, etc.) graphical frequency domain techniques are employed. Furthermore, to make comparisons possible, some of these examples are analyzed by several of the results presented. For example, the model of the coupled multicore reactor is treated in Chapters V and VII, deterministic and stochastic versions of the indirect control problem are considered in Chapters II and IV, respectively, continuous time and discrete time versions of a system with two nonlinearities are analyzed in Chapters II and III, respectively, and so forth.

Before proceeding further, some comments concerning the labeling of items is in order. Sections are assigned numerals which reflect the chapter and section numbers. For example, Section 2.3 signifies the third section in the second chapter. Extensive sections are sometimes divided into subsections identified by upper case common letters A, B, C, etc. Equations, definitions, theorems, corollaries, lemmas, examples, and special remarks are assigned monotonically increasing numerals which identify the chapter, section, and item number. For example, Theorem 3.3.6 denotes the sixth identified item in the third section of Chapter III. This theorem is followed by Eq. (3.3.7), the seventh identified item in the same section. Figures are identified separately. Thus, Fig. 1.2 denotes the second figure in Chapter I. Finally, the end of a proof is signified by the symbol ∎.

CHAPTER II

Systems Described by
Ordinary Differential Equations

In this chapter we consider the qualitative analysis of large scale systems described by ordinary differential equations. Our exposition will intentionally be detailed because (a) most of the existing results on large dynamical systems involve systems described by such equations, (b) ordinary differential equations play a crucial role in technology, and (c) many extensions and results not explicitly stated in the subsequent chapters (for systems described by other types of equations) will become apparent from results of the present chapter.

The necessary notation is established in the first section while in the second section we present selected results from the Lyapunov theory which serve as essential background material. In the third section we introduce several classes of large scale systems which are analyzed in the fourth section by the use of scalar Lyapunov functions. This is followed by a presentation of several results from the theory of Minkowski matrices (M-matrices) which constitute background material used throughout this book. We then apply M-matrices in the fifth section to the qualitative analysis of large systems. We begin the sixth section with a summary of several standard comparison theorems which usually go under the heading of "comparison principle." Next, we show how this principle can be applied to vector Lyapunov functions in the analysis of large scale systems. In the seventh section we present a method of determining estimates of trajectory behavior and trajectory bounds. To illustrate how the results can be applied and to demonstrate the usefulness of the method of analysis advanced herein, we consider several specific examples in the eighth section. To point out relative advantages and disadvantages of the present procedure, examples from diverse areas are included which were treated by previous workers using methods that differ significantly from the present approach. In the last section a brief discussion of the literature cited is presented.

2.1 Notation

Let V and W be arbitrary sets. Then $V \cup W$, $V \cap W$, $V-W$, and $V \times W$ denote the union, intersection, difference, and Cartesian product of V and W, respectively. If V is a subset of W we write $V \subset W$ and if x is an element of V we write $x \in V$. If f is a function or mapping of V into W we write $f: V \to W$ and we identify the domain of f and the range of f by $D(f)$ and $Ra(f)$, respectively.

Let $\varnothing$ denote the empty set, let R denote the real numbers, let $R^+ = [0, \infty)$, and let $J = [t_0, \infty)$ where $t_0 \geq 0$.

We let R^n be the Euclidean n-space and we let $|\cdot|$ represent the Euclidean norm. If $x \in R^n$, then $x^T = (x_1, \ldots, x_n)$ denotes the transpose of x. If $x, y \in R^n$, then $x \leq y$ signifies $x_i \leq y_i$, $x < y$ signifies $x_i < y_i$, and $x > 0$ signifies $x_i > 0$ for all $i = 1, \ldots, n$. Let $Y \subset R^n$. Then $\overline{Y}$ and ∂Y represent the closure and the boundary of Y, respectively. Also, $B(r) = \{x \in R^n : |x| < r\}$ and $\overline{B(r)} = \{x \in R^n : |x| \leq r\}$ for some $r > 0$.

Unless otherwise specified, matrices are usually assumed to be real. If $A = [a_{ij}]$ is an arbitrary matrix, then A^T denotes the transpose of A, $A > 0$ indicates that $a_{ij} > 0$, and $A \geq 0$ signifies that $a_{ij} \geq 0$ for all i, j.

Now let A be a square matrix. If A is nonsingular, then A^{-1} denotes the inverse of A. An eigenvalue of A is identified as $\lambda(A)$ and $\text{Re}\,\lambda(A)$ denotes the real part of $\lambda(A)$. If all eigenvalues of A happen to be real we write $\lambda_M(A)$ and $\lambda_m(A)$ to denote the largest and smallest eigenvalues of A, respectively. Matrix A is said to be stable if all its eigenvalues have negative real parts and unstable if at least one of its eigenvalues has positive real part. The determinant of an $n \times n$ matrix A is denoted by

$$\det A = \begin{vmatrix} a_{11} & \cdots & a_{1n} \\ \cdot & & \cdot \\ \cdot & \cdots & \cdot \\ \cdot & & \cdot \\ a_{n1} & \cdots & a_{nn} \end{vmatrix}.$$

If A is a diagonal matrix we write $A = \text{diag}[a_1, ..., a_n]$. The identity matrix is denoted by I.

Finally, the norm of an arbitrary matrix A, induced by the Euclidean norm, is given by

$$\|A\| = \min\{\alpha \in R^+ : \alpha|x| \geq |Ax|, \ x \in R^n\} = \sqrt{\lambda_M(A^{\mathrm{T}}A)}.$$

2.2 Lyapunov Stability and Related Results

In this section we present essential background material concerned with the stability analysis of dynamical systems described by ordinary differential equations. Since this material can be found in several standard texts dealing with the direct method of Lyapunov (also called the second method of Lyapunov) and related topics, we will not present proofs for any of the results presented in this section.

We consider systems which can appropriately be described by ordinary differential equations of the form

$$\dot{x} = g(x, t) \tag{I}$$

where $x \in R^n$, $t \in J$, $\dot{x} = dx/dt$, and $g: B(r) \times J \to R^n$ for some $r > 0$. Henceforth we assume that g is sufficiently smooth so that Eq. (I) possesses for every $x_0 \in B(r)$ and for every $t_0 \in R^+$ one and only one solution $x(t; x_0, t_0)$ for all $t \in J$, where $x_0 = x(t_0; x_0, t_0)$. We call x_0 an initial point, we refer to t as "time," and we call t_0 initial time. Henceforth we also assume that Eq. (I) admits the **trivial solution** $x = 0$ so that $g(0, t) = 0$ for all $t \in J$. This solution is also called an **equilibrium point** or a **singular point** of (I). In addition, we also assume that $x = 0$ is an **isolated** equilibrium, i.e., there exists $r' > 0$ so that $g(x', t) = 0$ for all $t \in J$ holds for no nonzero $x' \in B(r')$.

The preceding formulation pertains to local results. When discussing global results, we always assume that $g: R^n \times J \to R^n$ and that g is sufficiently smooth so that Eq. (I) possesses for every $x_0 \in R^n$ and for every $t_0 \in R^+$ a unique solution $x(t; x_0, t_0)$ for all $t \in J$. In this case we also assume that $x = 0$ is the only equilibrium of Eq. (I).

Since Eq. (I) can generally not be solved analytically in closed form, the qualitative properties of the equilibrium are of great practical interest. This motivates the following stability definitions in the sense of Lyapunov.

2.2.1. Definition. The equilibrium $x = 0$ of Eq. (I) is **stable** if for every $\mathscr{E} > 0$ and any $t_0 \in R^+$ there exists a $\delta(\mathscr{E}, t_0) > 0$ such that

$$|x(t; x_0, t_0)| < \mathscr{E} \quad \text{for all} \quad t \geq t_0$$

whenever $|x_0| < \delta(\mathscr{E}, t_0)$.

In the above definition, δ depends on $\mathscr{E}$ and t_0. If δ is independent of t_0, i.e., $\delta = \delta(\mathscr{E})$, then the equilibrium $x = 0$ of Eq. (I) is said to be **uniformly stable**.

2.2.2. Definition. The equilibrium $x = 0$ of Eq. (I) is **asymptotically stable** if (i) it is stable, and (ii) there exists an $\eta(t_0) > 0$ such that $\lim_{t \to \infty} x(t; x_0, t_0) = 0$ whenever $|x_0| < \eta$.

The set of all $x_0 \in R^n$ such that condition (ii) of Definition 2.2.2 is satisfied is called the **domain of attraction** of the equilibrium $x = 0$ of Eq. (I).

2.2.3. Definition. The equilibrium $x = 0$ of Eq. (I) is **uniformly asymptotically stable** if (i) it is uniformly stable, and (ii) for every $\mathscr{E} > 0$ and any $t_0 \in R^+$ there exists a $\delta_0 > 0$, independent of t_0 and $\mathscr{E}$, and a $T(\mathscr{E}) > 0$, independent of t_0, such that $|x(t; x_0, t_0)| < \mathscr{E}$ for all $t \geq t_0 + T(\mathscr{E})$ whenever $|x_0| < \delta_0$.

Of special interest in applications is the following special case of uniform asymptotic stability.

2.2.4. Definition. The equilibrium $x = 0$ of Eq. (I) is **exponentially stable** if there exists an $\alpha > 0$, and for every $\mathscr{E} > 0$ there exists a $\delta(\mathscr{E}) > 0$ such that

$$|x(t; x_0, t_0)| \leq \mathscr{E} e^{-\alpha(t - t_0)} \quad \text{for all} \quad t \geq t_0$$

whenever $|x_0| < \delta(\mathscr{E})$.

2.2.5. Definition. The equilibrium $x = 0$ of Eq. (I) is **unstable** if it is not stable. (In this case there exists a sequence $\{x_{0_n}\}$ of initial points and a sequence $\{t_n\}$ such that $|x(t_0 + t_n; x_{0_n}, t_0)| \geq \mathscr{E}$ for all n.)

When $g: R^n \times J \to R^n$ and Eq. (I) possesses unique solutions for all $x_0 \in R^n$ and every $t_0 \in R^+$, the following global characterizations are of interest.

2.2.6. Definition. A solution $x(t; x_0, t_0)$ of Eq. (I) is **bounded** if there exists a $\beta > 0$ such that $|x(t; x_0, t_0)| < \beta$ for all $t \geq t_0$, where β may depend on each solution.

2.2.7. Definition. The solutions of Eq. (I) are **uniformly bounded** if for any $\alpha > 0$ and $t_0 \in R^+$, there exists a $\beta = \beta(\alpha) > 0$ (independent of t_0) such that if $|x_0| < \alpha$, then $|x(t; x_0, t_0)| < \beta$ for all $t \geq t_0$.

2.2.8. Definition. The solutions of Eq. (I) are **uniformly ultimately bounded** (with bound B) if there exists a $B > 0$ and if corresponding to any $\alpha > 0$ and $t_0 \in R^+$, there exists a $T = T(\alpha) > 0$ (independent of t_0) such that $|x_0| < \alpha$ implies that $|x(t; x_0, t_0)| < B$ for all $t \geq t_0 + T$.

2.2.9. Definition. The equilibrium $x = 0$ of Eq. (I) is **asymptotically stable in the large** if it is stable and if every solution of Eq. (I) tends to zero as $t \to \infty$. (In this case the domain of attraction of the equilibrium of Eq. (I) is all of R^n.)

2.2.10. Definition. The equilibrium $x = 0$ of Eq. (I) is **uniformly asymptotically stable in the large** if (i) it is uniformly stable, (ii) the solutions of Eq. (I) are uniformly bounded, and (iii) for any $\alpha > 0$, any $\mathscr{E} > 0$, and $t_0 \in R^+$, there exists $T(\mathscr{E}, \alpha) > 0$, independent of t_0, such that if $|x_0| < \alpha$ then $|x(t; x_0, t_0)| < \mathscr{E}$ for all $t \geq t_0 + T(\mathscr{E}, \alpha)$.

2.2.11. Definition. The equilibrium $x = 0$ of Eq. (I) is **exponentially stable in the large** if there exists $\alpha > 0$ and for any $\beta > 0$, there exists $k(\beta) > 0$ such that

$$|x(t; x_0, t_0)| \leq k(\beta)|x_0|e^{-\alpha(t - t_0)} \quad \text{for all} \quad t \geq t_0$$

whenever $|x_0| < \beta$.

Results which yield conditions for stability, instability and boundedness in the sense of the above definitions involve the existence of functions $v: D \to R$, where in the case of local results $D = B(r) \times J$ for some $r > 0$, while in the case of global results $D = R^n \times J$. Henceforth we *always assume that such v-functions are continuous on their respective domains of definition and that they satisfy locally a Lipschitz condition with respect to x. Also, unless otherwise stated, we assume henceforth that* $v(0, t) = 0$ *for all* $t \in J$. The upper right-hand derivative of v with respect to t along solutions of Eq. (I) is given by

$$Dv_{(I)}(x, t) = \lim_{h \to 0^+} \sup(1/h)\{v[x(t+h; x, t), t+h] - v(x, t)\}$$

$$= \lim_{h \to 0^+} \sup(1/h)\{v[x + h \cdot g(x, t), t+h] - v(x, t)\}. \quad (2.2.12)$$

If v is continuously differentiable with respect to all of its arguments, then the total derivative of v with respect to t along solutions of Eq. (I) is given by

$$Dv_{(I)}(x, t) = \nabla v(x, t)^\mathsf{T} g(x, t) + \partial v(x, t)/\partial t, \quad (2.2.13)$$

where $\nabla v(x, t)$ denotes the gradient vector of the scalar function v and $\partial v/\partial t$ represents the partial derivative of v with respect to t. Whether v is continuous or continuously differentiable will either be clear from context or it will be specified. In the former case $Dv_{(I)}$ is specified by Eq. (2.2.12) while in the latter case $Dv_{(I)}$ is given by Eq. (2.2.13).

We now characterize several properties of v-functions in terms of special types of comparison functions.

2.2.14. Definition. A continuous function $\varphi: [0, r_1] \to R^+$ (or a continuous function $\varphi: [0, \infty) \to R^+$) is said to belong to **class K**, i.e., $\varphi \in K$, if $\varphi(0) = 0$ and if φ is strictly increasing on $[0, r_1]$ (or on $[0, \infty)$). If $\varphi: R^+ \to R^+$, if $\varphi \in K$, and if $\lim_{r \to \infty} \varphi(r) = \infty$, then φ is said to belong to **class KR**.

2.2.15. Definition. Two functions $\varphi_1, \varphi_2 \in K$ defined on $[0, r_1]$ (or on $[0, \infty)$) are said to be of the **same order of magnitude** if there exist positive constants

k_1 and k_2 such that $k_1 \varphi_1(r) \le \varphi_2(r) \le k_2 \varphi_1(r)$ for all $r \in [0, r_1]$ (or for all $r \in [0, \infty)$).

2.2.16. Definition. A function v is said to be **positive definite** if there exists $\varphi \in K$ such that $v(x, t) \ge \varphi(|x|)$ for all $t \in J$ and for all $x \in B(r)$ for some $r > 0$. (Recall the assumption $v(0, t) = 0$ for all $t \in J$.)

2.2.17. Definition. A function v is said to be **negative definite** if $-v$ is positive definite.

2.2.18. Definition. A function v, defined on $R^n \times J$, is said to be **radially unbounded** if there exists $\varphi \in KR$ such that $v(x, t) \ge \varphi(|x|)$ for all $x \in R^n$ and $t \in J$. (Recall the assumption $v(0, t) = 0$ for all $t \in J$.)

2.2.19. Definition. A function v is said to be **decrescent** if there exists $\varphi \in K$ such that $|v(x, t)| \le \varphi(|x|)$ for all $t \in J$ and for all $x \in B(r)$ for some $r > 0$. In this case v is also said "to admit an infinitely small upper bound" or "to become uniformly small."

2.2.20. Definition. A function v is said to be **positive (negative) semidefinite** if $v(x, t) \ge 0$ $(v(x, t) \le 0)$ for all $t \in J$ and $x \in B(r)$ for some $r > 0$. (Recall the assumption $v(0, t) = 0$ for all $t \in J$.)

For alternate equivalent definitions of the above concepts (positive definite, negative definite, etc.) the reader is referred to several standard texts dealing with the Lyapunov theory cited in Section 2.9.

We are now in a position to summarize several well-known stability and instability results. In the first four of these, which are local results, we assume that v is defined and continuous on $B(r) \times J$ for some $r > 0$.

2.2.21. Theorem. If there exists a positive definite function v with a negative semidefinite derivative $Dv_{(I)}$, then the equilibrium $x = 0$ of Eq. (I) is **stable**.

2.2.22. Theorem. If there exists a positive definite, decrescent function v with a negative semidefinite derivative $Dv_{(I)}$, then the equilibrium $x = 0$ of Eq. (I) is **uniformly stable**.

2.2.23. Theorem. If there exists a positive definite, decrescent function v with a negative definite derivative $Dv_{(I)}$, then the equilibrium $x = 0$ of Eq. (I) is **uniformly asymptotically stable**. In this case there exist $\varphi_1, \varphi_2, \varphi_3 \in K$ such that

$$\varphi_1(|x|) \le v(x, t) \le \varphi_2(|x|), \qquad Dv_{(I)}(x, t) \le -\varphi_3(|x|)$$

for all $x \in B(r)$ and $t \in J$.

2.2.24. Theorem. If in Theorem 2.2.23 $\varphi_1, \varphi_2, \varphi_3 \in K$ are of the same order of magnitude, then the equilibrium $x = 0$ of Eq. (I) is **exponentially stable**.

Theorem 2.2.24 is true if in particular there exist three constants $c_1 > 0$, $c_2 > 0$, and $c_3 > 0$ such that

$$c_1 |x|^2 \le v(x, t) \le c_2 |x|^2, \qquad Dv_{(\mathrm{I})}(x, t) \le -c_3 |x|^2$$

for all $x \in B(r)$ and $t \in J$.

In the next three theorems, which are of great practical importance, we assume that v is defined and continuous over $R^n \times J$ and that Eq. (I) possesses unique solutions for all $x_0 \in R^n$ and all $t_0 \in R^+$.

2.2.25. Theorem. If there exists a positive definite, decrescent, and radially unbounded function v with a negative definite derivative $Dv_{(\mathrm{I})}$, then the equilibrium $x = 0$ of Eq. (I) is **uniformly asymptotically stable in the large.** In this case there exist $\varphi_1, \varphi_2 \in KR$ and $\varphi_3 \in K$ such that

$$\varphi_1(|x|) \le v(x, t) \le \varphi_2(|x|), \qquad Dv_{(\mathrm{I})} \le -\varphi_3(|x|)$$

for all $x \in R^n$ and for all $t \in J$.

2.2.26. Theorem. If in Theorem 2.2.25 $\varphi_1, \varphi_2, \varphi_3 \in KR$ and if $\varphi_1, \varphi_2, \varphi_3$ are of the same order of magnitude, then the equilibrium $x = 0$ of Eq. (I) is **exponentially stable in the large.**

Theorem 2.2.26 is true if in particular there exist three constants $c_1 > 0$, $c_2 > 0$, and $c_3 > 0$ such that

$$c_1 |x|^2 \le v(x, t) \le c_2 |x|^2, \qquad Dv_{(\mathrm{I})}(x, t) \le -c_3 |x|^2$$

for all $x \in R^n$ and $t \in J$.

2.2.27. Theorem. If there exists a function v defined on $|x| \ge R$ (where R may be large) and $0 \le t < \infty$, and if there exist $\psi_1, \psi_2 \in KR$ such that
  (i) $\psi_1(|x|) \le v(x, t) \le \psi_2(|x|)$,
  (ii) $Dv_{(\mathrm{I})}(x, t) \le 0$,
for all $|x| \ge R$ and $0 \le t < \infty$, then the solutions of Eq. (I) are **uniformly bounded.** If in addition there exists $\psi_3 \in K$ (defined on R^+) and if (ii) is replaced by
  (iii) $Dv_{(\mathrm{I})}(x, t) \le -\psi_3(|x|)$
then the solutions of Eq. (I) are **uniformly ultimately bounded.**

The next theorem, which yields conditions for instability, is a local result. We assume once more that for some $r > 0$, v is defined and continuous on $B(r) \times J$.

2.2.28. Theorem. Assume there exists a function v having the following properties.
  (i)   For every $\mathscr{E} > 0$ and for every $t \ge t_0$ there exist points x' such that $v(x', t) < 0$ and such that $|x'| < \mathscr{E}$. The set of all points (x, t) such that $|x| < r$

and $v(x, t) < 0$ will be called the "domain $v < 0$." It is bounded by the hyper-surfaces $|x| = r$ and $v = 0$ and may consist of several component domains.

(ii)   In at least one of the component domains D of the domain $v < 0$, v is bounded from below and $0 \in \partial D$.

(iii)   In the domain D, $Dv_{(I)} \leq -\varphi(v)$ where $\varphi \in K$.

Then the equilibrium $x = 0$ of Eq. (I) is **unstable**.

If in particular there exists a positive definite function v (a negative definite function v) such that $Dv_{(I)}(x, t)$ is positive definite (negative definite), then the equilibrium $x = 0$ of Eq. (I) is unstable. In fact, in this case the equilibrium is said to be **completely unstable**.

Equation (I) is called a **nonautonomous ordinary differential equation**. Many systems of practical interest can appropriately be described by **autonomous ordinary differential equations** given by

$$\dot{x} = g(x) \tag{I'}$$

where $x \in R^n$, $\dot{x} = dx/dt$, and $g: B(r) \to R^n$ for some $r > 0$, or in the case of a global setting, $g: R^n \to R^n$. For the unique solutions of Eq. (I') we have $x(t; x_0, t_0) = x(t - t_0; x_0, 0)$ which allows us to assume $t_0 = 0$ without loss of generality. Furthermore, since Eq. (I') is a special case of Eq. (I), all preceding statements made for (I) hold equally as well for (I'), with obvious modifications. In particular, since Eq. (I') is invariant under translations of time, it makes sense to consider only uniform stability, uniform asymptotic stability, uniform asymptotic stability in the large, and so forth. Conditions for stability, in-stability, boundedness, and the like, involve in this case the existence of functions $v: B(r) \to R$ for some $r > 0$ or $v: R^n \to R$. Such functions are characterized as being positive definite, negative definite, or radially un-bounded as was done before using functions of class K and class KR and deleting all reference to $t \in J$. For Eq. (I'), the preceding theorems are modified by replacing $v(x, t)$ by $v(x)$. In addition, in Theorems 2.2.22, 2.2.23, and 2.2.25, all references to the word "decrescent" are deleted.

For system (I') there is a significant extension for asymptotic stability, which we want to consider. First we require the following concept, given here in a global setting.

2.2.29. Definition. A set Γ of points in R^n is **invariant** with respect to Eq. (I') if every solution of Eq. (I') starting in Γ remains in Γ for all time, i.e., if $x_0 \in \Gamma$ then $x(t; x_0, 0) \in \Gamma$ for all $t \in R$.

This concept has been used to prove the results summarized in the next theorem.

2.2.30. Theorem. If there exists a continuously differentiable function $v: R^n \to R$ (recall the assumption $v(0) = 0$) and $\psi \in KR$ such that

(i) $v(x) \geq \psi(|x|)$ for all $x \in R^n$,
(ii) $Dv_{(I')}(x) \leq 0$ for all $x \in R^n$,
(iii) the origin $x = 0$ is the only invariant subset of the set

$$E = \{x \in R^n : Dv_{(I')}(x) = 0\},$$

then the equilibrium $x = 0$ is **asymptotically stable in the large**. If hypothesis
(iii) is deleted then all solutions $x(t; x_0, 0)$ of Eq. (I') are **bounded** for $t \geq t_0$.

The preceding results are phrased in terms of the Euclidean norm $|\cdot|$. They
are also true with respect to any other equivalent norm defined on R^n. In this
case, convergence needs to be interpreted relative to the particular norm that
is used.

Henceforth we refer to results such as those given in this section as
Lyapunov-type theorems and we call a function v satisfying any theorem of
this type a **Lyapunov function**.

The Lyapunov theorems are very powerful. However, in general, great
difficulties arise in applying these results to high-dimensional systems with
complicated structure. The reason for this lies in the fact that there is no
universal and systematic procedure available which tells us how to find the
required Lyapunov functions. Although converse Lyapunov theorems have
been established, these results provide no clue (except in the case of linear
equations) for the construction of Lyapunov functions. For this reason we
will pursue an approach which allows us to analyze the stability of high-
dimensional systems with intricate structure in terms of simpler system components, which we shall call subsystems and interconnecting structure. This
viewpoint makes it often possible to circumvent many of the difficulties
associated with the Lyapunov method.

First we need to consider the description of large scale systems, also called
interconnected systems or composite systems. This is the topic of the next
section.

2.3   Large Scale Systems

Since it simplifies matters considerably, our exposition in the present
section is in a global setting. A development in a local setting involves obvious
modifications.

To fix some of the subsequent ideas, we begin with a specific example. In
particular, we consider systems described by the set of equations

$$\dot{z}_i = A_i z_i + D_i u_i \tag{2.3.1}$$

$$y_i = H_i z_i \tag{2.3.2}$$

where $z_i \in R^{n_i}$, $\dot{z}_i = dz_i/dt$, A_i is an $n_i \times n_i$ matrix, D_i is an $n_i \times m_i$ matrix, $u_i \in R^{m_i}$, H_i is a $p_i \times n_i$ matrix, and $y_i \in R^{p_i}$. Equations (2.3.1) and (2.3.2) describe the input–output characteristics of a linear time-invariant system described by ordinary differential equations. We call this system the ith **transfer system**, where u_i is interpreted as the **input** and y_i as the **output**. Associated with Eq. (2.3.1) is the system described by the linear equation

$$\dot{z}_i = A_i z_i, \tag{2.3.3}$$

which we call the ith **isolated subsystem** or the ith **free subsystem**.

Next, let us consider l transfer systems (i.e., $i = 1, ..., l$) and let us inter-connect these by means of the equations

$$u_i = \sum_{j=1}^{l} B_{ij} y_j + G_i u, \tag{2.3.4}$$

$i = 1, ..., l$, to form a composite system or interconnected system. Here B_{ij} is an $m_i \times p_j$ matrix, G_i is an $m_i \times q$ matrix and $u \in R^q$. In this case B_{ij} represents a linear, time-invariant connection from the output of the jth transfer system to the input of the ith transfer system. Frequently we may assume B_{ii} to be zero, since feedback around any transfer system can often (for purposes of analysis) be combined with the matrix A_i.

Combining Eqs. (2.3.1), (2.3.2), and (2.3.4), we obtain

$$\dot{z}_i = A_i z_i + \sum_{j=1}^{l} C_{ij} z_j + K_i u, \tag{2.3.5}$$

$i = 1, ..., l$, where $C_{ij} = D_i B_{ij} H_j$ and $K_i = D_i G_i$ are $n_i \times n_j$ and $n_i \times q$ matrices, respectively. Letting $\sum_{i=1}^{l} n_i = n$, $x^T = (z_1^T, ..., z_l^T) \in R^n$,

$$
A = \begin{bmatrix} A_1 & & 0 \\ & \cdot & \\ & & \cdot \\ 0 & & A_l \end{bmatrix}, \quad
C = \begin{bmatrix} C_{11} & \cdots & C_{1l} \\ \cdot & & \cdot \\ \cdot & \cdots & \cdot \\ \cdot & & \cdot \\ C_{l1} & \cdots & C_{ll} \end{bmatrix}, \quad
K = \begin{bmatrix} K_1 \\ \cdot \\ \cdot \\ \cdot \\ K_l \end{bmatrix}
$$

we can rewrite Eq. (2.3.5) as

$$\dot{x} = Ax + Cx + Ku. \tag{2.3.6}$$

Finally, let the output of this interconnected system be represented by

$$y = Hx \tag{2.3.7}$$

where $y \in R^p$ and H is a $p \times n$ matrix.

Equations (2.3.6) and (2.3.7) describe the input–output characteristics of the composite system considered, where u denotes the input and y the output. This system may be viewed as a linear interconnection of l isolated subsystems described by Eq. (2.3.3) with interconnecting structure specified by Eqs. (2.3.1), (2.3.2), and (2.3.4) and output characterized by Eq. (2.3.7).

Systems described by equations of the form (2.3.6) and (2.3.7) are examples of **large scale systems**. For obvious reasons, the terms **interconnected system** and **composite system** are more descriptive. In control theory, such systems are sometimes also called **multi-loop feedback systems, decentralized systems,** and the like.

In the Lyapunov stability analysis of the equilibrium $x = 0$ of the above system, the output equation (2.3.7) is of no concern whatsoever. Furthermore, in such an analysis it is the free or unforced dynamical system that is of interest. Thus, when investigating the Lyapunov stability of the above system, we consider the set of differential equations

$$\dot{z}_i = A_i z_i + \sum_{j=1}^{l} C_{ij} z_j, \tag{2.3.8}$$

$i = 1, ..., l$, or equivalently, the differential equation

$$\dot{x} = Ax + Cx \triangleq Fx, \tag{2.3.9}$$

which is a special case of Eq. (I) of Section 2.2. On the other hand, when studying the qualitative input–output properties of the above system, Eqs. (2.3.6) and (2.3.7) are of interest. We will devote Chapters VI and VII in their entirety to such investigations. Meanwhile, we primarily concern ourselves with Lyapunov stability and related concepts for large scale systems.

Many systems of practical interest (e.g., power systems, aerospace systems, circuits, economic systems, etc.) are appropriately described by nonlinear time varying ordinary differential equations and may often be viewed as interconnected or composite systems. In the following we consider several classes of ordinary differential equations which may be used to characterize such systems.

We consider systems described by equations of the form

(Σ_i) $\qquad\qquad\qquad \dot{z}_i = f_i(z_i, t) + g_i(z_1, ..., z_l, t), \qquad\qquad (2.3.10)$

$i = 1, ..., l$, where $z_i \in R^{n_i}$, $t \in J$, $f_i: R^{n_i} \times J \to R^{n_i}$, and $g_i: R^{n_1} \times \cdots \times R^{n_l} \times J \to R^{n_i}$. Henceforth we assume that $f_i(z_i, t) = 0$ for all $t \in J$ if and only if $z_i = 0$. Letting $\sum_{j=1}^{l} n_j = n$,

$$x^T = (z_1^T, ..., z_l^T) \in R^n$$

$$f(x, t)^T = [f_1(z_1, t)^T, ..., f_l(z_l, t)^T]$$

$$g(x, t)^T = [g_1(z_1, ..., z_l, t)^T, ..., g_l(z_1, ..., z_l, t)^T]$$

and

$$g_i(z_1, ..., z_l, t) \triangleq g_i(x, t), \qquad i = 1, ..., l,$$

so that $g(x,t)^{\mathrm{T}} = [g_1(x,t)^{\mathrm{T}},...,g_l(x,t)^{\mathrm{T}}]$, we can represent Eq. (2.3.10) equivalently as

$$(\mathscr{S}) \qquad\qquad \dot{x} = f(x,t) + g(x,t) \triangleq h(x,t). \qquad\qquad (2.3.11)$$

Clearly, $f: R^n \times J \to R^n$, $g: R^n \times J \to R^n$, and $h: R^n \times J \to R^n$. We always assume that $h(x,t) = 0$ for all $t \in J$ if and only if $x = 0$.

A system described by Eq. (2.3.11) may be viewed as a nonlinear and time varying interconnection of l systems represented by equations of the form

$$(\mathscr{S}_i) \qquad\qquad \dot{z}_i = f_i(z_i,t). \qquad\qquad (2.3.12)$$

Henceforth we assume that for every $t_0 \in R^+$ and every $x_0 \in R^n$, Eq. (2.3.11) possesses a unique solution $x(t;x_0,t_0)$ for $t \geq t_0$ with $x_0 = x(t_0;x_0,t_0)$. Also, we assume that for every $t_0 \in R^+$ and every $z_{i_0} \in R^{n_i}$, Eq. (2.3.12) has a unique solution $z_i(t;z_{i_0},t_0)$ for $t \geq t_0$ with $z_i(t_0;z_{i_0},t_0) = z_{i_0}$.

Subsequently we refer to Eq. (2.3.11) as **composite system** ($\mathscr{S}$), or **interconnected system** ($\mathscr{S}$), or **large scale system** ($\mathscr{S}$) **with decomposition** (Σ_i) (described by Eq. (2.3.10)). We refer to Eq. (2.3.12) as the ith **isolated subsystem** ($\mathscr{S}_i$) or as the ith **free subsystem** ($\mathscr{S}_i$) or as the ith **unforced subsystem** ($\mathscr{S}_i$). We call x in Eq. (2.3.11) a **hyper vector**. Finally note that Eqs. (2.3.11) and (2.3.12) are of the same form as Eq. (I) and as such, all results of Section 2.2 are applicable to composite system ($\mathscr{S}$) and to isolated subsystem ($\mathscr{S}_i$).

In specific applications, more information concerning system structure is usually available than indicated in Eq. (2.3.10). The following two classes of systems are examples of special cases of Eq. (2.3.10) encountered in practice. Let

$$g_i(x,t) = \sum_{j=1, i \neq j}^{l} C_{ij} z_j \qquad\qquad (2.3.13)$$

where C_{ij} is a constant $n_i \times n_j$ matrix. Then (Σ_i) assumes the form

$$\dot{z}_i = f_i(z_i,t) + \sum_{j=1, i \neq j}^{l} C_{ij} z_j, \qquad\qquad (2.3.14)$$

$i = 1,...,l$. This equation represents a system consisting of l isolated subsystems ($\mathscr{S}_i$) which are linearly interconnected. Its structural properties are depicted in the block diagram of Fig. 2.1.

Next let

$$g_i(x,t) = \sum_{j=1, i \neq j}^{l} g_{ij}(z_j,t) \qquad\qquad (2.3.15)$$

where $g_{ij}: R^{n_j} \times J \to R^{n_i}$. Then ($\Sigma_i$) assumes the form

$$\dot{z}_i = f_i(z_i,t) + \sum_{j=1, i \neq j}^{l} g_{ij}(z_j,t), \qquad\qquad (2.3.16)$$

$i = 1,...,l$.

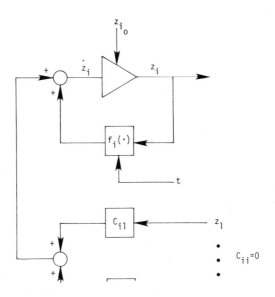

Figure 2.1   Interconnected system (2.3.14) where $i = 1, ..., l$.

Equations (2.3.10), (2.3.11), and (2.3.12) characterize a class of composite systems which can appropriately be described by nonautonomous ordinary differential equations. Deleting $t \in J$ in these equations, we also consider time-invariant interconnected systems described by the corresponding autonomous ordinary differential equations given by

(Σ_i') $\qquad\qquad \dot{z}_i = f_i(z_i) + g_i(z_1, ..., z_l), \qquad i = 1, ..., l,$ $\qquad$ (2.3.17)

$(\mathscr{S}')$ $\qquad\qquad \dot{x} = f(x) + g(x) \triangleq h(x),$ $\qquad\qquad\qquad$ (2.3.18)

$(\mathscr{S}_i')$ $\qquad\qquad \dot{z}_i = f_i(z_i),$ $\qquad\qquad\qquad\qquad\qquad$ (2.3.19)

respectively.

Note that in all cases considered thus far, the interconnecting structure enters additatively into the system description. As such, composite system $(\mathscr{S})$ with decomposition (Σ_i) is actually a special case of systems described by equations of the form

(Σ_i'') $\quad \dot{z}_i = f_i(z_i, g_i(z_1, ..., z_l, t), t) = f_i(z_i, g_i(x, t), t) \triangleq h_i(x, t),$ $\quad$ (2.3.20)

$i = 1, ..., l,$ where $z_i \in R^{n_i}$, $t \in J$, $\sum_{j=1}^{l} n_j = n$, $x^T = (z_1^T, ..., z_l^T) \in R^n$, $g_i: R^n \times J \to R^{r_i}, f_i: R^{n_i} \times R^{r_i} \times J \to R^{n_i}$, and $h_i: R^n \times J \to R^{n_i}$. Letting $h^T(x, t) = [h_1(x, t)^T, ..., h_l(x, t)^T]$, Eq. (2.3.20) can equivalently be represented by

$(\mathscr{S}'')$ $\qquad\qquad\qquad\qquad \dot{x} = h(x, t).$ $\qquad\qquad\qquad\qquad$ (2.3.21)

We assume that Eq. (2.3.21) possesses only one equilibrium, $x = 0$, and that for every $t_0 \in R^+$ and every $x_0 \in R^n$, Eq. (2.3.21) has a unique solution $x(t; x_0, t_0)$ for all $t \geq t_0$. In the present case we speak of composite system ($\mathscr{S}''$) with decomposition (Σ_i'') and leave the notion of isolated subsystem undefined.

In the present chapter, as well as in subsequent ones, we advance a method of qualitative analysis at different hierarchical levels, involving the following general steps.

Step 1. A large scale system ($\mathscr{S}$) (see Eq. (2.3.11)) is decomposed into l isolated subsystems ($\mathscr{S}_i$) (see Eq. (2.3.12)) which when interconnected in an appropriate fashion (see Eq. (2.3.10)) yield the original composite or interconnected system.

Step 2. The qualitative properties of the lower order (and hopefully simpler) free subsystems ($\mathscr{S}_i$) are characterized in terms of Lyapunov functions v_i, using standard and well-established techniques involving the results of Section 2.2.

Step 3. Qualitative properties of the overall system ($\mathscr{S}$) are deduced from the qualitative properties of the interconnecting structure and the individual free subsystems.

For composite system ($\mathscr{S}''$) with decomposition (Σ_i''), the above procedure has to be modified somewhat, for in this case free subsystems ($\mathscr{S}_i$) are not defined. Nevertheless, as will be demonstrated, even in this case a method of analysis at different hierarchical levels is still possible.

The process of decomposing a large system (Step 1) into an appropriate form is by no means a trivial task. We will not address ourselves explicitly to the problem of "tearing," pioneered by Kron [1]. However, we would like to point to two general approaches. In the first of these, the structural properties of the process being modeled usually dictate a natural decomposition. In the second approach, the decomposition is usually influenced by mathematical convenience to overcome technical difficulties.

In connection with Step 2 we note that for isolated subsystems of sufficiently low order and simplicity, many well-known results from the Lyapunov theory are available. Note also that in this step the converse Lyapunov theorems can play a crucial role. Thus, if the stability properties of the free subsystems are known a priori, we are usually in a position to search for Lyapunov functions with certain general properties.

We will use two methods of implementing Step 3. In the first approach, which is developed in Sections 2.4 and 2.5, scalar Lyapunov functions (consisting of weighted sums of Lyapunov functions for the isolated subsystems) are constructed and applied to the results of Section 2.2. In the second approach, vector Lyapunov functions (i.e., l-vectors whose components are

Lyapunov functions) are employed and stability is deduced from an lth-order differential inequality involving these vector Lyapunov functions by invoking an appropriate comparison principle.

The method outlined above frequently enables us to circumvent difficulties which arise when the Lyapunov approach is applied to high-dimensional systems with complex structure.

2.4 Analysis by Scalar Lyapunov Functions

In this section, which consists of five parts, we develop qualitative results for the systems considered in the previous section, making use of scalar Lyapunov functions. First we concern ourselves with uniform stability and uniform asymptotic stability. Next, we consider exponential stability. This is followed by instability and complete instability results. Uniform boundedness and uniform ultimate boundedness are treated in the fourth part. The section is concluded with a discussion of the results.

A. Uniform Stability and Uniform Asymptotic Stability

In characterizing the qualitative properties of the free subsystem $(\mathscr{S}_i)$, we will find it convenient to use the following convention.

2.4.1. Definition. Isolated subsystem $(\mathscr{S}_i)$ possesses **Property A** if there exists a continuously differentiable function $v_i: R^{n_i} \times J \to R$, functions $\psi_{i1}, \psi_{i2} \in KR$, $\psi_{i3} \in K$, and a constant $\sigma_i \in R$ such that the inequalities

$$\psi_{i1}(|z_i|) \leq v_i(z_i, t) \leq \psi_{i2}(|z_i|), \qquad Dv_{i(\mathscr{S}_i)}(z_i, t) \leq \sigma_i \psi_{i3}(|z_i|)$$

hold for all $z_i \in R^{n_i}$ and for all $t \in J$.

Clearly, if $\sigma_i < 0$, the equilibrium $z_i = 0$ of $(\mathscr{S}_i)$ is uniformly asymptotically stable in the large. If $\sigma_i = 0$, the equilibrium is uniformly stable. If $\sigma_i > 0$, the equilibrium of $(\mathscr{S}_i)$ may be unstable.

2.4.2. Theorem. The equilibrium $x = 0$ of composite system $(\mathscr{S})$ with decomposition (Σ_i) is **uniformly asymptotically stable in the large** if the following conditions are satisfied.

(i)   Each isolated subsystem $(\mathscr{S}_i)$ possesses Property A;

(ii)   given v_i and ψ_{i3} of hypothesis (i), there exist constants $a_{ij} \in R$ such that

$$\nabla v_i(z_i, t)^{\mathsf{T}} g_i(z_1, \ldots, z_l, t) \leq [\psi_{i3}(|z_i|)]^{1/2} \sum_{j=1}^{l} a_{ij} [\psi_{j3}(|z_j|)]^{1/2}$$

for all $z_i \in R^{n_i}$, $i = 1, \ldots, l$, and $t \in J$; and

(iii) given σ_i of hypothesis (i) there exists an l-vector $\alpha^T = (\alpha_1, ..., \alpha_l) > 0$ such that the test matrix $S = [s_{ij}]$ specified by

$$s_{ij} = \begin{cases} \alpha_i(\sigma_i + a_{ii}), & i = j \\ (\alpha_i a_{ij} + \alpha_j a_{ji})/2, & i \neq j \end{cases}$$

is negative definite.

Proof. For composite system $(\mathscr{S})$ we choose a Lyapunov function

$$v(x,t) = \sum_{i=1}^{l} \alpha_i v_i(z_i, t) \tag{2.4.3}$$

where the functions v_i are given in hypothesis (i) and where $\alpha_i > 0$, $i = 1, ..., l$ are constants given in hypothesis (iii). Clearly, $v(x,t)$ is continuously differentiable and $v(0,t) = 0$ for all $t \in J$, since each $v_i(z_i, t)$ satisfies these conditions. Since each isolated subsystem $(\mathscr{S}_i)$ possesses Property A, it follows that

$$\sum_{i=1}^{l} \alpha_i \psi_{i1}(|z_i|) \leq v(x,t) \leq \sum_{i=1}^{l} \alpha_i \psi_{i2}(|z_i|)$$

for all $x \in R^n$, $t \in J$, and $z_i \in R^{n_i}$, $i = 1, ..., l$. Since by assumption $\psi_{i1}, \psi_{i2} \in KR$, it follows that $v(x,t)$ is positive definite, decrescent, and radially unbounded. Indeed, there exist $\psi_1, \psi_2 \in KR$ such that

$$\psi_1(|x|) \leq \sum_{i=1}^{l} \alpha_i \psi_{i1}(|z_i|) \quad \text{and} \quad \psi_2(|x|) \geq \sum_{i=1}^{l} \alpha_i \psi_{i2}(|z_i|),$$

so that

$$\psi_1(|x|) \leq v(x,t) \leq \psi_2(|x|) \tag{2.4.4}$$

for all $x \in R^n$ and $t \in J$.

It remains to be shown that along solutions of composite system $(\mathscr{S})$ the derivative $Dv_{(\mathscr{S})}(x,t)$ is negative definite for all $x \in R^n$ and $t \in J$. We have

$$Dv_{(\mathscr{S})}(x,t) = \sum_{i=1}^{l} \{\alpha_i[(\partial v_i(z_i,t)/\partial t) + \nabla v_i(z_i,t)^T f_i(z_i,t)]$$

$$+ \alpha_i[\nabla v_i(z_i,t)^T g_i(z_1, ..., z_l, t)]\}$$

$$= \sum_{i=1}^{l} \{\alpha_i Dv_{i(\mathscr{S}_i)}(z_i,t) + \alpha_i[\nabla v_i(z_i,t)^T g_i(z_1, ..., z_l, t)]\}. \tag{2.4.5}$$

From hypotheses (i) and (ii) it now follows that

$$Dv_{(\mathscr{S})}(x,t) \leq \sum_{i=1}^{l} \{\alpha_i \sigma_i \psi_{i3}(|z_i|) + \alpha_i[\psi_{i3}(|z_i|)]^{1/2} \sum_{j=1}^{l} a_{ij}[\psi_{j3}(|z_j|)]^{1/2}\}.$$

Now let

$$w^{\mathrm{T}} = [\psi_{13}(|z_1|)^{1/2}, \ldots, \psi_{13}(|z_l|)^{1/2}] \tag{2.4.6}$$

and let $R = [r_{ij}]$ be the $l \times l$ matrix specified by

$$r_{ij} = \begin{cases} \alpha_i[\sigma_i + a_{ii}], & i = j \\ \alpha_i a_{ij}, & i \neq j. \end{cases}$$

We now have

$$Dv_{(\mathcal{S})}(x, t) \leq w^{\mathrm{T}} R w = w^{\mathrm{T}}((R + R^{\mathrm{T}})/2) w = w^{\mathrm{T}} S w,$$

where $S = [s_{ij}]$ is the test matrix given in hypothesis (iii). Since S is symmetric, all its eigenvalues are real. Also since by hypothesis (iii), S is negative definite, all its eigenvalues are negative so that $\lambda_M(S) < 0$. We thus have

$$Dv_{(\mathcal{S})}(x, t) \leq \lambda_M(S) w^{\mathrm{T}} w = \lambda_M(S) \sum_{i=1}^{l} \psi_{i3}(|z_i|). \tag{2.4.7}$$

Therefore, $Dv_{(\mathcal{S})}(x, t)$ is negative definite for all $x \in R^n$ and $t \in J$. In fact, there exists a function $\psi_3 \in K$ such that $\psi_3(|x|) \leq \sum_{i=1}^{l} \psi_{i3}(|z_i|)$, so that

$$Dv_{(\mathcal{S})}(x, t) \leq \lambda_M(S) \psi_3(|x|), \qquad \lambda_M(S) < 0, \tag{2.4.8}$$

for all $x \in R^n$ and $t \in J$.

Inequalities (2.4.4) and (2.4.8) show that all hypotheses of Theorem 2.2.25 are satisfied. Therefore, the equilibrium $x = 0$ of composite system $(\mathcal{S})$ is uniformly asymptotically stable in the large. ∎

Before proceeding further, we note that the test matrix $S = [s_{ij}]$ of the above theorem is negative definite if and only if

$$(-1)^k \begin{vmatrix} s_{11} & \cdots & s_{1k} \\ \cdot & & \cdot \\ \cdot & \cdots & \cdot \\ \cdot & & \cdot \\ s_{k1} & \cdots & s_{kk} \end{vmatrix} > 0, \qquad k = 1, 2, \ldots, l. \tag{2.4.9}$$

2.4.10. Corollary. If the test matrix S in Theorem 2.4.2 is negative semi-definite, then the equilibrium $x = 0$ of composite system $(\mathcal{S})$ with decomposition (Σ_i) is **uniformly stable**. (In this case we require that the main determinants (2.4.9) be nonnegative.)

Proof. Choose the positive definite, decrescent, and radially unbounded Lyapunov function given in Eq. (2.4.3). Since by assumption matrix S is negative semidefinite, we have $\lambda(S) \leq 0$. Thus, inequality (2.4.7) yields $Dv_{(\mathcal{S})}(x, t) \leq 0$ for all $x \in R^n$ and $t \in J$. Therefore, the hypotheses of Theorem 2.2.22 are satisfied and the equilibrium of system $(\mathcal{S})$ is uniformly stable. ∎

We can generalize Therorem 2.4.2 somewhat.

2.4.11. Theorem. The equilibrium $x = 0$ of composite system $(\mathscr{S})$ with decomposition (Σ_i) is **uniformly asymptotically stable in the large** if the following conditions hold.

(i)   Each isolated subsystem $(\mathscr{S}_i)$ possesses Property A;

(ii)   given v_i and ψ_{i3} of hypothesis (i), there exist continuous functions $a_{ij} \colon R^n \times J \to R$ such that

$$\nabla v_i(z_i, t)^{\mathrm{T}} g_i(z_1, \ldots, z_l, t) \le [\psi_{i3}(|z_i|)]^{1/2} \sum_{j=1}^{l} a_{ij}(x, t) [\psi_{j3}(|z_j|)]^{1/2}$$

for all $z_i \in R^{n_i}$, $i = 1, \ldots, l$, $x \in R^n$, and $t \in J$;

(iii)   there exists an l-vector $\alpha^{\mathrm{T}} = (\alpha_1, \ldots, \alpha_l) > 0$ and $\mathscr{E} > 0$, such that for all $x \in R^n$ and $t \in J$, the test matrix $S(x, t) + \mathscr{E}I$ is negative definite, where I denotes the $l \times l$ identity matrix and $S(x, t) = [s_{ij}(x, t)]$ is defined by

$$s_{ij}(x, t) = \begin{cases} \alpha_i[\sigma_i + a_{ii}(x, t)], & i = j \\ [\alpha_i a_{ij}(x, t) + \alpha_j a_{ji}(x, t)]/2, & i \ne j. \end{cases}$$

Proof. Choose the positive definite, decrescent, and radially unbounded Lyapunov function given in Eq. (2.4.3). Let $R(x, t) = [r_{ij}(x, t)]$ be the $l \times l$ matrix specified by

$$r_{ij}(x, t) = \begin{cases} \alpha_i[\sigma_i + a_{ii}(x, t)], & i = j \\ \alpha_i a_{ij}(x, t), & i \ne j \end{cases}$$

and let w be defined as in Eq. (2.4.6). Using Eq. (2.4.5) and hypotheses (i), (ii), and (iii), we obtain

$$Dv_{(\mathscr{S})}(x, t) \le w^{\mathrm{T}} R(x, t) w = w^{\mathrm{T}}[(R(x, t) + R(x, t)^{\mathrm{T}})/2]w$$

$$= w^{\mathrm{T}} S(x, t) w \le -\mathscr{E} w^{\mathrm{T}} w = -\mathscr{E} \sum_{i=1}^{l} \psi_{i3}(|z_i|).$$

Therefore, $Dv_{(\mathscr{S})}(x, t)$ is negative definite for all $x \in R^n$ and $t \in J$. Indeed there exists a function $\psi_3 \in K$ such that $\psi_3(|x|) \le \sum_{i=1}^{l} \psi_{i3}(|z_i|)$, so that

$$Dv_{(\mathscr{S})}(x, t) \le -\mathscr{E}\psi_3(|x|).$$

Thus, the hypotheses of Theorem 2.2.25 are satisfied and the equilibrium of system $(\mathscr{S})$ is uniformly asymptotically stable in the large. ∎

Deleting all references to $t \in J$, Theorems 2.4.2, 2.4.11, and Corollary 2.4.10 are also applicable to autonomous composite system $(\mathscr{S}')$ with decomposition (Σ_i'). For system $(\mathscr{S}')$ an extension involving invariant sets is possible.

2.4.12. Definition. Isolated subsystem $(\mathscr{S}_i')$ possesses **Property A'** if there exists a continuously differentiable function $v_i: R^{n_i} \to R$, $\psi_{i1} \in KR$, $\psi_{i3} \in K$, and a constant $\sigma_i \in R$ such that

$$\psi_{i1}(|z_i|) \le v_i(z_i), \qquad v_i(0) = 0, \qquad Dv_{i(\mathscr{S}_i')}(z_i) \le \sigma_i \psi_{i3}(|z_i|)$$

for all $z_i \in R^{n_i}$.

If $\sigma_i < 0$, the equilibrium $z_i = 0$ of system $(\mathscr{S}_i')$ is asymptotically stable in the large and if $\sigma_i > 0$, the equilibrium may be unstable. If $\sigma_i = 0$, all solutions of $(\mathscr{S}_i')$ are bounded for all $t \ge t_0 = 0$, and also the equilibrium of $(\mathscr{S}_i')$ is stable.

2.4.13. Theorem. Assume that for composite system $(\mathscr{S}')$ with decomposition (Σ_i') the following conditions hold.

(i)   Each isolated subsystem $(\mathscr{S}_i')$ possesses Property A';
(ii)   given v_i and ψ_{i3} of hypothesis (i), there exist constants $a_{ij} \in R$ such that

$$\nabla v_i(z_i)^{\mathrm{T}} g_i(z_1, ..., z_l) \le [\psi_{i3}(|z_i|)]^{1/2} \sum_{j=1}^{l} a_{ij} [\psi_{j3}(|z_j|)]^{1/2}$$

for all $z_i \in R^{n_i}$, $i = 1, ..., l$;
(iii)   given σ_i of hypothesis (i), there exists an l-vector $\alpha^{\mathrm{T}} = (\alpha_1, ..., \alpha_l) > 0$ such that the matrix $S = [s_{ij}]$ specified by

$$s_{ij} = \begin{cases} \alpha_i(\sigma_i + a_{ii}), & i = j \\ (\alpha_i a_{ij} + \alpha_j a_{ji})/2, & i \ne j \end{cases}$$

is negative semidefinite.

Then the following statements are true.

(a)   All solutions of $(\mathscr{S}')$ are **bounded**;
(b)   the equilibrium $x = 0$ of $(\mathscr{S}')$ is **stable**; and
(c)   if the origin $x = 0$ is the only invariant subset of the set $E = \{x \in R^n : Dv_{(\mathscr{S}')}(x) = 0\}$, where $v(x) = \sum_{i=1}^{l} \alpha_i v_i(z_i)$, then the equilibrium $x = 0$ of composite system $(\mathscr{S}')$ is **asymptotically stable in the large**.

Proof. For composite system $(\mathscr{S}')$ we choose the Lyapunov function

$$v(x) = \sum_{i=1}^{l} \alpha_i v_i(z_i)$$

where the functions v_i are given in hypothesis (i) and where $\alpha_i > 0$, $i = 1, ..., l$ are constants given in hypothesis (iii). Clearly $v(x)$ is continuously differentiable, $v(0) = 0$, and $v(x)$ is positive definite and radially unbounded. Thus, there exists $\psi_1 \in KR$ such that

$$v(x) \ge \psi_1(|x|) \tag{2.4.14}$$

for all $x \in R^n$. Modifying the proof of Theorem 2.4.2 in an obvious way, we see that

$$Dv_{(\mathscr{S}')}(x) \le \lambda_M(S) \sum_{i=1}^{l} \psi_{i3}(|z_i|)$$

for all $x \in R^n$. Since S is by assumption negative semidefinite, we have $\lambda_M(S) \le 0$, so that

$$Dv_{(\mathscr{S}')}(x) \le 0 \tag{2.4.15}$$

for all $x \in R^n$. Clearly, the equilibrium $x = 0$ of system $(\mathscr{S}')$ is stable. Furthermore, it follows from Theorem 2.2.30 that all solutions of $(\mathscr{S}')$ are bounded for $t \ge t_0 = 0$. If in addition the origin $x = 0$ is the only invariant subset of the set $E = \{x \in R^n : Dv_{(\mathscr{S}')}(x) = 0\}$, then it follows from Theorem 2.2.30 that the equilibrium of $(\mathscr{S}')$ is asymptotically stable in the large. ∎

Next we consider composite system $(\mathscr{S}'')$ with decomposition (Σ_i'').

2.4.16. Theorem. Assume that for composite system $(\mathscr{S}'')$ with decomposition (Σ_i'') the following conditions hold.

(i)   There exist continuously differentiable functions $v_i: R^{n_i} \times J \to R$ and $\psi_{i1}, \psi_{i2} \in KR$, $i = 1, ..., l$, such that

$$\psi_{i1}(|z_i|) \le v_i(z_i, t) \le \psi_{i2}(|z_i|)$$

for all $z_i \in R^{n_i}$ and $t \in J$;

(ii)   given v_i in hypothesis (i), there exist constants $a_{ij} \in R$ and $\psi_{i4} \in K$, $i, j = 1, ..., l$, such that

$$\nabla v_i(z_i, t)^T h_i(x, t) + \partial v_i(z_i, t)/\partial t \le \psi_{i4}(|z_i|) \sum_{j=1}^{l} a_{ij} \psi_{j4}(|z_j|)$$

for all $z_i \in R^{n_i}$, $x \in R^n$, $t \in J$, and $i = 1, ..., l$; and

(iii)   there exists an l-vector $\alpha^T = (\alpha_1, ..., \alpha_l) > 0$ such that the test matrix $S = [s_{ij}]$ specified by

$$s_{ij} = \begin{cases} \alpha_i a_{ii}, & i = j \\ (\alpha_i a_{ij} + \alpha_j a_{ji})/2, & i \ne j \end{cases}$$

is either negative semidefinite or negative definite.

(a)   If S is negative semidefinite, the equilibrium $x = 0$ of composite system $(\mathscr{S}'')$ is **uniformly stable**.

(b)   If S is negative definite, the equilibrium of $(\mathscr{S}'')$ is **uniformly asymptotically stable in the large**.

Proof. Given the functions v_i of hypothesis (i), we choose for composite system $(\mathscr{S}'')$ the Lyapunov function

$$v(x, t) = \sum_{i=1}^{l} \alpha_i v_i(z_i, t) \tag{2.4.17}$$

where $\alpha^T = (\alpha_1, ..., \alpha_l) > 0$ is given in hypothesis (iii). Clearly $v(x,t)$ is continuously differentiable and

$$\sum_{i=1}^{l} \alpha_i \psi_{i1}(|z_i|) \leq v(x,t) \leq \sum_{i=1}^{l} \alpha_i \psi_{i2}(|z_i|)$$

for all $z_i \in R^{n_i}$, $i = 1, ..., l$, $x \in R^n$, and $t \in J$. Since by assumption $\psi_{i1}, \psi_{i2} \in KR$, it follows that $v(x,t)$ is positive definite, decrescent, and radially unbounded, and there exist $\psi_1, \psi_2 \in KR$ such that

$$\psi_1(|x|) \leq \sum_{i=1}^{l} \alpha_i \psi_{i1}(|z_i|) \quad \text{and} \quad \psi_2(|x|) \geq \sum_{i=1}^{l} \alpha_i \psi_{i2}(|z_i|).$$

Hence,

$$\psi_1(|x|) \leq v(x,t) \leq \psi_2(|x|) \tag{2.4.18}$$

for all $x \in R^n$ and $t \in J$.

Along solutions of $(\mathscr{S}'')$ we have

$$Dv_{(\mathscr{S}'')}(x,t) = \sum_{i=1}^{l} \alpha_i \{\nabla v_i(z_i, t)^T h_i(x, t) + (\partial v_i(z_i, t)/\partial t)\}$$

$$\leq \sum_{i=1}^{l} \alpha_i \left\{ \psi_{i4}(|z_i|) \sum_{j=1}^{l} a_{ij} \psi_{j4}(|z_j|) \right\}$$

$$= w^T R w = w^T ((R + R^T)/2) w = w^T S w \leq \lambda_M(S) \sum_{i=1}^{l} \psi_{i4}(|z_i|)^2,$$

where $w^T = [\psi_{14}(|z_1|), ..., \psi_{14}(|z_l|)]$, $R = [r_{ij}] \triangleq [\alpha_i a_{ij}]$, and S is given in hypothesis (iii). Therefore, if S is negative semidefinite, then so is $Dv_{(\mathscr{S}'')}(x,t)$ and the equilibrium $x = 0$ of composite system $(\mathscr{S}'')$ is uniformly stable. Similarly, if S is negative definite then $Dv_{(\mathscr{S}'')}(x,t)$ is negative definite for all $x \in R^n$ and $t \in J$ and the equilibrium of $(\mathscr{S}'')$ is uniformly asymptotically stable in the large. ∎

B. Exponential Stability

In studying the exponential stability of composite system $(\mathscr{S})$ we will find it useful to employ the following convention.

2.4.19. Definition. Isolated subsystem $(\mathscr{S}_i)$ possesses **Property B** if there exists a continuously differentiable function $v_i: R^{n_i} \times J \to R$, $\psi_{i1}, \psi_{i2}, \psi_{i3} \in KR$ *which are of the same order of magnitude*, and a constant $\sigma_i \in R$, such that

$$\psi_{i1}(|z_i|) \leq v_i(z_i, t) \leq \psi_{i2}(|z_i|), \qquad Dv_{i(\mathscr{S}_i)}(z_i, t) \leq \sigma_i \psi_{i3}(|z_i|)$$

for all $z_i \in R^{n_i}$ and $t \in J$.

Isolated subsystem $(\mathscr{S}_i)$ possesses **Property B'** if it possesses Property B with $\psi_{i1}(|z_i|) = c_{i1}|z_i|^2$, $\psi_{i2}(|z_i|) = c_{i2}|z_i|^2$, and $\psi_{i3}(|z_i|) = |z_i|^2$, where $c_{i2} \geq c_{i1} > 0$ are constants.

If isolated subsystem $(\mathscr{S}_i)$ possesses either Property B or Property B', and if $\sigma_i < 0$, then the equilibrium $z_i = 0$ is exponentially stable in the large. If $\sigma_i = 0$, the equilibrium of $(\mathscr{S}_i)$ is uniformly stable and if $\sigma_i > 0$, the equilibrium of $(\mathscr{S}_i)$ may be unstable.

2.4.20. Theorem. The equilibrium $x = 0$ of composite system $(\mathscr{S})$ with decomposition (Σ_i) is **exponentially stable in the large** if the following conditions are satisfied.

(i) Each isolated subsystem $(\mathscr{S}_i)$ possesses Property B;

(ii) all comparison functions ψ_{ij}, $i = 1, ..., l$, $j = 1, 2, 3$ of hypothesis (i) are of the same order of magnitude;

(iii) given v_i and ψ_{i3} of hypothesis (i), there exist constants $a_{ij} \in R$ such that

$$\nabla v_i(z_i, t)^{\mathsf{T}} g_i(z_1, ..., z_l, t) \leq [\psi_{i3}(|z_i|)]^{1/2} \sum_{j=1}^{l} a_{ij} [\psi_{j3}(|z_j|)]^{1/2}$$

for all $z_i \in R^{n_i}$, $z_j \in R^{n_j}$, $i, j = 1, ..., l$, and $t \in J$; and

(iv) given σ_i in hypothesis (i), there exists an l-vector $\alpha^{\mathsf{T}} = (\alpha_1, ..., \alpha_l) > 0$ such that the matrix $S = [s_{ij}]$ specified by

$$s_{ij} = \begin{cases} \alpha_i(\sigma_i + a_{ii}), & i = j \\ (\alpha_i a_{ij} + \alpha_j a_{ji})/2, & i \neq j \end{cases}$$

is negative definite.

Proof. First we note that if the above hypotheses are satisfied, then all hypotheses of Theorem 2.4.2 are also satisfied. It follows that the equilibrium $x = 0$ of composite system $(\mathscr{S})$ is uniformly asymptotically stable in the large. As in the proof of Theorem 2.4.2, we choose a Lyapunov function

$$v(x, t) = \sum_{i=1}^{l} \alpha_i v_i(z_i, t)$$

and conclude that

$$\sum_{i=1}^{l} \alpha_i \psi_{i1}(|z_i|) \leq v(x, t) \leq \sum_{i=1}^{l} \alpha_i \psi_{i2}(|z_i|)$$

and

$$Dv_{(\mathscr{S})}(x, t) \leq \lambda_M(S) \sum_{i=1}^{l} \psi_{i3}(|z_i|)$$

for all $x \in R^n$, $t \in J$, and $z_i \in R^{n_i}$, $i = 1, ..., l$, where $\lambda_M(S) < 0$ is the largest eigenvalue of test matrix S.

To complete the proof we first show that there are $\psi_1, \psi_2, \psi_3 \in KR$ which are of the same order of magnitude (see Definition 2.2.15) such that

$$\psi_1(\|x\|_\infty) \leq v(x, t) \leq \psi_2(\|x\|_\infty) \tag{2.4.21}$$

and

$$Dv_{(\mathscr{S})}(x, t) \leq \lambda_M(S)\psi_3(\|x\|_\infty) \tag{2.4.22}$$

for all $x \in R^n$ and $t \in J$, where

$$\|x\|_\infty = \max_i |z_i|, \qquad i = 1, ..., l.$$

By hypothesis (ii), there is a function $\psi \in KR$, e.g., $\psi = \psi_{11}$, and positive constants k_{ij} such that

$$\psi_{i1}(r) \geq k_{i1} \psi(r), \qquad \psi_{i2}(r) \leq k_{i2} \psi(r), \qquad \psi_{i3}(r) \geq k_{i3} \psi(r)$$

for all $r \in [0, \infty)$ and $i = 1, ..., l$. Define

$$\psi_1(r) = \min_i (\alpha_i k_{i1}) \psi(r), \qquad \psi_2(r) = \left(\sum_{i=1}^{l} \alpha_i k_{i2} \right) \psi(r),$$

and

$$\psi_3(r) = \min_i (k_{i3}) \psi(r).$$

Clearly, $\psi_1, \psi_2, \psi_3 \in KR$ are of the same order of magnitude. Moreover, since ψ is strictly increasing, we have

$$\sum_{i=1}^{l} \alpha_i \psi_{i1}(|z_i|) \geq \sum_{i=1}^{l} \alpha_i k_{i1} \psi(|z_i|) \geq \min_i (\alpha_i k_{i1}) \sum_{i=1}^{l} \psi(|z_i|)$$

$$\geq \min_i (\alpha_i k_{i1}) \max_i \psi(|z_i|) = \min_i (\alpha_i k_{i1}) \psi \left(\max_i |z_i| \right)$$

$$= \min_i (\alpha_i k_{i1}) \psi(\|x\|_\infty) = \psi_1(\|x\|_\infty).$$

Also,

$$\sum_{i=1}^{l} \alpha_i \psi_{i2}(|z_i|) \leq \sum_{i=1}^{l} \alpha_i k_{i2} \psi(|z_i|) \leq \left(\sum_{i=1}^{l} \alpha_i k_{i2} \right) \max_i \psi(|z_i|)$$

$$= \left(\sum_{i=1}^{l} \alpha_i k_{i2} \right) \psi \left(\max_i |z_i| \right) = \psi_2(\|x\|_\infty),$$

and

$$\sum_{i=1}^{l} \psi_{i3}(|z_i|) \geq \sum_{i=1}^{l} k_{i3} \psi(|z_i|) \geq \left(\min_i k_{i3} \right) \sum_{i=1}^{l} \psi(|z_i|)$$

$$\geq \left(\min_i k_{i3} \right) \max_i \psi(|z_i|) = \left(\min_i k_{i3} \right) \psi \left(\max_i |z_i| \right) = \psi_3(\|x\|_\infty).$$

Thus, inequalities (2.4.21) and (2.4.22) are true. Since the Lyapunov theorems of Section 2.2 are valid for any norm on R^n, such as $\|\cdot\|_\infty$, it follows from Theorem 2.2.26 (with $|\cdot|$ replaced by $\|\cdot\|_\infty$) that the equilibrium $x = 0$ of composite system $(\mathcal{S})$ is exponentially stable in the large in the norm $\|\cdot\|_\infty$. Furthermore, since

$$|x(t; x_0, t_0)| \leq l^{1/2} \|x(t; x_0, t_0)\|_\infty, \qquad t \in J,$$

it follows that the equilibrium $x = 0$ of composite system $(\mathcal{S})$ is also exponentially stable in the large in the norm $|\cdot|$. ∎

In practice it is often convenient to use quadratic comparison functions $\psi_{ij}(r)$, $i = 1, ..., l$, $j = 1, 2, 3$. In this case, the above theorem assumes the following form.

2.4.23. Corollary. The equilibrium $x = 0$ of composite system $(\mathscr{S})$ with decomposition (Σ_i) is **exponentially stable in the large** if the following conditions are satisfied.

(i)   Each isolated subsystem $(\mathscr{S}_i)$ possesses Property B$'$;

(ii) given v_i of hypothesis (i), there exist constants $a_{ij} \in R$ such that

$$\nabla v_i(z_i, t)^{\mathsf{T}} g_i(z_1, ..., z_l, t) \leq |z_i| \sum_{j=1}^{l} a_{ij} |z_j|$$

for all $z_i \in R^{n_i}$, $z_j \in R^{n_j}$, $i, j = 1, ..., l$, $t \in J$; and

(iii) given σ_i in hypothesis (i), there exists an l-vector $\alpha^{\mathsf{T}} = (\alpha_1, ..., \alpha_l) > 0$ such that the matrix $S = [s_{ij}]$ specified by

$$s_{ij} = \begin{cases} \alpha_i(\sigma_i + a_{ii}), & i = j \\ (\alpha_i a_{ij} + \alpha_j a_{ji})/2, & i \neq j \end{cases}$$

is negative definite.

Proof. Since each isolated subsystem $(\mathscr{S}_i)$ possesses Property B$'$, we have $\psi_{i1}(|z_i|) = c_{i1}|z_i|^2$, $\psi_{i2}(|z_i|) = c_{i2}|z_i|^2$, and $\psi_{i3}(|z_i|) = |z_i|^2$, $i = 1, ..., l$. Each of these comparison functions is of the same order of magnitude. Thus, all hypotheses of Theorem 2.4.20 are satisfied. Hence, the equilibrium $x = 0$ of composite system $(\mathscr{S})$ is exponentially stable in the large.   ∎

If composite system $(\mathscr{S})$ is decomposed into n scalar subsystems $(\mathscr{S}_i)$, it is possible to reduce the conservatism of Corollary 2.4.23 by eliminating norms in hypothesis (ii). The price paid for this improvement is an increase of the order of the test matrix S. We have

2.4.24. Corollary. The equilibrium $x = 0$ of composite system $(\mathscr{S})$ with decomposition (Σ_i) is **exponentially stable in the large** if the following conditions are satisfied.

(i)   Each isolated subsystem $(\mathscr{S}_i)$ possesses Property B$'$ with $n_i = 1$ so that $z_i = x_i$;

(ii) given v_i of hypothesis (i), there exist constants $a_{ij} \in R$ such that

$$[\partial v_i(x_i, t)/\partial x_i][g_i(x_1, ..., x_n, t)] \leq x_i \sum_{j=1}^{n} a_{ij} x_j$$

for all $x_i \in R$, $i = 1, ..., n$, and $t \in J$; and

(iii) hypothesis (iii) of Corollary 2.4.23 holds with $l = n$.

Proof. Choose as a Lyapunov function

$$v(x, t) = \sum_{i=1}^{n} \alpha_i v_i(x_i, t).$$

Following along the lines of the proof of Theorem 2.4.2, we see that

$$c_1|x|^2 \leq v(x,t) \leq c_2|x|^2$$

and

$$Dv_{(\mathscr{S})}(x,t) \leq c_3|x|^2$$

for all $x \in R^n$ and $t \in J$, where $c_1 = \min_i\{\alpha_i c_{i1}\}$, $c_2 = \max_i\{\alpha_i c_{i2}\}$, and $c_3 = \lambda_M(S) < 0$. The conclusion of the corollary follows from the remark following Theorem 2.2.26.   ■

For composite system $(\mathscr{S})$ with decomposition specified by Eq. (2.3.16), we have the following result.

2.4.25. Corollary. The equilibrium $x = 0$ of composite system $(\mathscr{S})$ with decomposition (2.3.16) is **exponentially stable in the large** if the following conditions are satisfied.
   (i) Each isolated subsystem $(\mathscr{S}_i)$ possesses Property B′;
   (ii) given v_i of hypothesis (i), there exists a positive constant c_{i4} such that

$$|\nabla v_i(z_i,t)| \leq c_{i4}|z_i|$$

for all $z_i \in R^{n_i}$;
   (iii) for each $i, j = 1, \dots, l$, $i \neq j$, there exists a constant $k_{ij} > 0$ such that

$$|g_{ij}(z_j,t)| \leq k_{ij}|z_j|$$

for all $z_j \in R^{n_j}$ and $t \in J$; and
   (iv) given σ_i in hypothesis (i), there exists an l-vector $\alpha^T = (\alpha_1, \dots, \alpha_l)^T > 0$ such that the test matrix $S = [s_{ij}]$ specified by

$$s_{ij} = \begin{cases} \alpha_i\sigma_i, & i = j \\ (\alpha_i c_{i4}k_{ij} + \alpha_j c_{j4}k_{ji})/2, & i \neq j \end{cases}$$

is negative definite.

Proof. Given v_i in hypothesis (i) and $\alpha > 0$ in hypothesis (iv), we choose for system $(\mathscr{S})$ the Lyapunov function

$$v(x,t) = \sum_{i=1}^{l} \alpha_i v_i(z_i,t).$$

From hypothesis (i) it follows that

$$c_1|x|^2 \leq v(x,t) \leq c_2|x|^2 \tag{2.4.26}$$

for all $x \in R^n$ and $t \in J$, where $c_1 = \min_i\{\alpha_i c_{i1}\}$ and $c_2 = \max_i\{\alpha_i c_{i2}\}$.
Along solutions of system $(\mathscr{S})$ we have, taking hypotheses (i)–(iv) into

account

$$Dv_{(\mathscr{S})}(x,t) = \sum_{i=1}^{l} \left\{ \alpha_i \, Dv_{i(\mathscr{S}_i)}(z_i,t) + \alpha_i \, \nabla v_i(z_i,t)^{\mathrm{T}} \sum_{j=1,\,i\neq j}^{l} g_{ij}(z_j,t) \right\}$$

$$\leq \sum_{i=1}^{l} \left\{ \alpha_i \sigma_i |z_i|^2 + \alpha_i c_{i4} |\dot{z}_i| \sum_{j=1,\,i\neq j}^{l} k_{ij}|z_j| \right\}$$

$$= w^{\mathrm{T}} R w = w^{\mathrm{T}}((R+R^{\mathrm{T}})/2)\,w = w^{\mathrm{T}} S w \leq \lambda_M(S)|x|^2$$

$$(2.4.27)$$

for all $x \in R^n$ and $t \in J$. Here, $w^{\mathrm{T}} = (|z_1|, ..., |z_l|)$, $\lambda_M(S) < 0$ is the largest eigenvalue of test matrix S given in hypothesis (iv), and $R = [r_{ij}]$ is specified by

$$r_{ij} = \begin{cases} \alpha_i \sigma_i, & i = j \\ \alpha_i c_{i4} k_{ij}, & i \neq j. \end{cases}$$

The conclusion of the theorem follows now from Theorem 2.2.26 and inequalities (2.4.26) and (2.4.27). ■

For composite system $(\mathscr{S})$ with decomposition specified by Eq. (2.3.14), the above result assumes the following form.

2.4.28. Corollary. The equilibrium $x = 0$ of composite system $(\mathscr{S})$ with decomposition (2.3.14) is **exponentially stable in the large** if the following conditions are satisfied.
(i)   Hypotheses (i) and (ii) of Corollary 2.4.25 hold;
(ii)   given σ_i of hypothesis (i), there exists an l-vector $\alpha^{\mathrm{T}} = (\alpha_1, ..., \alpha_l) > 0$ such that the test matrix $S = [s_{ij}]$ specified by

$$s_{ij} = \begin{cases} \alpha_i \sigma_i, & i = j \\ [\alpha_i c_{i4} \|C_{ij}\| + \alpha_j c_{j4} \|C_{ji}\|]/2, & i \neq j \end{cases}$$

is negative definite.

To prove Corollary 2.4.28, replace in hypothesis (iii) of Corollary 2.4.25 $g_{ij}(z_j,t)$ by $C_{ij} z_j$ and k_{ij} by $\|C_{ij}\|$.
We now consider system $(\mathscr{S}'')$.

2.4.29. Theorem. The equilibrium $x = 0$ of composite system $(\mathscr{S}'')$ with decomposition (Σ_i'') is **exponentially stable in the large** if the following conditions hold.
(i)   There exist continuously differentiable functions $v_i \colon R^{n_i} \times J \to R$ and constants $c_{i2} \geq c_{i1} > 0$, $i = 1, ..., l$, such that

$$c_{i1}|z_i|^2 \leq v_i(z_i,t) \leq c_{i2}|z_i|^2$$

for all $z_i \in R^{n_i}$ and $t \in J$;

(ii)   given v_i of hypothesis (i), there exist constants $a_{ij} \in R$, $i, j = 1, ..., l$, such that

$$\nabla v_i(z_i, t)^{\mathrm{T}} h_i(x, t) + \partial v_i(z_i, t)/\partial t \leq |z_i| \sum_{j=1}^{l} a_{ij} |z_j|$$

for all $z_i \in R^{n_i}$, $i = 1, ..., l$, $x \in R^n$, and $t \in J$;

(iii)   there exists a vector $\alpha^{\mathrm{T}} = (\alpha_1, ..., \alpha_l) > 0$ such that the matrix $S = [s_{ij}]$ specified by

$$s_{ij} = \begin{cases} \alpha_i a_{ii}, & i = j \\ (\alpha_i a_{ij} + \alpha_j a_{ji})/2, & i \neq j \end{cases}$$

is negative definite.

The proof of this theorem follows along similar lines as the proof of Theorem 2.4.16.

C. Instability and Complete Instability

In our next definition we let $B_i(r_i) = \{z_i \in R^{n_i} : |z_i| < r_i\}$ for some $r_i > 0$.

2.4.30. Definition. For isolated subsystem $(\mathscr{S}_i)$, let there exist a continuously differentiable function $v_i : B_i(r_i) \times J \to R$, three functions $\psi_{i1}, \psi_{i2}, \psi_{i3} \in K$, and constants $\delta_{i1}, \delta_{i2}, \sigma_i \in R$ such that

$$\delta_{i1} \psi_{i1}(|z_i|) \leq v_i(z_i, t) \leq \delta_{i2} \psi_{i2}(|z_i|), \qquad Dv_{i(\mathscr{S}_i)}(z_i, t) \leq \sigma_i \psi_{i3}(|z_i|)$$

for all $z_i \in B_i(r_i)$ and $t \in J$. If $\delta_{i1} = \delta_{i2} = -1$ we say that $(\mathscr{S}_i)$ possesses **Property C**. If $\delta_{i1} = \delta_{i2} = 1$ we say that $(\mathscr{S}_i)$ possesses **Property A″**.

If $(\mathscr{S}_i)$ has Property C with $\sigma_i < 0$, then the equilibrium $z_i = 0$ is completely unstable. If $(\mathscr{S}_i)$ has Property A″ and $\sigma_i < 0$, the equilibrium of $(\mathscr{S}_i)$ is uniformly asymptotically stable.

2.4.31. Theorem. Let $N \neq \varnothing$, $N \subset L = \{1, ..., l\}$. Assume that for composite system $(\mathscr{S})$ with decomposition (Σ_i) the following conditions hold.

(i)   If $i \in N$, then isolated subsystem $(\mathscr{S}_i)$ has Property C and if $i \notin N$, $i \in L$, then $(\mathscr{S}_i)$ has Property A″;

(ii)   given v_i of hypothesis (i), there exist constants $a_{ij} \in R$ such that

$$\nabla v_i(z_i, t)^{\mathrm{T}} g_i(z_1, ..., z_l, t) \leq [\psi_{i3}(|z_i|)]^{1/2} \sum_{j=1}^{l} a_{ij} [\psi_{j3}(|z_j|)]^{1/2}$$

for all $z_i \in B_i(r_i)$, $z_j \in B_j(r_j)$, $i, j \in L$, and $t \in J$;

(iii)   given σ_i of hypothesis (i), there exists an l-vector $\alpha^{\mathrm{T}} = (\alpha_1, ..., \alpha_l) > 0$

such that the test matrix $S = [s_{ij}]$ specified by

$$s_{ij} = \begin{cases} \alpha_i(\sigma_i + a_{ii}), & i = j \\ (\alpha_i a_{ij} + \alpha_j a_{ji})/2, & i \neq j \end{cases}$$

is negative definite.

(a) If $N \neq L$, then the equilibrium $x = 0$ of composite system $(\mathscr{S})$ is **unstable**.

(b) If $N = L$, the equilibrium of $(\mathscr{S})$ is **completely unstable**.

Proof. Given v_i of hypothesis (i) and α of hypothesis (iii) we choose

$$v(x, t) = \sum_{i=1}^{l} \alpha_i v_i(z_i, t).$$

Proceeding as in the proof of Theorem 2.4.2, we obtain

$$Dv_{(\mathscr{S})}(x, t) \leq \lambda_M(S) \sum_{i=1}^{l} \psi_{i3}(|z_i|)$$

for all $z_i \in B_i(r_i)$, $i \in L$, and $t \in J$. Since test matrix S is negative definite, we have $\lambda_M(S) < 0$ and $Dv_{(\mathscr{S})}(x, t)$ is negative definite.

Now consider the set

$$D = \{(x, t) \in R^n \times J : z_i \in B_i(r) \text{ whenever } i \in N, \text{ where } r < \min r_i,$$
$$\text{and } z_i = 0 \text{ whenever } i \notin N, \text{ and } t \in J\}.$$

For $(x, t) \in D$ we have

$$-\sum_{i \in N} \alpha_i \psi_{i1}(|z_i|) \leq v(x, t) \leq -\sum_{i \in N} \alpha_i \psi_{i2}(|z_i|)$$

i.e., in every neighborhood of the origin $x = 0$, there is at least one point $x' \neq 0$ for which $v(x', t) < 0$ for all $t \in J$. Furthermore, on the set D, $v(x, t)$ is bounded from below. Thus, all conditions of Theorem 2.2.28 are satisfied. If $N \neq L$, then the equilibrium $x = 0$ of composite system $(\mathscr{S})$ is unstable. If $N = L$, then $v(x, t)$ is negative definite and the equilibrium of $(\mathscr{S})$ is completely unstable. ∎

In the next result we consider composite system $(\mathscr{S}'')$.

2.4.32. Theorem. Let $L = \{1, ..., l\}$, $N \subset L$, and $N \neq \varnothing$. Assume that for composite system $(\mathscr{S}'')$ with decomposition (Σ_i'') the following conditions hold.

(i) There exist continuously differentiable functions $v_i : B_i(r_i) \times J \to R$ and $\psi_{i1}, \psi_{i2} \in K$, $i \in L$, such that

$$\psi_{i1}(|z_i|) \leq v_i(z_i, t) \leq \psi_{i2}(|z_i|), \quad i \notin N, \quad i \in L,$$

and

$$-\psi_{i1}(|z_i|) \le v_i(z_i, t) \le -\psi_{i2}(|z_i|), \qquad i \in N,$$

for all $z_i \in B_i(r_i)$ for some $r_i > 0$, $i \in L$, and all $t \in J$;

(ii)   given v_i of hypothesis (i), there exist $a_{ij} \in R$ and $\psi_{i4} \in K$, $i, j \in L$, such that

$$\nabla v_i(z_i, t)^T h_i(x, t) + \partial v_i(z_i, t)/\partial t \le \psi_{i4}(|z_i|) \sum_{j=1}^{l} a_{ij} \psi_{j4}(|z_j|)$$

for all $z_i \in B_i(r_i)$, $z_j \in B_j(r_j)$, $i, j \in L$, and $t \in J$; and

(iii)   there exists a vector $\alpha^T = (\alpha_1, \ldots, \alpha_l) > 0$ such that the matrix $S = [s_{ij}]$ given by

$$s_{ij} = \begin{cases} \alpha_i a_{ii}, & i = j \\ (\alpha_i a_{ij} + \alpha_j a_{ji})/2, & i \ne j \end{cases}$$

is negative definite.

(a)   If $N \ne L$, then the equilibrium $x = 0$ of composite system $(\mathscr{S}'')$ is **unstable**.

(b)   If $N = L$, the equilibrium of $(\mathscr{S}'')$ is **completely unstable**.

The proof of this theorem involves obvious modifications of the proofs of Theorems 2.4.16 and 2.4.31.

D. Uniform Boundedness and Uniform Ultimate Boundedness

And now we consider the boundedness of solutions of system $(\mathscr{S})$.

2.4.33. Definition. Isolated subsystem $(\mathscr{S}_i)$ possesses **Property D** if there exists a continuously differentiable function $v_i: R^{n_i} \times J \to R$, $\psi_{i1}, \psi_{i2}, \psi_{i3} \in KR$ and a constant $\sigma_i \in R$ such that

$$\psi_{i1}(|z_i|) \le v_i(z_i, t) \le \psi_{i2}(|z_i|), \qquad Dv_{i(\mathscr{S}_i)}(z_i, t) \le \sigma_i \psi_{i3}(|z_i|)$$

for all $t \in J$ and for all $|z_i| \ge R_i$ (where R_i may be large) and such that $v_i(z_i, t)$ and $Dv_{i(\mathscr{S}_i)}(z_i, t)$ are bounded on $B_i(R_i) \times J$.

If $\sigma_i \le 0$, then the solutions of subsystem $(\mathscr{S}_i)$ are uniformly bounded and if $\sigma_i < 0$, then the solutions of $(\mathscr{S}_i)$ are uniformly ultimately bounded.

2.4.34. Theorem. The solutions of composite system $(\mathscr{S})$ with decomposition (Σ_i) are **uniformly bounded**, in fact, **uniformly ultimately bounded** if the following conditions are satisfied.

(i)   Each isolated subsystem $(\mathscr{S}_i)$ possesses Property D;

(ii)   given v_i and ψ_{i3} of hypothesis (i), there exist constants $a_{ij} \in R$ such that

$$\nabla v_i(z_i, t)^T g_i(z_1, \ldots, z_l, t) \le [\psi_{i3}(|z_i|)]^{1/2} \sum_{j=1}^{l} a_{ij} [\psi_{j3}(|z_j|)]^{1/2}$$

for all $z_i \in R^{n_i}$, $z_j \in R^{n_j}$, $i, j = 1, ..., l$, and $t \in J$; and

(iii)   given σ_i of hypothesis (i), there exists an *l*-vector $\alpha^T = (\alpha_1, ..., \alpha_l) > 0$ such that the matrix $S = [s_{ij}]$ specified by

$$s_{ij} = \begin{cases} \alpha_i(\sigma_i + a_{ii}), & i = j \\ (\alpha_i a_{ij} + \alpha_j a_{ji})/2, & i \neq j \end{cases}$$

is negative definite.

Proof. Given the functions v_i of hypothesis (i) and the vector α of hypothesis (iii), we choose

$$v(x, t) = \sum_{i=1}^{l} \alpha_i v_i(z_i, t).$$

Following the proof of Theorem 2.4.2 it is clear that there exist $\psi_1, \psi_2, \psi_3 \in KR$ such that

$$\psi_1(|x|) \leq v(x, t) \leq \psi_2(|x|), \qquad Dv_{(\mathscr{S})}(x, t) \leq \lambda_M(S)\psi_3(|x|)$$

where $\lambda_M(S) < 0$, whenever $x \in R^n - B_1(R_1) \times \cdots \times B_l(R_l)$ and $t \in J$.

To complete the proof we need to consider the situation where some of the z_i are such that $|z_i| < R_i$.

First consider the case where $|z_i| \geq R_i$, $i = 1, ..., r$, and $|z_i| < R_i$, $i = r+1, ..., l$. Then

$$\sum_{i=r+1}^{l} \alpha_i v_i(z_i, t) + \sum_{i=1}^{r} \alpha_i \psi_{i1}(|z_i|) \leq v(x, t) \leq \sum_{i=1}^{r} \alpha_i \psi_{i2}(|z_i|) + \sum_{i=r+1}^{l} \alpha_i v_i(z_i, t).$$

Since v_i is continuous on $R^{n_i} \times J$ and bounded on $B_i(R_i) \times J$, $i = 1, ..., l$, it follows that there are $\varphi_1, \varphi_2 \in KR$ such that $\varphi_1(|x|) \leq v(x, t) \leq \varphi_2(|x|)$ for all $t \in J$ and for all $x \in R^n$ such that $|z_i| < R_i$, $i = r+1, ..., l$, and such that $|z_i|$, $i = 1, ..., r$, are sufficiently large. Along solutions of $(\mathscr{S})$ we have

$$Dv_{(\mathscr{S})}(x, t) = \sum_{i=1}^{r} \alpha_i \{ Dv_{i(\mathscr{S}_i)}(z_i, t) + \nabla v_i(z_i, t)^T g_i(z_1, ..., z_l, t) \}$$

$$+ \sum_{i=r+1}^{l} \alpha_i \{ Dv_{i(\mathscr{S}_i)}(z_i, t) + \nabla v_i(z_i, t)^T g_i(z_1, ..., z_l, t) \}$$

$$\leq \sum_{i=1}^{r} \alpha_i \sigma_i \psi_{i3}(|z_i|) + \sum_{i=1}^{r} \alpha_i [\psi_{i3}(|z_i|)]^{1/2} \sum_{j=1}^{r} a_{ij} [\psi_{j3}(|z_j|)]^{1/2}$$

$$+ \sum_{i=1}^{r} \alpha_i [\psi_{i3}(|z_i|)]^{1/2} \sum_{j=r+1}^{l} a_{ij} [\psi_{j3}(|z_j|)]^{1/2}$$

$$+ \sum_{i=r+1}^{l} \alpha_i Dv_{i(\mathscr{S}_i)}(z_i, t)$$

$$+ \sum_{i=r+1}^{l} \alpha_i [\psi_{i3}(|z_i|)]^{1/2} \sum_{j=1}^{r} a_{ij} [\psi_{j3}(|z_j|)]^{1/2}$$

$$+ \sum_{i=r+1}^{l} \alpha_i [\psi_{i3}(|z_i|)]^{1/2} \sum_{j=r+1}^{l} a_{ij} [\psi_{j3}(|z_j|)]^{1/2}.$$

Now for all $|z_i| < R_i$, $i = r+1, \ldots, l$, there exist constants K_1, K_2, K_3, and M_i such that

$$K_1 \geq \sum_{j=r+1}^{l} |a_{ij}| [\psi_{j3}(|z_j|)]^{1/2}, \qquad M_i \geq |Dv_{i(\mathcal{S}_i)}(z_i, t)|,$$

$$K_2 \geq \sum_{i=r+1}^{l} \alpha_i [\psi_{i3}(|z_i|)]^{1/2},$$

$$K_3 \geq \sum_{i=r+1}^{l} \alpha_i [\psi_{i3}(|z_i|)]^{1/2} \sum_{j=r+1}^{l} a_{ij} [\psi_{j3}(|z_j|)]^{1/2}.$$

Let $w^T = [\psi_{13}(|z_1|), \ldots, \psi_{r3}(|z_r|)]$, let the $r \times r$ matrix $R = [r_{ij}]$ be specified by

$$r_{ij} = \begin{cases} \alpha_i(\sigma_i + a_{ii}), & i = j \\ \alpha_i a_{ij}, & i \neq j, \end{cases}$$

and let $\tilde{S} = (R + R^T)/2$. Then

$$Dv_{(\mathcal{S})}(x, t) \leq w^T \tilde{S} w + K_1 \sum_{i=1}^{r} \alpha_i [\psi_{i3}(|z_i|)]^{1/2} + \sum_{i=r+1}^{l} \alpha_i M_i$$

$$+ K_2 \sum_{j=1}^{r} |a_{ij}| [\psi_{j3}(|z_j|)]^{1/2} + K_3.$$

Since the test matrix S is negative definite by hypothesis (iii) so is $\tilde{S}$ and $\lambda_M(\tilde{S}) < 0$. We have

$$Dv_{(\mathcal{S})}(x, t) \leq \lambda_M(\tilde{S}) \sum_{i=1}^{r} \psi_{i3}(|z_i|) + K_1 \sum_{i=1}^{r} \alpha_i [\psi_{i3}(|z_i|)]^{1/2}$$

$$+ \sum_{i=r+1}^{l} \alpha_i M_i + K_2 \sum_{j=1}^{r} |a_{ij}| [\psi_{j3}(|z_j|)]^{1/2} + K_3.$$

Therefore, if the $|z_i|$ for $i = 1, \ldots, r$ are sufficiently large, the sign of $Dv_{(\mathcal{S})}(x, t)$ is determined by $\lambda_M(\tilde{S}) \sum_{i=1}^{r} \psi_{i3}(|z_i|)$, and $Dv_{(\mathcal{S})}(x, t) < 0$ for all $t \in J$.

Above we have assumed that $|z_i| \geq R_i$, $i = 1, \ldots, r$, and $|z_i| < R_i$, $i = r+1, \ldots, l$. For any other combination of indices, the proof is similar. ∎

E. Discussion

At this point, a few remarks concerning the preceding results are in order.

In all results for composite systems $(\mathcal{S})$ and $(\mathcal{S}')$, the analysis is accomplished in terms of the qualitative properties of the lower order isolated subsystems $(\mathscr{S}_i)$ and $(\mathscr{S}_i')$, respectively, and in terms of the properties of the interconnecting structure expressed by $g_i(x, t)$ and $g_i(x)$, respectively. These results involve generally a reduction in dimension (from n to l) and are in such a form as to reveal qualitative trade-off effects between subsystems and interconnecting structure. In the case of composite system $(\mathscr{S}'')$, a reduction in dimension is also realized, however, the results do not yield qualitative trade-off information between isolated subsystems and interconnecting structure, for in this case the notion of isolated subsystem is undefined.

For systems with a fine decomposition (i.e., a decomposition into many subsystems) it is as a rule easy to find Lyapunov functions for the lower order subsystems; however, the resulting stability (or instability or bounded- ness) conditions tend to be conservative. (Note however, that in the case of Corollary 2.4.24 involving the finest possible decomposition, the conserva- tism can be reduced appreciably.) On the other hand, in the case of systems with a coarse decomposition (i.e., a decomposition into few subsystems) it is usually more difficult to find Lyapunov functions for the subsystems (which may no longer be of low order); however, the resulting stability conditions are usually less conservative. This points to the following advantages (items (a) and (b) below) and disadvantages (item (c)) of the present method.

(a) It is often possible to circumvent difficulties that arise when the Lyapunov method is applied to systems of high dimension and intricate interconnecting structure.

(b) In the case of systems $(\mathscr{S})$ and $(\mathscr{S}')$ the analysis is accomplished in terms of system components and system structure.

(c) If a system is decomposed into many subsystems, the results may be conservative.

Theorems 2.4.16, 2.4.29, and 2.4.32 are of course also applicable to com- posite system $(\mathscr{S})$, since this system is a special case of system $(\mathscr{S}'')$. This demonstrates that the decomposition of large scale systems into *isolated sub- systems* is a conceptual convenience and not a necessity. Indeed, Theorems 2.4.16, 2.4.29, and 2.4.32 may in principle yield less conservative conditions than corresponding results involving isolated subsystems, when applied to a given problem, for the former set of results involves fewer majorizations (estimates by comparison functions) than the latter set. However, it is usually more difficult to apply the former than the latter.

Of all the preceding results, Corollaries 2.4.25 and 2.4.28 are perhaps the easiest to apply. The reason being that these results pertain to systems for which a great deal of information of the interconnecting structure is available. This suggests that the method of analysis advanced herein is in a sense much more important than the specific results presented. That is to say, given a specific system with special structure, it may often be more desirable to arrive at a result tailored to this particular system, using the present method of analysis, rather than try to force this system into one of the specific forms considered in Section 2.3. We will demonstrate this further in Section 2.8.

In all results presented thus far we require the existence of a vector $\alpha^T = (\alpha_1, ..., \alpha_l) > 0$ which arises in the choice of Lyapunov functions employed. These v-functions are always of the form

$$v(\cdot) = \sum_{i=1}^{l} \alpha_i v_i(\cdot).$$

The motivation for choosing functions of this type is as follows. We may think of v as representing a measure of energy associated with a given composite system and we may view v_i as providing a measure of energy for isolated subsystem $(\mathscr{S}_i)$. Once the isolated subsystems are interconnected to form the composite system, it is reasonable to assume that the qualitative effects of different subsystems on the overall system may vary in importance. Therefore it seems reasonable to assign different *weights* $\alpha_i > 0$ to different functions v_i, thus forming the indicated *weighted sum* of Lyapunov functions which serves as a Lyapunov function for the overall interconnected system. It is of course possible to combine the weighting factors α_i with the respective functions v_i. In this case the test matrices S are modified, replacing each α_i by 1. However, there are the following two good reasons for not doing this.

(a)   In applications to specific problems, the presence of α provides us with an added degree of flexibility, and in addition, judicious choice of α enables us frequently to reduce the conservatism of the results.

(b)   In the *special case* when the off-diagonal elements of S are non-negative, the presence of α will enable us to obtain results which are equivalent to the present ones, however, are much easier to apply. This will be accomplished in the next section.

In physical large systems, it is usually true that a given subsystem $(\mathscr{S}_k)$ is not connected to every other subsystem $(\mathscr{S}_i)$, $i = 1, ..., l$, $i \neq k$. Thus, in applications, matrix S is usually sparse and it may be possible to apply linear programming methods (see Dantzig [1]) and sparse matrix results (see Tewarson [1]) to determine the optimal choice of α. On the other hand, when the order of S is small, the optimal choice of α is usually obvious.

The concept which gave rise to the qualitative analysis of dynamical systems at different hierarchical levels is the notion of **vector Lyapunov function**, introduced by Bellman [2]. In a certain sense, the results of this section may be viewed as analysis via vector Lyapunov functions. Specifically, if $V^T = (v_1, ..., v_l) \in R^l$ and $\alpha^T = (\alpha_1, ..., \alpha_l) \in R^l$, $\alpha_i > 0$, $i = 1, ..., l$, then

$$v(\cdot) = \alpha^T V = \sum_{i=1}^{l} \alpha_i v_i(\cdot).$$

Nevertheless, $v(\cdot)$ is a scalar Lyapunov function. We shall reserve the term "vector Lyapunov approach" to cases when an appropriate comparison principle is applied to a differential inequality (of order l) involving vector Lyapunov functions. This type of approach will be considered in Section 2.6.

To simplify matters, we consider in the following discussion only Theorem 2.4.2. However, similar statements apply to the remaining results as well. We call the parameter σ_i, introduced in Definition 2.4.1, a **degree** or a **margin of stability**. In order to satisfy hypothesis (iii) of Theorem 2.4.2, it is necessary that $(\sigma_i + a_{ii}) < 0$, $i = 1, ..., l$. Thus, if $\sigma_i > 0$, so that $(\mathscr{S}_i)$ may be unstable, we require that $a_{ii} < 0$ and $|a_{ii}| > \sigma_i$. In other words, this theorem is applicable to composite systems with unstable subsystems, provided that there exists a sufficient amount of stabilizing feedback (i.e., **local negative feedback**) associated with $(\mathscr{S}_i)$, which is not an integral part of $(\mathscr{S})$. Stability results of the type presented herein for composite systems with unstable subsystems and *without* local stabilizing feedback have apparently not been established at this time. This problem is of great practical importance and needs to be pursued further. (Results for systems with unstable subsystems are given in Thompson [2]. However, they do not involve a reduction in dimension, and as such, this problem may be regarded as being essentially unsolved.) Generally speaking, the greater the margin of stability associated with each subsystem (i.e., the more negative the terms $(\sigma_i + a_{ii})$, $i = 1, ..., l$), the easier it is to satisfy the negative definiteness requirement of matrix S. Note that as in the case of weights α_i, it is of course possible to combine σ_i with ψ_{i3}, replacing σ_i by $-1, 0,$ or 1, when $(\mathscr{S}_i)$ is uniformly asymptotically stable, uniformly stable, or possibly unstable, respectively. We emphasize that the off-diagonal terms of test matrix S can have arbitrary signs. In the next section we consider additional results for which the off-diagonal terms of the test matrix are required to possess the same sign. Note also that in this section the test matrices are always symmetric. This will in general not be required in the results of the next section. Finally note that inequalities of the type encountered in hypothesis (ii) of Theorem 2.4.2, which express restraints on the interaction among the subsystems, are more easily satisfied than may appear at first glance. For example, hypothesis (ii) of Theorem 2.4.2 can be satisfied for appropriately chosen a_{ij} if $g_i(x, t)$ are linear functions and $v_i(z_i, t)$ are quadratic, $i = 1, ..., l$.

The results of this section can often be utilized in compensation and stabilization procedures at different hierarchical levels. To simplify matters, we consider once more Theorem 2.4.2. Similar statements can be made for the remaining results. We choose in hypothesis (iii) of this theorem $\alpha_i = 1$, $i = 1, ..., l$. It can be shown (see, e.g., Taussky [1]) that all eigenvalues of matrix S are negative if the **diagonal dominance conditions**

$$-(\sigma_{ii} + a_{ii}) > \sum_{j=1, i \neq j}^{l} |a_{ij} + a_{ji}|/2, \qquad i = 1, ..., l,$$

hold. Now suppose we are able to ascertain by some means (such as Theorem 2.4.31) that a specific system $(\mathscr{S})$ is unstable. Or suppose we are able to determine by Theorem 2.4.2 that $(\mathscr{S})$ is uniformly asymptotically stable, however, the degree of stability of $(\mathscr{S})$, as expressed by $\lambda_M(S)$, is unsatisfactory. Using appropriate local feedback associated with the subsystems and the interconnecting structure, it is frequently possible to stabilize $(\mathscr{S})$, or to enhance the stability of $(\mathscr{S})$. In this process, the local compensators are chosen so as to increase the terms $-(\sigma_{ii}+a_{ii})$ and decrease the terms $|a_{ij}+a_{ji}|$. However, it must be noted that this procedure may often yield overly conservative results. Furthermore, it may physically be impossible or undesirable to use this type of an approach..

We conclude our discussion by noting that all preceding results involving test matrices with off-diagonal terms having arbitrary signs are applicable to **strongly coupled systems** as well as **weakly coupled systems**.

2.5   Application of M-Matrices

In the *special case* when the off-diagonal elements of the test matrices S in the results of Section 2.4 are nonnegative, we can utilize the properties of **Minkowski matrices,** called **M-matrices,** to establish results which are easier to apply because they do not involve usage of weighting vectors α. However, we are obliged to emphasize at the outset that (a) in the case the off-diagonal elements of S are nonnegative, the results of the preceding section and corresponding results of this section are equivalent, (b) whereas it is always possible to obtain the results of this section from those of the preceding section, the converse is in general not true, and (c) the results of the present section are generally applicable only to weakly coupled systems.

This section consists of six parts. In the first of these we give a summary of selected results from the theory of M-matrices. Since this material is well covered in several references, all proofs of M-matrix results are omitted. In the remaining parts we establish several Lyapunov results for composite systems $(\mathscr{S})$ and $(\mathscr{S}'')$, most of which have corresponding counterparts in Section 2.4.

A. M-Matrices

We begin with the following definition.

2.5.1. Definition. A real $l \times l$ matrix $D = [d_{ij}]$ is said to be an **M-matrix** if $d_{ij} \le 0$, $i \ne j$ (i.e., all off-diagonal elements of D are nonpositive), and if all principal minors of D are positive (i.e., all principal minor determinants of D are positive).

Let $D = [d_{ij}]$ be an $l \times l$ matrix. If the determinants D_k, given by

$$D_k = \begin{vmatrix} d_{11} & \cdots & d_{1k} \\ \cdot & & \cdot \\ \cdot & \cdots & \cdot \\ \cdot & & \cdot \\ d_{k1} & \cdots & d_{kk} \end{vmatrix}, \qquad k = 1, \dots, l,$$

are all positive, we say that the "successive principal minors of D are positive," or, the "leading principal minors of D are positive."

2.5.2. Theorem. Let $D = [d_{ij}]$ be a real $l \times l$ matrix such that $d_{ij} \leq 0$ for all $i \neq j$. Then the following statements are equivalent.

(i)   The principal minors of D are all positive (i.e., D is an M-matrix).

(ii)   The successive principal minors of D are all positive.

(iii)   There is a vector $u \in R^l$ such that $u > 0$ and such that $Du > 0$.

(iv)   There is a vector $v \in R^l$ such that $v > 0$ and such that $D^Tv > 0$.

(v) D is nonsingular and all elements of D^{-1} are nonnegative (in fact, all diagonal elements of D^{-1} are positive).

(vi)   The real parts of all eigenvalues of D are positive.

A direct consequence of parts (iii) and (iv) of Theorem 2.5.2 is the following.

2.5.3. Corollary. Let $D = [d_{ij}]$ be an $l \times l$ matrix with nonpositive off-diagonal elements. Then the following statements are true.

(i) D is an M-matrix if and only if there exist positive constants λ_j, $j = 1, \dots, l$, such that

$$\sum_{j=1}^{l} \lambda_j d_{ij} > 0, \qquad i = 1, \dots, l. \tag{2.5.4}$$

(ii) D is an M-matrix if and only if there exist positive constants η_j, $j = 1, \dots, l$, such that

$$\sum_{j=1}^{l} \eta_j d_{ji} > 0, \qquad i = 1, \dots, l. \tag{2.5.5}$$

Additional useful properties of M-matrices which we will require and which are a consequence of Theorem 2.5.2 are the following.

2.5.6. Corollary. Let $D = [d_{ij}]$ be an $l \times l$ matrix with nonpositive off-diagonal elements. Then D is an M-matrix if and only if there exists a diagonal matrix $A = \text{diag}[\alpha_1, \dots, \alpha_l]$, $\alpha_i > 0$, $i = 1, \dots, l$, such that the matrix

$$B = AD + D^T A \tag{2.5.7}$$

is positive definite.

2.5.8. Corollary. Let $C = \text{diag}[c_1, \dots, c_l] > 0$ be a diagonal $l \times l$ matrix and let $Q = [q_{ij}] \geq 0$ be an $l \times l$ matrix. Then $C - Q$ is an M-matrix if and only if there is a diagonal $l \times l$ matrix $A = \mathrm{diag}[\alpha_1, ..., \alpha_l] > 0$ such that the matrix

$$B = CAC - Q^{\mathrm{T}}AQ \qquad (2.5.9)$$

is positive definite.

2.5.10. Corollary. If D is an M-matrix, then $D - \mu I$ is an M-matrix if and only if $\mu < \min_{i \in L} \mathrm{Re}[\lambda_i(D)]$, $L = \{1, ..., l\}$, where I is the $l \times l$ identity matrix and $\mathrm{Re}[\lambda_i(D)]$ denotes the real part of the ith eigenvalue of D.

B. Uniform Asymptotic Stability

Our first result is as follows.

2.5.11. Theorem. Assume that for composite system $(\mathscr{S})$ with decomposition (Σ_i) the following conditions hold.

(i)   Each isolated subsystem $(\mathscr{S}_i)$ possesses Property A (see Definition 2.4.1); and

(ii) given v_i and ψ_{i3} of hypothesis (i), there exist constants $a_{ij} \in R$ such that

(a) $a_{ij} \geq 0$ for all $i \neq j$, and

(b) $\nabla v_i(z_i, t)^{\mathrm{T}} g_i(z_1, ..., z_l, t) \leq [\psi_{i3}(|z_i|)]^{1/2} \sum_{j=1}^{l} a_{ij} [\psi_{j3}(|z_j|)]^{1/2}$ for all $z_i \in R^{n_i}$, $z_j \in R^{n_j}$, $i, j = 1, ..., l$, and $t \in J$.

The equilibrium $x = 0$ of system $(\mathscr{S})$ is **uniformly asymptotically stable in the large** if *any one* of the following conditions hold.

(iii) given σ_i of hypothesis (i), the successive principal minors of the $l \times l$ test matrix $D = [d_{ij}]$ are all positive, where

$$d_{ij} = \begin{cases} -(\sigma_i + a_{ii}), & i = j \\ -a_{ij}, & i \neq j. \end{cases}$$

(iv)   The real parts of the eigenvalues of D are all positive.

(v)   There exist positive constants λ_i, $i = 1, ..., l$, such that

$$-(\sigma_i + a_{ii}) - \sum_{j=1, i \neq j}^{l} (\lambda_j/\lambda_i) a_{ij} > 0, \qquad i = 1, ..., l. \qquad (2.5.12)$$

(vi)   There exist positive constants η_i, $i = 1, ..., l$, such that

$$-(\sigma_i + a_{ii}) - \sum_{j=1, i \neq j}^{l} (\eta_j/\eta_i) a_{ji} > 0, \qquad i = 1, ..., l. \qquad (2.5.13)$$

Proof. Since by assumption $a_{ij} \geq 0$ for all $i \neq j$, and since the successive principal minors of matrix D are all positive, it follows from part (ii) of Theorem 2.5.2 that D is an M-matrix. In view of Corollary 2.5.6 there exists a diagonal matrix $A = \text{diag}[\alpha_1, ..., \alpha_l] > 0$ such that the matrix

$$-2S \triangleq AD + D^{\mathrm{T}}A$$

is positive definite. Here, $-2S = -2[s_{ij}]$ is specified by

$$s_{ij} = \begin{cases} \alpha_i(\sigma_i + a_{ii}), & i = j \\ (\alpha_i a_{ij} + \alpha_j a_{ji})/2, & i \neq j. \end{cases}$$

But if $-2S$ is positive definite, then the matrix $S = [s_{ij}]$ is negative definite. Therefore, if hypotheses (i)–(iii) of the present theorem are satisfied, then all hypotheses of Theorem 2.4.2 are also satisfied. Hence, the equilibrium $x = 0$ of system $(\mathscr{S})$ is uniformly asymptotically stable in the large. The proof is now complete, for hypotheses (iii)–(vi) are equivalent statements (see Corollary 2.5.3 and part (vi) of Theorem 2.5.2). ∎

Henceforth we shall refer to inequalities (2.5.12) and (2.5.13) as **row dominance condition** and **column dominance condition**, respectively.

In all results of the previous and present section, we have used Lyapunov functions which are continuously differentiable. In our next theorem, we make use of continuous Lyapunov functions which need not be continuously differentiable. In doing so, we employ the following convention.

2.5.14. Definition. Isolated subsystem $(\mathscr{S}_i)$ possesses **Property Ã** if there exists a continuous function $v_i: R^{n_i} \times J \to R$, $\psi_{i1}, \psi_{i2} \in KR$, $\psi_{i3} \in K$, constants $L_i, \sigma_i \in R$, $L_i > 0$, such that

$$\psi_{i1}(|z_i|) \leq v_i(z_i, t) \leq \psi_{i2}(|z_i|),$$

$$Dv_{i(\mathscr{S}_i)}(z_i, t) = \lim_{h \to 0^+} \sup(1/h)\{v_i[z_i(t+h; z_i, t), t+h] - v_i(z_i, t)\}$$

$$\leq \sigma_i \psi_{i3}(|z_i|),$$

and

$$|v_i(z_i', t) - v_i(z_i'', t)| \leq L_i|z_i' - z_i''|$$

for all $z_i, z_i', z_i'' \in R^{n_i}$ and $t \in J$.

2.5.15. Theorem. The equilibrium $x = 0$ of composite system $(\mathscr{S})$ with decomposition (Σ_i) is **uniformly asymptotically stable in the large** if the following conditions are satisfied.

(i)   Each isolated subsystem $(\mathscr{S}_i)$ possesses Property Ã;

(ii)   given ψ_{i3} of hypothesis (i), there exist constants $a_{ij} \geq 0$, $i, j = 1, ..., l$, such that

$$|g_i(z_1, ..., z_l, t)| \leq \sum_{j=1}^{l} a_{ij} \psi_{j3}(|z_j|), \qquad i = 1, ..., l,$$

for all $z_i \in R^{n_i}$, $i = 1, ..., l$, and $t \in J$; and

(iii)   given σ_i and L_i of hypothesis (i), the successive principal minors of the $l \times l$ test matrix $D = [d_{ij}]$ are all positive, where

$$d_{ij} = \begin{cases} -(\sigma_i + L_i a_{ii}), & i = j \\ -L_i a_{ij}, & i \neq j. \end{cases}$$

Proof. Given v_i of hypothesis (i), let $\alpha^T = (\alpha_1, \ldots, \alpha_l) > 0$ be an arbitrary constant vector and choose as a Lyapunov function for system $(\mathscr{S})$,

$$v(x, t) = \sum_{i=1}^{l} \alpha_i v_i(z_i, t).$$

From hypothesis (i) it follows that v is continuous, positive definite, decrescent, and radially unbounded.

Along solutions of system $(\mathscr{S})$ we have

$$Dv_{(\mathscr{S})}(x, t) = \lim_{h \to 0^+} \sup(1/h) \left\{ \sum_{i=1}^{l} \alpha_i [v_i(z_i(t + h; z_i, t), t + h) - v_i(z_i, t)] \right\}$$

$$= \lim_{h \to 0^+} \sup(1/h) \sum_{i=1}^{l} \alpha_i \{ v_i[z_i + h \cdot f_i(z_i, t) + h \cdot g_i(x, t) + o(h), t + h]$$

$$- v_i(z_i, t) \}$$

$$= \lim_{h \to 0^+} \sup(1/h) \sum_{i=1}^{l} \alpha_i \{ v_i[z_i + h \cdot f_i(z_i, t) + o(h), t + h] - v_i(z_i, t)$$

$$+ v_i[z_i + h \cdot f_i(z_i, t) + h \cdot g_i(x, t) + o(h), t + h]$$

$$- v_i[z_i + h \cdot f_i(z_i, t) + o(h), t + h] \}$$

where $o(h)$ denotes higher order terms so that $o(h)/h \to 0$ as $h \to 0$. In view of hypotheses (i) and (ii) we now have

$$Dv_{(\mathscr{S})}(x, t) \leq \sum_{i=1}^{l} \alpha_i Dv_{i(\mathscr{S}_i)}(z_i, t) + \sum_{i=1}^{l} \alpha_i L_i |g_i(x, t)|$$

$$\leq \sum_{i=1}^{l} \alpha_i \left\{ \sigma_i \psi_{i3}(|z_i|) + L_i \sum_{j=1}^{l} a_{ij} \psi_{j3}(|z_j|) \right\}.$$

Now let $w^T = [\psi_{13}(|z_1|), \ldots, \psi_{13}(|z_l|)]$ and note that $w = 0$ if and only if $x = 0$. We have

$$Dv_{(\mathscr{S})}(x, t) \leq -\alpha^T Dw = -y^T w,$$

where $\alpha^T D = y^T$ and matrix D is defined in hypothesis (iii). Since D has positive successive principal minors and since $d_{ij} \leq 0$ for all $i \neq j$, it follows from parts (ii) and (v) of Theorem 2.5.2 that D is an M-matrix, that D^{-1}

exists and that $D^{-1} \geq 0$. Thus,

$$\alpha = (D^{-1})^{\mathsf{T}} y.$$

Since each row and column of D^{-1} must contain at least one nonzero element (in fact, the diagonal elements of D^{-1} are positive), we can always choose y in such a fashion that $y > 0$, so that $\alpha > 0$. Therefore,

$$Dv_{(\mathscr{S})}(x, t) \leq -y^{\mathsf{T}} w < 0, \qquad x \neq 0.$$

Hence, $Dv_{(\mathscr{S})}(x, t)$ is negative definite for all $x \in R^n$ and $t \in J$. It follows that the equilibrium $x = 0$ of composite system $(\mathscr{S})$ is uniformly asymptotically stable in the large.   ∎

By taking advantage of Corollary 2.5.3 and following the proof of Theorem 2.5.15, we can establish our next result.

2.5.16. Corollary. Assume that hypothesis (i) of Theorem 2.5.15 is satisfied with $\sigma_i \in R$ replaced by a continuous function $\sigma_i \colon R^{n_i} \times J \to R$. Assume that hypothesis (ii) of Theorem 2.5.15 is satisfied with $a_{ij} \in R$ replaced by a continuous function $a_{ij} \colon R^{n_j} \times J \to R$ with the property $a_{ij}(z_j, t) \geq 0$ for all $z_j \in R^{n_j}$ and all $t \in J$. Let $D(x, t) = [d_{ij}(z_j, t)]$ be determined by

$$d_{ij}(z_j, t) = \begin{cases} -[\sigma_i(z_i, t) + L_i a_{ii}(z_i, t)], & i = j \\ -L_i a_{ij}(z_j, t), & i \neq j. \end{cases}$$

If there exist constants $\lambda_i > 0$, $i = 1, \ldots, l$, and a constant $\gamma > 0$ such that

$$d_{jj}(z_j, t) - \sum_{i = 1, i \neq j}^{l} (\lambda_i / \lambda_j) |d_{ij}(z_j, t)| \geq \gamma, \qquad j = 1, \ldots, l,$$

for all $z_j \in R^{n_j}$, $j = 1, \ldots, l$, $t \in J$, then the equilibrium $x = 0$ of composite system $(\mathscr{S})$ is **uniformly asymptotically stable in the large**.

For system $(\mathscr{S}'')$ we have the following result.

2.5.17. Theorem. Assume that hypotheses (i) and (ii) of Theorem 2.4.16 are true. The equilibrium $x = 0$ of composite system $(\mathscr{S}'')$ with decomposition (Σ_i'') is **uniformly asymptotically stable in the large** if $a_{ij} \geq 0$ for all $i \neq j$ and if the successive principal minors of the test matrix $D = [d_{ij}]$, where $d_{ij} = -a_{ij}$, are all positive.

Proof. Using an argument similar to that of Theorem 2.5.11, the proof follows from Theorem 2.4.16 and Corollary 2.5.6.   ∎

C. Exponential Stability

Next, we consider several exponential stability results for composite systems $(\mathscr{S})$ and $(\mathscr{S}'')$.

2.5.18. Theorem. Assume that hypotheses (i)–(iii) of Theorem 2.4.20 are true. Then the equilibrium $x = 0$ of composite system $(\mathscr{S})$ with decomposition (Σ_i) is **exponentially stable in the large** if $a_{ij} \geq 0$ for all $i \neq j$ and if the successive principal minors of the test matrix $D = [d_{ij}]$ are all positive, where

$$d_{ij} = \begin{cases} -(\sigma_i + a_{ii}), & i = j \\ -a_{ij}, & i \neq j. \end{cases}$$

Proof. The proof follows from Theorem 2.4.20 and Corollary 2.5.6, following a similar procedure as in the proof of Theorem 2.5.11. ∎

2.5.19. Corollary. Assume that hypotheses (i) and (ii) of Corollary 2.4.23 are true and assume that $a_{ij} \geq 0$ for all $i \neq j$. The equilibrium $x = 0$ of system $(\mathscr{S})$ is **exponentially stable in the large** if the successive principal minors of the test matrix $D = [d_{ij}]$ are all positive, where

$$d_{ij} = \begin{cases} -(\sigma_i + a_{ii}), & i = j \\ -a_{ij}, & i \neq j. \end{cases}$$

Proof. The proof follows from Corollaries 2.4.23 and 2.5.6. ∎

The next two results are concerned with composite system $(\mathscr{S})$ having decompositions determined by Eqs. (2.3.16) and (2.3.14), respectively.

2.5.20. Corollary. Assume that hypotheses (i)–(iii) of Corollary 2.4.25 are true. The equilibrium $x = 0$ of system $(\mathscr{S})$ with decomposition (2.3.16) is **exponentially stable in the large** if the successive principal minors of the test matrix $D = [d_{ij}]$ are all positive, where

$$d_{ij} = \begin{cases} -\sigma_i, & i = j \\ -c_{i4} k_{ij}, & i \neq j. \end{cases}$$

Proof. The proof follows from Corollaries 2.4.25 and 2.5.6. ∎

2.5.21. Corollary. Assume that hypothesis (i) of Corollary 2.4.28 is true. Then the equilibrium $x = 0$ of system $(\mathscr{S})$ with decomposition (2.3.14) is **exponentially stable in the large** if the successive principal minors of the test matrix $D = [d_{ij}]$ are all positive, where

$$d_{ij} = \begin{cases} -\sigma_i, & i = j \\ -c_{i4} \| C_{ij} \|, & i \neq j. \end{cases}$$

Proof. This proof follows from Corollaries 2.4.28 and 2.5.6. ∎

In the next theorem we consider system $(\mathscr{S}'')$.

2.5.22. Theorem. Assume that hypotheses (i) and (ii) of Theorem 2.4.29 are true. Then the equilibrium $x = 0$ of composite system $(\mathscr{S}'')$ with decomposition (Σ_i'') is **exponentially stable in the large** if $a_{ij} \geq 0$ for all $i \neq j$ and if the successive principal minors of the test matrix $D = [d_{ij}]$ are all positive, where $d_{ij} = -a_{ij}$.

Proof. The proof follows from Theorem 2.4.29 and Corollary 2.5.6. ∎

The next result provides another interesting application of M-matrices.

2.5.23. Corollary. Let D denote the test matrix of Theorem 2.5.11 (or Theorem 2.5.18). Assume that composite system $(\mathscr{S})$ with decomposition (Σ_i) has been shown to be uniformly asymptotically stable in the large (or exponentially stable in the large) using one of these results. Then any modification of the isolated subsystems $(\mathscr{S}_i)$, or their feedback as expressed by a_{ii}, $i = 1, \ldots, l$, which increases $(\sigma_i + a_{ii})$, $i = 1, \ldots, l$, by an amount less than $\mu = \min_k \operatorname{Re}[\lambda_k(D)]$, $k = 1, \ldots, l$, will leave system $(\mathscr{S})$ uniformly asymptotically stable in the large (exponentially stable in the large).

Proof. The proof follows directly from Theorem 2.5.11 (or Theorem 2.5.18) and from Corollary 2.5.10. ∎

The parameter μ in Corollary 2.5.23 may be interpreted as a **margin of stability** or as a **degree of stability** of the overall interconnected system $(\mathscr{S})$ and may be used to judge how sensitive the qualitative properties of system $(\mathscr{S})$ are with respect to structural changes.

D. Instability and Complete Instability

Next, we consider the instability of systems $(\mathscr{S})$ and $(\mathscr{S}'')$.

2.5.24. Theorem. Assume that hypotheses (i) and (ii) of Theorem 2.4.31 are true, that $a_{ij} \geq 0$ for all $i \neq j$, and that all principal minors of the test matrix $D = [d_{ij}]$ are positive, where

$$d_{ij} = \begin{cases} -(\sigma_i + a_{ii}), & i = j \\ -a_{ij}, & i \neq j. \end{cases}$$

(a) If $N \neq L$, then the equilibrium $x = 0$ of composite system $(\mathscr{S})$ is **unstable**.

(b) If $N = L$, the equilibrium of system $(\mathscr{S})$ is **completely unstable**.

Proof. The proof follows from Theorem 2.4.31 and Corollary 2.5.6. ∎

2.5.25. Theorem. Assume that hypotheses (i) and (ii) of Theorem 2.4.32 hold, that $a_{ij} \geq 0$ for all $i \neq j$, and that all principal minors of the test matrix $D = [d_{ij}]$ are positive, where $d_{ij} = -a_{ij}$. If $N \neq L$, then the equilibrium $x = 0$ of composite system $(\mathscr{S}'')$ is **unstable** and if $N = L$, the equilibrium of $(\mathscr{S}'')$ is **completely unstable**.

Proof. The proof follows from Theorem 2.4.32 and Corollary 2.5.6. ∎

E. Uniform Boundedness and Uniform Ultimate Boundedness

Our last result yields sufficient conditions for boundedness of solutions of composite system $(\mathscr{S})$.

2.5.26. Theorem. Assume that hypotheses (i) and (ii) of Theorem 2.4.34 are true, that $a_{ij} \geq 0$ for all $i \neq j$, and that the successive principal minors of the test matrix $D = [d_{ij}]$ are all positive, where

$$d_{ij} = \begin{cases} -(\sigma_i + a_{ii}), & i = j \\ -a_{ij}, & i \neq j. \end{cases}$$

Then the solutions of composite system $(\mathscr{S})$ with decomposition (Σ_i) are **uniformly bounded**, in fact, **uniformly ultimately bounded**.

Proof. The proof follows from Theorem 2.4.34 and Corollary 2.5.6. ∎

F. Discussion

We conclude this section with a few observations, phrased in the notation of Theorem 2.5.11. Similar statements can be made for the remaining results. To satisfy hypotheses (iii)–(vi) of this theorem, it is necessary that $(\sigma_i + a_{ii}) < 0$, $i = 1, \ldots, l$. Thus, if $\sigma_i > 0$, so that subsystem $(\mathscr{S}_i)$ may be unstable, we require that $a_{ii} < 0$ and $|a_{ii}| > \sigma_i$. In other words, as in the case of Theorem 2.4.2, Theorem 2.5.11 is applicable to composite systems with unstable subsystems, provided that a sufficient amount of stabilizing feedback is associated with each unstable subsystem, where the feedback is not an integral part of the subsystem structure. Clearly, the greater the degree of stability associated with each subsystem, as expressed by $\sigma_i + a_{ii}$, $i = 1, \ldots, l$, the easier it is to satisfy hypotheses (iii)–(vi) of Theorem 2.5.11. Also, the weaker the interconnections, i.e., the smaller the terms $a_{ij} \geq 0$, $i \neq j$, the easier it is to satisfy these hypotheses. Thus, Theorem 2.5.11 constitutes a set of **weak coupling conditions** for uniform asymptotic stability of system $(\mathscr{S})$. The same is true in the case of Theorem 2.4.2 when $a_{ij} \geq 0$, $i \neq j$. These statements are of course not surprising, for if $a_{ij} \geq 0$, $i \neq j$, then Theorems 2.4.2 and 2.5.11 are equivalent. However, it is emphasized that the test matrix S in Theorem 2.4.2 has no restrictions in sign for the off-diagonal elements (i.e., a_{ij}, $i \neq j$, can have any sign) and therefore Theorem 2.4.2 is a more general result than Theorem

2.5.11. Note however that the latter result is easier to apply than the former. Specifically, in Theorem 2.4.2 we utilize a symmetric test matrix S involving a weighting vector $\alpha > 0$, while in Theorem 2.5.11 we employ a test matrix D which in general need not be symmetric and which does not involve a weighting vector.

Note that hypothesis (iv) of Theorem 2.5.11 enables us to deduce the stability properties of the nonlinear, nonautonomous n-dimensional composite system $(\mathscr{S})$ from the linear autonomous system $\dot{y} = Dy$, where $l \leq n$. Note also that in the row and column dominance conditions (2.5.12), (2.5.13), there appear arbitrary sets of constants $\{\lambda_i\}$, $\{\eta_i\}$, $i = 1, ..., l$, respectively. Due to the simple form of these conditions, usage of linear programming methods appear to be attractive to determine the optimal choice of these constants in specific high-dimensional problems (to obtain the least conservative results). In the case of low-dimensional problems, this choice may often be determined by inspection.

The row and column dominance conditions are especially well suited for systematic stabilization and compensation procedures of large scale systems. Conceptually, such procedures involve the following steps.

(a)   Enhance the degree of stability of specific subsystems, using local feedback at the subsystem level structure (i.e., increase $-(\sigma_i + a_{ii})$); or

(b)   weaken coupling effects, using local feedback at the interconnecting structure level (i.e., decrease appropriate choices of $a_{ij} \geq 0$, $i \neq j$); or

(c)   combine items (a) and (b).

In this procedure Corollary 2.5.23 is useful in providing a measure for the margin of stability μ of the overall composite system $(\mathscr{S})$.

The constants $a_{ij} \geq 0$, $i \neq j$, in Theorem 2.5.11 provide a measure of dynamic interaction among subsystems. If in particular $a_{ij} = 0$, we view $(\mathscr{S}_j)$ as not being connected to $(\mathscr{S}_i)$ and if $a_{ji} = 0$, we view $(\mathscr{S}_i)$ as not being connected to $(\mathscr{S}_j)$. Suppose now that any one of the equivalent hypotheses (iii)–(vi) of Theorem 2.5.11 have been satisfied for system $(\mathscr{S})$ for a given set of constants $a_{ij} \geq 0$, $i \neq j$. Then it is clear that if any one (or all) of the $a_{ij} \geq 0$, $i \neq j$, are decreased or set equal to zero, the stability conditions (iii)–(vi) are preserved. (This has motivated some authors to speak of so-called "connective stability." Since results of the type considered above have this feature automatically built in, we decline to pursue this notion any further.) Notice further that if an increase of say a_{ik}, $i \neq k$, is accompanied by an appropriate decrease of a_{ip}, $p = 1, ..., l$, $i \neq p \neq k$, then stability condition (2.5.12) can be preserved.

The preceding discussion suggests that in certain cases one may want to pursue the following approach in the *planning of reliable large scale dynamical systems.*

(a)   If applicable, a given system is viewed as an interconnection of subsystems, also called "areas";

(b)   each area is planned in such a fashion so that it is endowed with a sufficient margin of stability;

(c)   the dynamic interaction among various areas is limited in such a fashion that either the row or the column dominance conditions are satisfied;

(d)   the overall system is planned in such a way that it exhibits a sufficient degree of stability;

(e)   if at a future point in time it is necessary to increase the interaction among certain areas, this will have to be accompanied by an appropriate increase in the margin of stability of the affected areas and/or by a suitable decrease in the strength of coupling elsewhere.

Note that a system planned in this fashion is reliable in the sense that subsystems can be disconnected and reconnected (intentionally or by accident) without affecting the basic stability properties of the main system and the disconnected subsystems.

2.6   Application of the Comparison Principle
to Vector Lyapunov Functions

In the present section we first give a brief overview of several comparison theorems which are the basis of the comparison principle. Next, we show how this principle can be applied in the analysis of large scale systems using vector Lyapunov functions. We will show that for most of the specific cases considered thus far in the literature, this method reduces to the scalar Lyapunov function approach considered in Sections 2.4 and 2.5, where usage of the comparison principle is not required. Furthermore, we will demonstrate that the method of the present section, when applied to interconnected systems, involves test matrices which are always required to be M-matrices. Since the comparison principle is treated in detail in several texts, proofs of the comparison theorems are omitted.

We begin by considering a scalar ordinary differential equation of the form

$$\dot{y} = G(y, t) \qquad\qquad \text{(C)}$$

where $y \in R$, $t \in J$, and $G: B(r) \times J \to R$ for some $r > 0$. Assume that G is continuous on $B(r) \times J$ and that $G(0, t) = 0$ for all $t \geq t_0$. Under these assumptions it is well known that Eq. (C) possesses solutions $y(t; y_0, t_0)$ for every $y_0 = y(t_0; y_0, t_0) \in B(r)$, which are not necessarily unique. These solutions either exist for all $t \in [t_0, \infty)$ or else must leave the domain of definition of G at some finite time $t_1 > t_0$. Also, under the above assumptions, Eq. (C) admits the trivial solution $y = 0$ for all $t \geq t_0$. As before, we assume that $y = 0$ is an isolated equilibrium. Finally, for the sake of brevity, we frequently write $y(t)$ in place of $y(t; y_0, t_0)$ to denote solutions, with $y(t_0) = y_0$.

2.6.1. Definition. Let $p(t)$ be a solution of Eq. (C) in the interval $[t_0, a)$. Then $p(t)$ is called a **maximal solution** of (C) if for any other solution $y(t)$ existing on $[t_0, a)$, such that $p(t_0) = y(t_0) = y_0$, we have $y(t) \le p(t)$ for all $t \in [t_0, a)$.

2.6.2. Definition. Let $q(t)$ be a solution of Eq. (C) on the interval $[t_0, a)$. Then $q(t)$ is called a **minimal solution** of (C) if for any other solution $y(t)$ existing on $[t_0, a)$, such that $q(t_0) = y(t_0) = y_0$, we have $y(t) \ge q(t)$ for all $t \in [t_0, a)$.

For maximal and minimal solutions of Eq. (C) we have the following existence theorem.

2.6.3. Theorem. If G is continuous on $B(r) \times J$ and if $y_0 \in B(r)$, then Eq. (C) has both a maximal solution $p(t)$ and a minimal solution $q(t)$ for any $p(t_0) = q(t_0) = y_0$. Each of these solutions either exists for all $t \in [t_0, \infty)$ or else must leave the domain of definition of G at some finite time $t_1 > t_0$.

The following *comparison theorem* is fundamental to the theory.

2.6.4. Theorem. Assume that G is continuous on $B(r) \times J$ and that $p(t)$ is the maximal solution of Eq. (C) on the interval $[t_0, a)$ with $p(t_0) = y_0$. If $r(t)$ is a continuous function such that $r(t_0) \le y_0$, if

$$Dr(t) = \lim_{h \to 0^+} \sup [r(t+h) - r(t)]/h$$

and if

$$Dr(t) \le G(r(t), t) \qquad \text{almost everywhere on} \quad [t_0, a),$$

then $r(t) \le p(t)$ on $[t_0, a)$.

We also have

2.6.5. Theorem. Assume that G is continuous on $B(r) \times J$ and that $q(t)$ is the minimal solution of Eq. (C) on the interval $[t_0, a)$ with $q(t_0) = y_0$. If $s(t)$ is a continuous function such that $s(t_0) \ge y_0$, and if

$$Ds(t) \ge G(s(t), t) \qquad \text{almost everywhere on} \quad [t_0, a),$$

then $s(t) \ge q(t)$ on $[t_0, a)$.

Theorem 2.6.4 (as well as Theorem 2.6.5), referred to as a **comparison principle**, is a very important tool in applications because it can be used to reduce the problem of determining the behavior of solutions of Eq. (I),

$$\dot{x} = g(x, t), \tag{I}$$

$x \in R^n$, $g: B(r) \times J \to R^n$, to the solution of a scalar equation (C). To be more specific, we have in mind the application of Theorem 2.6.4 (as well as Theorem

2.6.5) to the case where $r(t) = v(x(t), t)$ (or $s(t) = v(x(t), t)$), where $v: B(r) \times J \to R$ is a Lyapunov function and $x(t)$ is a solution of Eq. (I). (As in Section 2.2, we assume that $g(0, t) = 0$ for all $t \geq t_0$.) In particular, applying Theorem 2.6.4 to $v(x, t)$, one can easily see that the following results are true.

2.6.6. Theorem. Let g and G be continuous on their respective domains of definition. Let $v: B(r) \times J \to R$ be a continuous, positive definite function such that

$$Dv_{(1)}(x, t) \leq G(v(x, t), t). \tag{2.6.7}$$

Then the following statements are true.

(i) If the trivial solution of Eq. (C) is stable, then the trivial solution of Eq. (I) is stable;

(ii) if v is decrescent and if the trivial solution of Eq. (C) is uniformly stable, then the trivial solution of Eq. (I) is uniformly stable;

(iii) if v is decrescent and if the trivial solution of Eq. (C) is uniformly asymptotically stable, then the trivial solution of Eq. (I) is uniformly asymptotically stable;

(iv) if there are constants $a > 0$ and $b > 0$ such that $a|x|^b \leq v(x, t)$, if v is decrescent, and if the trivial solution of Eq. (C) is exponentially stable, then the trivial solution of Eq. (I) is exponentially stable;

(v) if $g: R^n \times J \to R^n$, $G: R \times J \to R$, $v: R^n \times J \to R$, and (2.6.7) holds for all $x \in R^n$, $t \in J$, $v \in R$, and if the solutions of Eq. (C) are uniformly bounded (uniformly ultimately bounded), then the solutions of Eq. (I) are uniformly bounded (uniformly ultimately bounded).

In practice, the special case $G(y, t) \equiv 0$ is most commonly used in parts (i) and (ii) and the special case $G(y, t) = -\alpha y$ for some constant $\alpha > 0$ is most commonly used in parts (iii) and (iv) of the above theorem.

Applying Theorem 2.6.5 to $v(x, t)$, we can also see that the next result is true.

2.6.8. Theorem. Let g and G be continuous on their respective domains of definition. Let $v: B(r) \times J \to R$ be a continuous, positive definite function such that $Dv_{(1)}(x, t) \geq G(v(x, t), t)$. If the trivial solution of Eq. (C) is unstable, then the trivial solution of Eq. (I) is also unstable.

The generality and effectiveness of the preceding comparison technique can be improved by considering *vector-valued comparison equations* and *vector Lyapunov functions* (see Section 2.4E). Here, the scalar case is included as a special case. Specifically, consider a system of *l* ordinary differential equations,

$$\dot{y} = H(y, t), \qquad y(t_0) = y_0 \tag{VC}$$

where $y \in R^l$, $t \in J$, $H: B(r) \times J \to R^l$ is continuous on $B(r) \times J$, and $H(0, t) = 0$ for all $t \geq t_0$. Under these assumptions Eq. (VC) possesses solutions $y(t)$ for every $y_0 = y(t_0) \in B(r)$ which again are not necessarily unique. These solutions either exist for all $t \in [t_0, \infty)$ or else must leave the domain of definition of H at some finite time $t_1 > t_0$. Furthermore, under the above assumptions Eq. (VC) admits the trivial solution $y = 0$ for all $t \geq t_0$. Once more we assume that $y = 0$ is an isolated equilibrium.

If we let as usual $y \leq z$ denote $y_i \leq z_i$, $i = 1, \ldots, l$, and $y < z$ denote $y_i < z_i$, $i = 1, \ldots, l$ then Definition 2.6.1 of maximal solution and Definition 2.6.2 of minimal solution still make sense. In order to extend Theorems 2.6.6 and 2.6.8 to the vector case, we require the following additional concept.

2.6.9. Definition. A function $H(y, t) = (H_1(y, t), \ldots, H_l(y, t))^T$ is said to be **quasimonotone** if for each component H_j, $j = 1, \ldots, l$, the inequality $H_j(y, t) \leq H_j(z, t)$ is true whenever $y_i \leq z_i$ for all $i \neq j$ and $y_j = z_j$.

The above property was used by Müller [1] and Kamke [1]. It is sometimes called the Wazewski condition with reference to the work by Wazewski [1].

2.6.10. Theorem. If $H(y, t)$ is continuous and quasimonotone and if $y_0 \in B(r)$, then Eq. (VC) has a maximal and minimal solution. Each of these solutions must either be defined for all $t \in [t_0, \infty)$ or else leave the domain of definition of H at some finite time $t_1 > t_0$.

Analogous to Theorem 2.6.4 we have the following comparison theorem.

2.6.11. Theorem. Assume that H is continuous on $B(r) \times J$, that H is quasimonotone, and let $p(t)$ be the maximal solution of Eq. (VC) on $[t_0, a)$ with $p(t_0) = y_0$. If $r(t)$ is a continuous function such that $r(t_0) \leq y_0$, if $Dr(t) = [\overline{\lim}_{h \to 0^+} (r_1(t+h) - r_1(t))/h, \ldots, \overline{\lim}_{h \to 0^+} (r_l(t+h) - r_l(t))/h]^T$, and if

$$Dr(t) \leq H(r(t), t) \qquad \text{almost everywhere on} \quad [t_0, a)$$

then $r(t) \leq p(t)$ on $[t_0, a)$.

We also have the following theorem.

2.6.12. Theorem. Assume that H is continuous on $B(r) \times J$, that H is quasimonotone, and let $q(t)$ be the minimal solution of Eq. (VC) on $[t_0, a)$ with $q(t_0) = y_0$. If $s(t)$ is a continuous function such that $s(t_0) \geq y_0$ and if

$$Ds(t) \geq H(s(t), t) \qquad \text{almost everywhere on} \quad [t_0, a)$$

then $s(t) \geq q(t)$ on $[t_0, a)$.

Now let $v_i(x, t)$, $i = 1, \ldots, l$, denote l continuous Lyapunov functions and let

$$V(x, t) = (v_1(x, t), \ldots, v_l(x, t))^T.$$

We call $V(x, t)$ a **vector Lyapunov function**. Such functions were originally introduced by Bellman [2]. Applying Theorem 2.6.11 to $V(x, t)$ the following result is easily established.

2.6.13. Theorem. Let g and H be continuous on their respective domains of definition and let H be quasimonotone. Let $V(x, t)$ be a continuous non-negative vector Lyapunov function (of dimension l) such that $|V(x, t)|$ is positive definite and such that

$$DV_{(I)}(x, t) \le H(V(x, t), t). \tag{2.6.14}$$

Then the following statements are true.

(i)   If the trivial solution of Eq. (VC) is stable, then the trivial solution of Eq. (I) is also stable;

(ii)   if $|V(x, t)|$ is decrescent and if the trivial solution of Eq. (VC) is uniformly stable, then the trivial solution of Eq. (I) is also uniformly stable;

(iii)   if $|V(x, t)|$ is decrescent and if the trivial solution of Eq. (VC) is uniformly asymptotically stable, then the trivial solution of Eq. (I) is also uniformly asymptotically stable;

(iv)   if there are constants $a > 0$ and $b > 0$ such that $a|x|^b \le |V(x, t)|$, if $|V(x, t)|$ is decrescent, and if the trivial solution of Eq. (VC) is exponentially stable, then the trivial solution of Eq. (I) is also exponentially stable;

(v)   if $g: R^n \times J \to R^n$ and $H: R^l \times J \to R^l$ and $V: R^n \times J \to R^l$ and if (2.6.14) is true for all $(x, t) \in R^n \times J$, and if the solutions of Eq. (VC) are uniformly bounded (uniformly ultimately bounded), then the solutions of Eq. (I) are also uniformly bounded (uniformly ultimately bounded).

Applying Theorem 2.6.12 to $V(x, t)$ we can also prove the following instability result.

2.6.15. Theorem. Let H and g be continuous on their respective domains of definition and let H be quasimonotone. Let $V(x, t)$ be a continuous non-negative vector Lyapunov function (of dimension l) such that $|V(x, t)|$ is positive definite and such that

$$DV_{(I)}(x, t) \ge H(V(x, t), t).$$

If the trivial solution of Eq. (VC) is unstable, then the trivial solution of Eq. (I) is also unstable.

Proofs of the preceding results can be found in several standard references, some of which we cite in Section 2.9. Let us now see how these results apply in the qualitative analysis of large scale systems. As Matrosov [1, 2] has pointed out, from the point of view of applications, the following special case of (VC),

$$\dot{y} = Py + m(y, t) \tag{VC'}$$

is particularly important. Here $P = [p_{ij}]$ is a real $l \times l$ matrix and the function $m: B(r) \times J \to R^l$ is assumed to consist of second or higher order terms, so that

$$\lim_{|y| \to 0} |m(y, t)| / |y| = 0, \qquad \text{uniformly in} \quad t \geq t_0.$$

Applying the **principle of stability in the first approximation** (see Hahn [2, p. 122]) to Eq. (VC'), we have the following result. If matrix P has either only eigenvalues with negative real parts or at least one eigenvalue with a positive real part, then the equilibrium of (VC') shows the same stability behavior as that of the corresponding linearized system,

$$\dot{y} = Py.$$

Using the principle of stability in the first approximation and Theorems 2.6.13 and 2.6.15, we immediately obtain the following result.

2.6.16. Corollary. Let g be continuous and let $V(x, t)$ be a nonnegative continuous vector Lyapunov function (of dimension l) such that $|V(x, t)|$ is positive definite and decrescent. Suppose there is an $l \times l$ matrix $P = [p_{ij}]$ and a function $m(V, t)$ such that

$$DV_{(l)}(x, t) \leq PV(x, t) + m(V(x, t), t) \qquad (2.6.17)$$

and

$$p_{ij} \geq 0 \qquad \text{if} \quad i \neq j, \qquad (2.6.18)$$

and $m(V, t)$ is quasimonotone in V, and

$$\lim_{|y| \to 0} |m(y, t)| / |y| = 0, \qquad \text{uniformly in} \quad t \geq t_0. \qquad (2.6.19)$$

Then the following statements are true.

(i) If matrix P has only eigenvalues with negative real part, the trivial solution of Eq. (I) is uniformly asymptotically stable;

(ii) if matrix P has only eigenvalues with negative real part and if in addition $|V(x, t)| \geq \delta |x|^2$ for some $\delta > 0$, then the trivial solution of Eq. (I) is exponentially stable;

(iii) if the inequality (2.6.17) is reversed and if P has at least one eigenvalue with positive real part, then the trivial solution of Eq. (I) is unstable;

(iv) if the inequality (2.6.17) is reversed and if the real parts of all eigenvalues of P are positive, then the trivial solution of Eq. (I) is completely unstable.

To simplify matters, we consider only parts (i) and (ii) of Corollary 2.6.16 in the following discussion. Similar statements can be made for the remaining parts. First we note that in (i) and (ii) of Corollary 2.6.16, the quasimonoticity condition (2.6.18) means that $-P$ is an M-matrix. Therefore, the condition that all eigenvalues of P have negative real parts, i.e., $\text{Re}[\lambda(P)] < 0$, is equivalent to the computable conditions

$$(-1)^k \begin{vmatrix} p_{11} & \cdots & p_{1k} \\ \cdot & & \cdot \\ \cdot & \cdots & \cdot \\ \cdot & & \cdot \\ p_{k1} & \cdots & p_{kk} \end{vmatrix} > 0, \qquad k = 1, \dots, l. \qquad (2.6.20)$$

Since $-P$ is an M-matrix, note also that the conditions $\mathrm{Re}[\lambda_k(P)] < 0$, $k = 1, \dots, l$, are equivalent to the existence of a column vector $\alpha^\mathrm{T} = (\alpha_1, \dots, \alpha_l) > 0$ such that $\alpha^\mathrm{T} P < 0$ (see Theorem 2.5.2, part (iv)). So in this case we can define the scalar Lyapunov function

$$v(x, t) = \sum_{i=1}^{l} \alpha_i v_i(x, t),$$

where $V(x, t) = (v_1(x, t), \dots, v_l(x, t))^\mathrm{T}$. Under the assumptions of Corollary 2.6.16, this function $v(x, t)$ is a positive definite and decrescent scalar Lyapunov function such that

$$Dv_{(1)}(x, t) \le \alpha^\mathrm{T} [PV(x, t) + m(V(x, t), t)]$$
$$= (\alpha^\mathrm{T} P)V(x, t) + \alpha^\mathrm{T} m(V(x, t), t).$$

Since $(\alpha^\mathrm{T} P)V(x, t)$ is negative definite and since $\alpha^\mathrm{T} m(V, t)$ consists of terms of second or higher order in $V(x, t)$, it follows that $Dv_{(1)}(x, t)$ is negative definite in a neighborhood of the origin $x = 0$. This argument shows that in the important special case of (2.6.17) the vector Lyapunov function $V(x, t)$ can be reduced to a scalar Lyapunov function $v(x, t)$ and because of the quasi-monoticity requirements, the seemingly more general approach of applying the comparison principle to vector Lyapunov functions is really equivalent to an approach utilizing scalar Lyapunov functions (as discussed, e.g., in Section 2.5). It would be interesting to know whether or not this same equivalence is true for systems of inequalities more general than (2.6.17). Furthermore, we note that whereas the present approach of applying the comparison principle to vector Lyapunov functions appears to require test matrices (e.g., matrix P) whose off-diagonal terms are of the same sign, this is not the case in the approach of Section 2.4.

Turning our attention to interconnected systems, let us consider once more composite system $(\mathscr{S})$ with decomposition (Σ_i) and isolated subsystems $(\mathscr{S}_i)$ (see Eqs. (2.3.10)–(2.3.12)). In particular, recall the equations for (Σ_i) and $(\mathscr{S}_i)$,

(Σ_i) $\qquad\qquad \dot{z}_i = f_i(z_i, t) + g_i(x, t), \qquad i = 1, \dots, l,$ $\qquad$ (2.6.21)

$(\mathscr{S}_i)$ $\qquad\qquad \dot{z}_i = f_i(z_i, t), \qquad i = 1, \dots, l.$ $\qquad\qquad$ (2.6.22)

For interconnected system $(\mathscr{S})$, the method originally proposed by Bailey

[1, 2] has been the most successful procedure of constructing vector Lyapunov functions (see Matrosov [1, 2]). Specifically, we consider isolated subsystems $(\mathscr{S}_i)$ for which we can find positive definite, decrescent, continuously differentiable Lyapunov functions v_i such that the derivative of v_i with respect to t along solutions of $(\mathscr{S}_i)$ satisfies

$$Dv_{i(\mathscr{S}_i)}(z_i, t) \leq -\beta_i v_i(z_i, t). \tag{2.6.23}$$

Systems for which (2.6.23) can be satisfied are often linear or nearly linear and the corresponding v_i is usually a quadratic function in z_i. Suppose now that we can find an $l \times l$ matrix $P = [p_{ij}]$ satisfying

$$p_{ij} \geq 0, \qquad i \neq j, \tag{2.6.24}$$

and a quasimonotone function $m: R^l \times J \to R^l$ satisfying

$$\lim_{|y| \to 0} |m(y, t)| / |y| = 0, \qquad \text{uniformly in} \quad t \geq t_0. \tag{2.6.25}$$

Assume also that the following inequalities hold,

$$-\nabla v_i(z_i, t)^{\mathrm{T}} g_i(x, t) + \sum_{j=1}^{l} (p_{ij} + \delta_{ij}\beta_i) v_j(z_j, t) + m_i(V(x, t), t) \geq 0,$$
$$\tag{2.6.26}$$

$i = 1, \dots, l$, where δ_{ij} is the Kronecker delta, $m_i(V, t)$ is the ith component of $m(V, t)$, and $V(x, t) = (v_1(z_1, t), \dots, v_l(z_l, t))^{\mathrm{T}}$. We now have

$$\begin{aligned} DV_{(\mathscr{S})}(x, t) &= [Dv_{1(\mathscr{S}_1)}(z_1, t) + \nabla v_1(z_1, t)^{\mathrm{T}} g_1(x, t), \dots, Dv_{l(\mathscr{S}_l)}(z_l, t) \\ &\quad + \nabla v_l(z_l, t)^{\mathrm{T}} g_l(x, t)]^{\mathrm{T}} \\ &\leq [-\beta_1 v_1(z_1, t) + \nabla v_1(z_1, t)^{\mathrm{T}} g_1(x, t), \dots, -\beta_l v_l(z_l, t) \\ &\quad + \nabla v_l(z_l, t)^{\mathrm{T}} g_l(x, t)]^{\mathrm{T}}. \end{aligned} \tag{2.6.27}$$

Combining (2.6.26) and (2.6.27) yields

$$DV_{(\mathscr{S})}(x, t) \leq PV(x, t) + m(V(x, t), t). \tag{2.6.28}$$

All conditions of Corollary 2.6.16 are now satisfied. In particular, if matrix P has only eigenvalues with negative real parts, then the equilibrium $x = 0$ of composite system $(\mathscr{S})$ is asymptotically stable. Parts (ii)–(iv) of Corollary 2.6.16 apply equally as well, with obvious modifications.

2.7 Estimates of Trajectory Behavior and Trajectory Bounds

In this section we obtain estimates of trajectory behavior and trajectory bounds of dynamical systems, using Lyapunov-type results. In our approach we define stability in terms of subsets of R^n which are *prespecified* in a given problem and which in general may be time varying. The properties of these sets yield information about the transient behavior and about trajectory bounds of such systems. Before considering interconnected systems, we need to develop essential preliminary results.

We begin by considering systems described by equations of the form

$$\dot{x} = g(x, t) \tag{I}$$

where $x \in R^n$, $t \in J$, and $g: R^n \times J \to R^n$. We assume that Eq. (I) possesses for every $x_0 \in R^n$ and for every $t_0 \in R^+$ a unique solution $x(t; x_0, t_0)$ for all $t \in J$. However, *in the present section we do not insist that $x = 0$ be an equilibrium of Eq.* (I).

Subsequently we employ time varying open subsets of R^n defined for all $t \in J$ and denoted by $S(t)$, $S_0(t)$, and $D(t)$, with appropriate subscripts or superscripts if necessary. The boundary of $S(t)$ is denoted by $\partial S(t)$ and its closure is denoted by $\overline{S(t)}$. Henceforth we assume that all such sets possess the following "well-behaved" property.

2.7.1. Definition. A time-varying open set $S(t)$ is said to possess **Property P** if, whenever $p: J \to R^n$ is continuous and $p(t_1) \in S(t_1)$ for some $t_1 \in J$, then either $p(t) \in S(t)$ for all $t \in [t_1, \infty)$ or there exists a $t_2 \in (t_1, \infty)$ for which $p(t_2) \in \partial S(t_2)$ and $p(t) \in S(t)$ for all $t \in [t_1, t_2]$.

For example, the set $S_1(t) = \{x \in R^n : |x| < ke^{-at}, \ k > 0, \ a > 0, \ t \in R^+\}$ possesses Property P. On the other hand, the set

$$S_2(t) = \begin{cases} \{x \in R^n : |x| < c_1 + c_2, \ c_1 > 0, \ c_2 > 0\}, & 0 \leq t < b \\ \{x \in R^n : |x| < c_1\}, & b \leq t < \infty \end{cases}$$

does not possess Property P.

In the next definition we let $x(t; x_i, t_i)$ be the solution of Eq. (I) which satisfies $x(t_i; x_i, t_i) = x_i$.

2.7.2. Definition. System (I) is called **stable** with respect to $\{S_0(t_0), S(t), t_0\}$ if $x_0 \in S_0(t_0)$ implies that $x(t; x_0, t_0) \in S(t)$ for all $t \in J$. System (I) is **uniformly stable** with respect to $\{S_0(t), S(t)\}$, if for all $t_i \in J$, $x_i \in S_0(t_i)$ implies that $x(t; x_i, t_i) \in S(t)$ for all $t \in [t_i, \infty)$. System (I) is **unstable** with respect to $\{S_0(t_0), S(t), t_0\}$, $S_0(t_0) \subset S(t_0)$, if there exists an $x_0 \in S_0(t_0)$ and a $t_c \in J$ such that $x(t_c; x_0, t_0) \in \partial S(t_c)$.

We emphasize that in Definition 2.7.2 the sets $S_0(t)$ and $S(t)$ are pre-specified in a given problem. In the literature the term *practical stability* is used for the special case $S_0(t) \equiv B(\alpha)$ and $S(t) \equiv B(\beta)$, $\alpha \leq \beta$. Also, the term *finite time stability* is used for the special case when J is replaced by the interval $[t_0, t_0 + T)$, $T > 0$.

2.7.3. Theorem. System (I) is **stable** with respect to $\{S_0(t_0), S(t), t_0\}$, $S(t_0) \supset S_0(t_0)$, if there exists a continuous function $v\colon R^n \times J \to R$ and a function G which satisfies the assumptions of Theorem 2.6.4 such that

(i) $Dv_{(I)}(x, t) \le G(v(x, t), t)$ for all $x \in S(t)$ and $t \in J$; and

(ii) $p(t; \sup_{x \in S_0(t_0)} v(x, t_0), t_0) < \inf_{x \in \partial S(t)} v(x, t)$,   for all $t \in J$,   where $p(t; y_0, t_0)$ is defined in Theorem 2.6.4.

System (I) is **uniformly stable** with respect to $\{S_0(t), S(t)\}$, $S(t) \supset S_0(t)$ for all $t \in J$, if (i) and (ii) are replaced by

(iii) $Dv_{(I)}(x, t) \le G(v(x, t), t)$ for all $x \in [S(t) - \overline{S_0(t)}]$ and $t \in J$; and

(iv) $p(t_2; \sup_{x \in \partial S_0(t_1)} v(x, t_1), t_1) < \inf_{x \in \partial S(t_2)} v(x, t_2)$ for all $t_2 > t_1$, $t_1, t_2 \in J$.

Proof. Since the proofs for stability and uniform stability are very similar, only the proof for stability is given.

The proof is by contradiction. Let $x_0 \in S_0(t_0)$ and assume there exists $t_2 \in (t_0, \infty)$ for which $x(t_2; x_0, t_0) \notin S(t_2)$. Since $S(t)$ possesses Property P, there exists $t_1 \in (t_0, t_2]$ such that $x(t; x_0, t_0) \in S(t)$ for all $t \in [t_0, t_1)$ and such that $x(t_1; x_0, t_0) \in \partial S(t_1)$. Define $r(t)$ as $r(t) = v[x(t; x_0, t_0), t]$ (refer to Theorem 2.6.4) and write

$$Dr(t) = Dv_{(I)}(x(t; x_0, t_0), t) \le G[v(x(t; x_0, t_0), t), t] = G(r(t), t).$$

Now $r(t_0) = v(x(t_0; x_0, t_0), t_0) = v(x_0, t_0) \le \sup_{x \in S_0(t_0)} v(x, t_0)$, and thus it follows from Theorem 2.6.4 that $v(x(t; x_0, t_0), t) \le p(t; \sup_{x \in S_0(t_0)} v(x, t_0), t_0)$ for all $t \in (t_0, t_1]$. In view of hypothesis (ii) we can now write

$$v(x(t_1; x_0, t_0), t_1) < \inf_{x \in \partial S(t_1)} v(x, t_1).$$

But the above inequality implies that $x(t_1; x_0, t_0) \notin \partial S(t_1)$, a contradiction. Hence, no t_2 as asserted exists and $x(t; x_0, t_0) \in S(t)$ for all $t \in J$. ∎

Note that from a computational point of view, the v-functions in Theorem 2.7.3 as well as in the remaining results of this section, may sometimes have less stringent requirements than those used in the usual Lyapunov theorems. Thus, in Theorem 2.7.3 there are in general no definiteness conditions on v and we do not require that $v(0, t) \equiv 0$.

2.7.4. Theorem. System (I) is **unstable** with respect to $\{S_0(t_0), S(t), t_0\}$, $S(t_0) \supset S_0(t_0)$, if there exists a continuous function $v\colon R^n \times J \to R$ and a function G which satisfies the assumptions of Theorem 2.6.5 and a $t_1 \in (t_0, \infty)$ such that

(i) $Dv_{(I)}(x, t) \ge G(v(x, t), t)$ for all $x \in S(t)$ and $t \in J$,

(ii) $q(t_1; \inf_{x \in S_0(t_0)} v(x, t_0), t_0) \ge \sup_{x \in \partial S(t_1)} v(x, t_1)$, and

(iii) $v(x, t_1) < \sup_{x \in \partial S(t_1)} v(x, t_1)$ for all $x \in S(t_1)$

where $q(t; y_0, t_0)$ is defined in Theorem 2.6.5.

Proof. The proof of this theorem is similar to that of Theorem 2.7.3. ∎

It is often convenient to view the right-hand side of Eq. (I) as the sum of two functions $f: R^n \times J \to R^n$ and $u: R^n \times J \to R^n$, i.e.,

$$\dot{x} = f(x, t) + u(x, t) \triangleq g(x, t). \tag{2.7.5}$$

Here we view $u(x, t)$ as (deterministic) "persistent perturbation terms" and we call

$$\dot{x} = f(x, t) \tag{2.7.6}$$

the "unperturbed system." System (2.7.5) is a special case of systems described by equations of the form

$$\dot{x} = f(x, u(x, t), t) \triangleq h(x, t) \tag{2.7.7}$$

where $u: R^n \times J \to R^r$, $f: R^n \times R^r \times J \to R^n$, and $h: R^n \times J \to R^n$.

2.7.8. Definition. System (2.7.7) is **totally stable** with respect to

$$\{S_0(t_0), S(t), D(t), t_0\},$$

if the conditions

(a) $x_0 \in S_0(t_0)$ and
(b) $u(x, t) \in D(t)$ whenever $x \in S(t)$ and $t \in J$

imply that $x(t; x_0, t_0) \in S(t)$ for all $t \in J$. System (2.7.7) is **uniformly totally stable** with respect to $\{S_0(t), S(t), D(t)\}$ if for all $t_i \in J$ the conditions

(a) $x_i \in S_0(t_i)$ and
(b) $u(x, t) \in D(t)$ whenever $x \in S(t)$ and $t \in [t_i, \infty)$

imply that $x(t; x_i, t_i) \in S(t)$ for all $t \in [t_i, \infty)$.

For system (2.7.5) we have the following result.

2.7.9. Theorem. System (2.7.5) is **totally stable** with respect to

$$\{S_0(t_0), S(t), D(t), t_0\}, \qquad S(t_0) \supset S_0(t_0),$$

if there exists a continuously differentiable function $v: R^n \times J \to R$ and two integrable functions $v: J \to R$, $\eta: J \to R$, such that
(i) $Dv_{(2.7.6)}(x, t) < v(t)$ for all $x \in S(t)$ and $t \in J$;
(ii) $\nabla v(x, t)^T u \leq \eta(t)$ for all $u \in D(t)$, $x \in S(t)$, $t \in J$; and
(iii) $\int_{t_0}^t [v(\tau) + \eta(\tau)] \, d\tau \leq \inf_{x \in \partial S(t)} v(x, t) - \sup_{x \in S_0(t_0)} v(x, t_0)$.
System (2.7.5) is **uniformly totally stable** with respect to $\{S_0(t), S(t), D(t)\}$, $S_0(t) \subset S(t)$ for all $t \in J$, if (i)–(iii) are replaced by
(iv) $Dv_{(2.7.6)}(x, t) < v(t)$ for all $x \in [S(t) - \overline{S_0(t)}]$ and $t \in J$;

(v) $\nabla v(x,t)^\mathsf{T} u \le \eta(t)$ for all $x \cdot \in [S(t) - \overline{S_0(t)}]$, $u \in D(t)$, $t \in J$; and

(vi) $\int_{t_1}^{t_2} [\nu(\tau) + \eta(\tau)]\, d\tau \le \inf_{x \in \partial S(t_2)} v(x, t_2) - \sup_{x \in \partial S_0(t_1)} v(x, t_1)$ for   all $t_2 > t_1, t_1, t_2 \in J$.

Proof. Since the proofs for total stability and uniform total stability are similar, only the proof for the former is given.

The proof is by contradiction. Let $x_0 \in S_0(t_0)$ and assume that $u(x,t) \in D(t)$ whenever $x \in S(t)$ and $t \in J$. Assume that there exists a $t_2 \in (t_0, \infty)$ for which $x(t_2; x_0, t_0) \notin S(t_2)$. Then, since $S(t)$ possesses Property P, there exists $t_1 \in (t_0, t_2]$ such that $x(t; x_0, t_0) \in S(t)$ for all $t \in [t_0, t_1)$ and such that $x(t_1; x_0, t_0) \in \partial S(t_1)$. Now

$$v(x(t_1; x_0, t_0), t_1) = v(x_0, t_0) + \int_{t_0}^{t_1} Dv_{(2.7.5)}(x(t; x_0, t_0), t)\, dt$$

$$= v(x_0, t_0)$$
$$+ \int_{t_0}^{t_1} [\nabla v(x(t; x_0, t_0), t)^\mathsf{T} f(x(t; x_0, t_0), t)$$
$$+ \partial v(x(t; x_0, t_0), t)/\partial t]\, dt$$
$$+ \int_{t_0}^{t_1} \nabla v(x(t; x_0, t_0), t)^\mathsf{T} u(x(t; x_0, t_0), t)\, dt.$$

In view of hypotheses (i)–(iii) we have

$$v(x(t_1; x_0, t_0), t_1) < \sup_{x \in S_0(t_0)} v(x, t_0) + \int_{t_0}^{t_1} [\nu(t) + \eta(t)]\, dt$$

$$\le \sup_{x \in S_0(t_0)} v(x, t_0) + \inf_{x \in \partial S(t_1)} v(x, t_1) - \sup_{x \in S_0(t_0)} v(x, t_0)$$

$$= \inf_{x \in \partial S(t_1)} v(x, t_1).$$

But this inequality implies that $x(t_1; x_0, t_0) \notin \partial S(t_1)$ which is a contradiction. Therefore no t_2 as asserted above exists and $x(t; x_0, t_0) \in S(t)$ for all $t \in J$. ■

For the special case when $u(t) \equiv 0$, Theorem 2.7.9 yields a stability result with respect to $\{S_0(t_0), S(t), t_0\}$, and a uniform stability result with respect to $\{S_0(t), S(t)\}$ for the unperturbed system (2.7.6). In this case we take $\eta(t) \equiv 0$ and delete hypothesis (ii) of Theorem 2.7.9 in its entirety.

Turning our attention to composite systems, let us consider once more interconnected system $(\mathscr{S}'')$ with decomposition (Σ_i''), i.e.,

(Σ_i'') $\dot{z}_i = f_i(z_i, g_i(z_1, \ldots, z_l, t), t)$

$$= f_i(z_i, g_i(x, t), t) = h_i(x, t), \qquad i = 1, \ldots, l, \tag{2.7.10}$$

$(\mathscr{S}'')$ $\dot{x} = h(x, t)$ $\tag{2.7.11}$

where all symbols in Eqs. (2.7.10) and (2.7.11) are as defined in Eqs. (2.3.20) and (2.3.21), respectively. However, in the present case we do not require that $x = 0$ be an equilibrium.

In the following, all time varying sets possess Property P. The notation $x \in \bigtimes_{j=1}^{l} S^j(t)$ signifies $z_j \in S^j(t)$, $j = 1, ..., l$, and $t \in J$. In particular, $\bigtimes_{j=1}^{l} B^j(r_j) = \{x \in R^n : |z_j| < r_j, j = 1, ..., l\}$.

2.7.12. Definition. Composite system $(\mathscr{S}'')$ with decomposition (Σ_i'') is **stable** with respect to $\{\bigtimes_{j=1}^{l} S_0^j(t_0), \bigtimes_{j=1}^{l} S^j(t), t_0\}$ if $x_0 \in \bigtimes_{j=1}^{l} S_0^j(t_0)$ implies that $x(t; x_0, t_0) \in \bigtimes_{j=1}^{l} S^j(t)$ for all $t \in J$. Composite system $(\mathscr{S}'')$ with decomposition (Σ_i'') is **uniformly stable** with respect to $\{\bigtimes_{j=1}^{l} S_0^j(t), \bigtimes_{j=1}^{l} S^j(t)\}$ if for all $t_i \in J$, $x_i \in \bigtimes_{j=1}^{l} S_0^j(t_i)$ implies that $x(t; x_i, t_i) \in \bigtimes_{j=1}^{l} S^j(t)$ for all $t \in [t_i, \infty)$.

We are now in a position to prove the following result.

2.7.13. Theorem. Composite system $(\mathscr{S}'')$ with decomposition (Σ_i'') is **stable** with respect to $\{\bigtimes_{j=1}^{l} S_0^j(t_0), \bigtimes_{j=1}^{l} S^j(t), t_0\}$, $S^j(t_0) \supset S_0^j(t_0)$, $j = 1, ..., l$, if the following conditions hold.

(i) $g_i(x, t) \in D_i(t) \subset R^{r_i}$ whenever $x \in \bigtimes_{j=1}^{l} \overline{S^j(t)}$ and $t \in J$;

(ii) each system described by

$$\dot{z}_i = f_i(z_i, g_i(x, t), t) \qquad (2.7.14)$$

is totally stable with respect to $\{S_0^i(t_0), S^i(t), D_i(t), t_0\}$, $S_0^i(t_0) \subset S^i(t_0)$, $i = 1, ..., l$.

Composite system $(\mathscr{S}'')$ is **uniformly stable** with respect to

$$\left\{ \bigtimes_{j=1}^{l} S_0^j(t), \bigtimes_{j=1}^{l} S^j(t) \right\}, \qquad S^j(t) \supset S_0^j(t), \quad j = 1, ..., l,$$

for all $t \in J$, if (i) is true and if (ii) is replaced by the following.

(iii) Each system described by Eq. (2.7.14) is uniformly totally stable with respect to $\{S_0^i(t), S^i(t), D_i(t)\}$ $S_0^i(t) \subset S^i(t)$, $i = 1, ..., l$, and $t \in J$.

Proof. The proofs for stability and uniform stability are similar. Only the proof for stability is given.

Let $x_0 \in \bigtimes_{j=1}^{l} S_0^j(t_0) \subset \bigtimes_{j=1}^{l} S^j(t_0)$. For purposes of contradicton, assume that at some $t_2 \in J$, $x(t_2; x_0, t_0) \notin \bigtimes_{j=1}^{l} S^j(t_2)$. Since $x(t; x_0, t_0)$ is continuous and since sets $S^j(t)$, $j = 1, ..., l$, possess Property P, there exists a finite first time $t_1 < t_2$ such that

$$x(t_1; x_0, t_0) \in \partial \left\{ \bigtimes_{j=1}^{l} S^j(t_1) \right\}.$$

Hence, $x(t; x_0, t_0) \in \bigtimes_{j=1}^{l} \overline{S^j(t)}$ for all $t \in [t_0, t_1]$. Now by hypothesis (i), $g_i(x, t) \in D_i(t)$ whenever $x \in \bigtimes_{j=1}^{l} \overline{S^j(t)}$ and $t \in J$. Thus, $g_i(x(t; x_0, t_0), t) \in$

$D_i(t)$, $i = 1, \ldots, l$, for all $t \in [t_0, t_1]$. Next, define

$$g_i^*(t) = \begin{cases} g_i(x(t; x_0, t_0), t), & t \in [t_0, t_1] \\ u_i(t), & t \in (t_1, \infty), \end{cases}$$

$i = 1, \ldots, l$, and choose $u_i(t)$ in such a fashion that $u_i(t) \in D_i(t)$ for all $t \in (t_1, \infty)$. Then clearly $g_i^*(t) \in D_i(t)$ for all $t \in J$. Corresponding to each equation describing (Σ_i''), consider

$$\dot{y}_i = f_i(y_i, g_i^*(t), t), \qquad i = 1, \ldots, l,$$

where the mappings f_i, $i = 1, \ldots, l$, are defined as in (Σ_i''). Let $y_{i_0} = z_{i_0}$. By hypothesis (ii), each system described by Eq. (2.7.14) is totally stable with respect to $\{S_0^i(t_0), S^i(t), D_i(t), t_0\}$, $S_0^i(t_0) \subset S^i(t_0)$, $i = 1, \ldots, l$. Thus, $y_i(t; y_{i_0}, t_0) \in S^i(t)$ for all $t \in J$. By causality (i.e., by uniqueness), $z_i(t; z_{i_0}, t_0) = y_i(t; y_{i_0}, t_0)$ for all $t \in [t_0, t_1]$. Therefore, $z_i(t; z_{i_0}, t_0) \in S^i(t)$ for all $t \in [t_0, t_1]$ and $x(t; x_0, t_0) \in \bigtimes_{j=1}^l S^j(t)$ for all $t \in [t_0, t_1]$. But above we assumed that

$$x(t_1; x_0, t_0) \in \partial \left\{ \bigtimes_{j=1}^l S^j(t_1) \right\}.$$

Thus, we have arrived at a contradiction and there exists no $t_1 \in J$ such that $x(t_1; x_0, t_0) \in \partial \{ \bigtimes_{j=1}^l S^j(t_1) \}$. This in turn implies that there is no $t_2 \in J$ such that $x(t_2; x_0, t_0) \notin \bigtimes_{j=1}^l S^j(t_2)$, because each $S^j(t)$ possesses Property P and because of the continuity of $x(t; x_0, t_0)$. Therefore $x(t; x_0, t_0) \in \bigtimes_{j=1}^l S^j(t)$ for all $t \in J$. ∎

2.8 Applications

In order to demonstrate the usefulness of the method of analysis advanced in the preceding sections, we consider several specific examples.

2.8.1. Example. (*Longitudinal Motion of an Aircraft.*) The controlled longi-tudinal motion of an aircraft may be represented by the set of equations (see Piontkovskii and Rutkovskaya [1])

$$\dot{x}_k = -\rho_k x_k + \sigma, \qquad k = 1, 2, 3, 4,$$

$$\dot{\sigma} = \sum_{k=1}^4 \beta_k x_k - rp\sigma - f(\sigma) \tag{2.8.2}$$

where $\rho_k > 0$, $r > 0$, $p > 0$, β_k are constants, where $x_k \in R$, $\sigma \in R$, and where $f: R \to R$ has the following properties: (a) it is continuous on R, (b) $f(\sigma) = 0$ if and only if $\sigma = 0$, and (c) $\sigma f(\sigma) > 0$ for all $\sigma \neq 0$. We call any function f

with these properties an *admissible nonlinearity*. If the equilibrium $x^T = (x_1, x_2, x_3, x_4, \sigma) = 0$ is asymptotically stable in the large for all admissible nonlinearities f, we call system (2.8.2) *absolutely stable* (see Aizerman and Gantmacher [1], Lefschetz [1]).

System (2.8.2) may be viewed as a linear interconnection of isolated subsystems $(\mathscr{S}_1)$ and $(\mathscr{S}_2)$ described by

$$\dot{x}_k = -\rho_k x_k, \qquad k = 1, 2, 3, 4, \qquad (\mathscr{S}_1)$$

and

$$\dot{\sigma} = -rp\sigma - f(\sigma). \qquad (\mathscr{S}_2)$$

Systems $(\mathscr{S}_1)$ and $(\mathscr{S}_2)$ are interconnected to form system (2.8.2) by means of the matrices $C_{12}^T = [1, 1, 1, 1]$ and $C_{21} = [\beta_1, \beta_2, \beta_3, \beta_4]$, where the notation of Eq. (2.3.14) is used.

Without loss of generality we assume in the following that $\rho_1 \leq \rho_2 \leq \rho_3 \leq \rho_4$. Let $z_1^T = (x_1, x_2, x_3, x_4)$ and $z_2 = \sigma$. For $(\mathscr{S}_1)$ and $(\mathscr{S}_2)$ choose $v_1(z_1) = c_1 z_1^T z_1$ and $v_2(z_2) = c_2 z_2^2$, where $c_1 > 0$ and $c_2 > 0$ are constants. Then

$$Dv_{1(\mathscr{S}_1)}(z_1) \leq -2c_1 \rho_1 |z_1|^2, \qquad |\nabla v_1(z_1)| \leq 2c_1 |z_1|$$

$$Dv_{2(\mathscr{S}_2)}(z_2) \leq -2rpc_2 |z_2|^2, \qquad |\nabla v_2(z_2)| \leq 2c_2 |z_2|$$

for all $z_1 \in R^4$ and $z_2 \in R$. The norms of C_{12} and C_{21} induced by the Euclidean norm are $\|C_{12}\| = 2$ and $\|C_{21}\| = (\sum_{i=1}^{4} \beta_i^2)^{1/2}$.

Hypothesis (i) of Corollary 2.4.28 is satisfied. The test matrix S of this corollary is specified by

$$s_{11} = -2\alpha_1 c_1 \rho_1, \qquad s_{22} = -2\alpha_2 rpc_2,$$

$$s_{12} = s_{21} = 2\alpha_1 c_1 + \alpha_2 c_2 \left(\sum_{i=1}^{4} \beta_i^2 \right)^{1/2}.$$

Choosing $\alpha_1 = 1/(4c_1)$ and $\alpha_2 = 1/[2c_2(\sum_{i=1}^{4} \beta_i^2)^{1/2}]$, matrix S assumes the form

$$S = \begin{bmatrix} -\rho_1/2 & 1 \\ 1 & -rp \Big/ \left(\sum_{i=1}^{4} \beta_i^2 \right)^{1/2} \end{bmatrix}.$$

This matrix is negative definite if and only if

$$\sum_{i=1}^{4} \xi_i^2 < 1, \qquad \xi_i = (2\beta_i)/(\rho_1 \, pr). \qquad (2.8.3)$$

It follows from Corollary 2.4.28 that the equilibrium $x = 0$ of Eq. (2.8.2) is exponentially stable in the large for any admissible function f if inequality (2.8.3) is satisfied.

We can also apply Corollary 2.5.21. The test matrix D of this corollary is given by

$$D = \begin{bmatrix} 2c_1\rho_1 & -4c_1 \\ -2c_2\left(\sum_{i=1}^{4}\beta_i^2\right)^{1/2} & 2rpc_2 \end{bmatrix}.$$

Matrix D has positive successive principal minors if and only if inequality (2.8.3) is satisfied. Thus, with the above choice of α_1 and α_2, Corollaries 2.4.28 and 2.5.21 yield the same result.

Utilizing an approach of the type discussed in Section 2.6, Piontkovskii and Rutkovskaya [1] obtain the stability condition

$$\sum_{i=1}^{4} 4\xi_i^2 < 1 \tag{2.8.4}$$

using the above Lyapunov functions $v_1(z_1)$ and $v_2(z_2)$ as components of a vector Lyapunov function. Condition (2.8.3) is clearly less conservative than condition (2.8.4).

2.8.5. Example. Let us reconsider system (2.8.2). For $(\mathscr{S}_1)$ choose $v_1(z_1) = |z_1|$ and for $(\mathscr{S}_2)$ choose $v_2(z_2) = |z_2|$. Then $Dv_{1(\mathscr{S}_1)}(z_1) \leq -\rho_1|z_1|$ and $Dv_{2(\mathscr{S}_2)}(z_2) \leq -rp|\sigma|$. Note that v_1 and v_2 are globally Lipschitzian with $L_1 = L_2 = 1$. Hypothesis (i) of Theorem 2.5.15 is clearly satisfied. Hypothesis (ii) of this theorem is also satisfied with $a_{11} = a_{22} = 0$, $a_{12} = 2$ and $a_{21} = (\sum_{i=1}^{4} \beta_i^2)^{1/2}$. The test matrix D of this theorem assumes the form

$$D = \begin{bmatrix} \rho_1 & -2 \\ -\left(\sum_{i=1}^{4}\beta_i^2\right)^{1/2} & rp \end{bmatrix}.$$

This matrix has positive successive principal minors if and only if the inequality

$$\sum_{i=1}^{4} \xi_i^2 < 1, \qquad \xi_i = (2\beta_i)/(\rho_1 pr) \tag{2.8.6}$$

is satisfied. It follows from Theorem 2.5.15 that the equilibrium $x = 0$ of system (2.8.2) is uniformly asymptotically stable in the large if inequality (2.8.6) is true. Thus, with the preceding choices of Lyapunov functions and weighting vector, Corollaries 2.4.28 and 2.5.21 and Theorem 2.5.15 yield the same stability condition for system (2.8.2).

2.8.7. Example. Let us alter system (2.8.2) by assuming that $f(\sigma)$ is given by

$$f(\sigma) = \sigma(\sigma^2 - a^2), \qquad a > 0, \tag{2.8.8}$$

which is not an admissible nonlinearity in the sense of Example 2.8.1. Specifically, in this case we have $\sigma f(\sigma) > 0$ for all $|\sigma| > a$, $\lim_{|\sigma| \to \infty} \sigma f(\sigma) = \infty$, $f(0) = 0$, $f(a) = 0$, $f(-a) = 0$, and $\sigma f(\sigma) < 0$ for all $0 < |\sigma| < a$.

Utilizing either Theorem 2.4.34 or 2.5.26 and proceeding in an identical fashion as in Example 2.8.1, we conclude that all solutions of system (2.8.2) with $f(\sigma)$ specified by Eq. (2.8.8) are uniformly bounded, and in fact, uniformly ultimately bounded if inequality (2.8.3) is satisfied.

2.8.9. Example. (*Indirect Control Problem.*) An important class of problems arising in automatic control theory is the indirect control problem (see, e.g., Aizerman and Gantmacher [1], Lefschetz [1]) characterized by the set of equations

$$\dot{x} = Ax + bf(\sigma)$$
$$\dot{\sigma} = -\rho\sigma - rf(\sigma) + a^{\mathrm{T}}x \tag{2.8.10}$$

where $x \in R^n$, $\sigma \in R$, A is a stable $n \times n$ matrix (i.e., all eigenvalues of A have negative real parts), $b \in R^n$, $\rho > 0$, $r > 0$, $a \in R^n$, and $f: R \to R$ has the following properties: (a) it is continuous on R, (b) $f(\sigma) = 0$ if and only if $\sigma = 0$, and (c) $0 < \sigma f(\sigma) < k\sigma^2$ for all $\sigma \neq 0$, where $k > 0$ is a constant. System (2.8.10) is called *absolutely stable* if its equilibrium $(x^{\mathrm{T}}, \sigma) = 0$ is asymptotically stable in the large for any *admissible nonlinearity f* with the above properties.

System (2.8.10) may be viewed as a nonlinear interconnection of isolated subsystems $(\mathscr{S}_1)$ and $(\mathscr{S}_2)$ described by

$$\dot{x} = Ax \tag{$\mathscr{S}_1$}$$

and

$$\dot{\sigma} = -\rho\sigma - rf(\sigma). \tag{$\mathscr{S}_2$}$$

Using the notation of Eq. (2.3.16), these subsystems are interconnected to form composite system (2.8.10) by means of the relations $g_{12}(\sigma) = f(\sigma)b$ and $g_{21}(x) = a^{\mathrm{T}}x$.

In the subsequent analysis, as well as in later examples, we utilize the following well-known result (see, e.g., Hahn [2, p. 117]).

2.8.11. Theorem. Let $y \in R^n$, let B be an $n \times n$ matrix, and consider the equation

$$\dot{y} = By. \tag{2.8.12}$$

If all eigenvalues of B have negative real parts or if at least one eigenvalue has positive real part, then there exists a Lyapunov function of the form

$$v(y) = y^{\mathrm{T}}Py, \qquad P^{\mathrm{T}} = P \tag{2.8.13}$$

such that

$$Dv_{(2.8.12)}(y) = -y^{\mathrm{T}}Cy \tag{2.8.14}$$

is definite, where

$$-C = B^T P + PB. \tag{2.8.15}$$

Thus, if the conditions of Theorem 2.8.11 are satisfied, then it is possible to construct a Lyapunov function $v(y)$ by assuming a definite matrix C and solving the *Lyapunov matrix equation* (2.8.15).

Returning to the problem on hand, since A of $(\mathscr{S}_1)$ is a stable matrix it follows from Theorem 2.8.11 that there exists a function $v_1: R^n \to R$ and constants $c_{1i} > 0$, $i = 1, 2, 3, 4$, such that

$$c_{11}|x|^2 \le v_1(x) \le c_{12}|x|^2, \quad Dv_{1(\mathscr{S}_1)}(x) \le -c_{13}|x|^2, \quad |\nabla v_1(x)| \le c_{14}|x|$$

for all $x \in R^n$.

For $(\mathscr{S}_2)$ we choose $v_2(\sigma) = \sigma^2/2$. Then $Dv_{2(\mathscr{S}_2)}(\sigma) \le -\rho|\sigma|^2$ and $|\nabla v_2(\sigma)| = |\sigma|$ for all $\sigma \in R$.

Hypotheses (i)–(iii) of Corollary 2.4.25 are now satisfied with $k_{12} = k|b|$ and $k_{21} = |a|$. Choosing $\alpha_1 = 1/(k|b|)$ and $\alpha_2 = c_{14}/|a|$, matrix S of Corollary 2.4.25 assumes the form

$$S = \begin{bmatrix} -c_{13}/(k|b|) & c_{14} \\ c_{14} & -c_{14}\rho/|a| \end{bmatrix}.$$

This matrix is negative definite if and only if

$$k < (\rho c_{13})/(|a||b|c_{14}). \tag{2.8.16}$$

It follows from Corollary 2.4.25 that the equilibrium $(x^T, \sigma) = 0$ of composite system (2.8.10) is asymptotically stable in the large (i.e., system (2.8.10) is absolutely stable) if inequality (2.8.16) is true.

We can apply Corollary 2.5.20 as well. The test matrix D of this corollary is given by

$$D = \begin{bmatrix} c_{13} & -c_{14}k|b| \\ -|a| & \rho \end{bmatrix}.$$

Matrix D has positive successive principal minors if and only if inequality (2.8.16) is satisfied. Therefore, with the above choice of α_1 and α_2, Corollaries 2.4.25 and 2.5.20 yield the same stability result.

Using an approach of the type discussed in Section 2.6 involving Lyapunov functions $v_1(x)$ and $v_2(\sigma)$ as components of a vector Lyapunov function, Piontkovskii and Rutkovskaya [1] obtain the stability condition

$$k < [(\rho c_{13})/(|a||b|c_{14})](c_{11}/c_{12})^{1/2}. \tag{2.8.17}$$

Since $(c_{11}/c_{12})^{1/2} \le 1$, it follows that condition (2.8.17) is in general more conservative than condition (2.8.16).

To determine the constants c_{13} and c_{14} in (2.8.16), let $v(x) = x^T P x$ with $P^T = P$, so that $Dv_{1(\mathscr{S}_1)}(x) = -x^T C x$. Given a positive definite matrix C, we can solve for P by solving the Lyapunov matrix equation $-C = A^T P + PA$. In so doing we obtain $\lambda_m(P)|x|^2 \le v_1(x) \le \lambda_M(P)|x|^2$, $Dv_{1(\mathscr{S}_1)}(x) \le -\lambda_m(C)|x|^2$, and $|\nabla v_1(x)| \le 2\lambda_M(P)|x|$. Inequality (2.8.16) assumes now the form

$$k < \tfrac{1}{2}[\lambda_m(C)/\lambda_M(P)][\rho/(|a|\,|b|)]. \tag{2.8.18}$$

To obtain the least conservative result, matrix C needs to be chosen in such a fashion as to maximize $\lambda_m(C)/\lambda_M(P)$.

2.8.19. Example. Choosing an appropriate nonsingular linear transformation, we can represent the indirect control problem of Example 2.8.9 equivalently by the set of equations

$$\dot{x}_1 = A_1 x_1 + b_1 f(\sigma)$$
$$\dot{x}_2 = A_2 x_2 + b_2 f(\sigma) \tag{2.8.20}$$
$$\dot{\sigma} = -\rho\sigma - rf(\sigma) + a_1{}^T x_1 + a_2{}^T x_2$$

where $x_1 \in R^{n_1}$, $x_2 \in R^{n_2}$, A_1 is an $n_1 \times n_1$ matrix, A_2 is an $n_2 \times n_2$ matrix, $b_1 \in R^{n_1}$, $b_2 \in R^{n_2}$, $a_1 \in R^{n_1}$, $a_2 \in R^{n_2}$, $n_1 + n_2 = n$, and the remaining symbols are as defined in Example 2.8.9.

System (2.8.20) may be viewed as a nonlinear interconnection of three isolated subsystems $(\mathscr{S}_1)$, $(\mathscr{S}_2)$, $(\mathscr{S}_3)$ of the form

$$\dot{x}_1 = A_1 x_1 \tag{$\mathscr{S}_1$}$$
$$\dot{x}_2 = A_2 x_2 \tag{$\mathscr{S}_2$}$$
$$\dot{\sigma} = -\rho\sigma - rf(\sigma). \tag{$\mathscr{S}_3$}$$

Using the notation of Eq. (2.3.16), these subsystems are interconnected to form composite system (2.8.20) by means of the relations $g_{12}(x_2) = g_{21}(x_1) = 0$, $g_{13}(\sigma) = f(\sigma)b_1$, $g_{23}(\sigma) = f(\sigma)b_2$, $g_{31}(x_1) = a_1{}^T x_1$, and $g_{32}(x_2) = a_2{}^T x_2$.

Assume that all eigenvalues of A_1 have positive real parts and that A_2 is stable. In accordance with Theorem 2.8.11, there exist $v_1 : R^{n_1} \to R$, $v_2 : R^{n_2} \to R$, and positive constants c_{ij}, $i = 1, 2$ and $j = 1, 2, 3, 4$, such that

$$-c_{11}|x_1|^2 \le v_1(x_1) \le -c_{12}|x|^2, \qquad c_{21}|x_2|^2 \le v_2(x_2) \le c_{22}|x_2|^2,$$
$$Dv_{1(\mathscr{S}_1)}(x_1) \le -c_{13}|x_1|^2, \qquad\qquad Dv_{2(\mathscr{S}_2)}(x_2) \le -c_{23}|x_2|^2,$$
$$|\nabla v_1(x_1)| \le c_{14}|x_1|, \qquad\qquad\quad |\nabla v_2(x_2)| \le c_{24}|x_2|,$$

for all $x_1 \in R^{n_1}$ and for all $x_2 \in R^{n_2}$.

For $(\mathscr{S}_3)$ choose $v_3(\sigma) = \tfrac{1}{2}\sigma^2$. Then $Dv_{3(\mathscr{S}_3)}(\sigma) \le -\rho|\sigma|^2$ and $|\nabla v_3(\sigma)| = |\sigma|$ for all $\sigma \in R$.

Using the notation of Theorems 2.4.31 and 2.5.24, we obtain $a_{13} = c_{14} k |b_1|$, $a_{23} = c_{24} k |b_2|$, $a_{31} = |a_1|$, $a_{32} = |a_2|$, and $a_{12} = a_{21} = a_{11} = a_{22} = a_{33} = 0$. Hypotheses (i) and (ii) of Theorem 2.4.31 are thus clearly satisfied. The test matrix D of Theorem 2.5.24 assumes the form

$$D = \begin{bmatrix} c_{13} & 0 & -c_{14}k|b_1| \\ 0 & c_{23} & -c_{24}k|b_2| \\ -|a_1| & -|a_2| & \rho \end{bmatrix}.$$

This matrix has positive successive principal minors if and only if the inequality

$$k < \frac{c_{13} c_{23} \rho}{c_{23} c_{14} |a_1| |b_1| + c_{13} c_{24} |a_2| |b_2|} \tag{2.8.21}$$

is satisfied. It follows from Theorem 2.5.24 that the equilibrium $x^T = (x_1^T, x_2^T, \sigma) = 0$ is unstable (for all admissible nonlinearities f) if inequality (2.8.21) holds.

2.8.22. Example. (*System with Two Nonlinearities.*) Consider the system described by the set of equations

$$\dot{z}_1 = A_1 z_1 + b_1 f_1(\sigma_1), \qquad \dot{z}_2 = A_2 z_2 + b_2 f_2(\sigma_2),$$
$$\sigma_1 = c_1^T z_2, \qquad \sigma_2 = c_2^T z_1, \tag{2.8.23}$$

where for $i = 1, 2$, $z_i \in R^{n_i}$, A_i is a stable $n_i \times n_i$ matrix, $c_1 \in R^{n_2}$, $c_2 \in R^{n_1}$, and $f_i: R \to R$ has the following properties: (a) it is continuous on R, (b) $f_i(\sigma_i) = 0$ if and only if $\sigma_i = 0$, and (c) $0 < \sigma_i f_i(\sigma_i) < k_i \sigma_i^2$ for all $\sigma_i \neq 0$, where $k_i > 0$ is a constant.

System (2.8.23) may be viewed as a nonlinear interconnection of two linear isolated subsystems $(\mathscr{S}_1)$ and $(\mathscr{S}_2)$,

$$\dot{z}_1 = A_1 z_1, \tag{$\mathscr{S}_1$}$$
$$\dot{z}_2 = A_2 z_2, \tag{$\mathscr{S}_2$}$$

which are interconnected to form composite system (2.8.23) by means of the relations $g_{12}(z_2) = b_1 f_1(\sigma_1)$ and $g_{21}(z_1) = b_2 f_2(\sigma_2)$, where the notation of Eq. (2.3.16) is used.

Since A_1 and A_2 are stable matrices there exist $v_i: R^{n_i} \to R$ and constants $c_{ij} > 0$, $i = 1, 2, j = 1, 2, 3, 4$, such that

$$c_{i1} |z_i|^2 \leq v_i(z_i) \leq c_{i2} |z_i|^2, \qquad Dv_{i(\mathscr{S}_i)}(z_i) \leq -c_{i3} |z_i|^2,$$
$$|\nabla v_i(z_i)| \leq c_{i4} |z_i|$$

for all $z_i \in R^{n_i}$. Thus, hypotheses (i) and (ii) of Corollary 2.4.25 are satisfied.

Hypothesis (iii) is satisfied with $k_{12} = k_1 |b_1| |c_1|$ and $k_{21} = k_2 |b_2| |c_2|$. Choosing $\alpha_1 = c_{24}/(k_1 |c_1| |b_1|)$ and $\alpha_2 = c_{14}/(k_2 |c_2| |b_2|)$, matrix S of Corollary 2.4.25 assumes the form

$$S = \begin{bmatrix} -c_{13} c_{24}/(k_1 |b_1| |c_1|) & c_{14} c_{24} \\ c_{14} c_{24} & -c_{23} c_{14}/(k_2 |b_2| |c_2|) \end{bmatrix}.$$

This matrix is negative definite if and only if

$$k_1 k_2 < (c_{13} c_{23})/(c_{14} c_{24} |b_1| |b_2| |c_1| |c_2|). \tag{2.8.24}$$

It follows from Corollary 2.4.25 that the equilibrium $x^T = (x_1{}^T, x_2{}^T) = 0$ of system (2.8.23) is exponentially stable in the large if (2.8.24) holds.

Following an approach of the type discussed in Section 2.6 and using the above Lyapunov functions $v_1(z_1)$ and $v_2(z_2)$ as components of a vector Lyapunov function, Piontkovskii and Rutkovskaya [1] obtain the inequality

$$k_1 k_2 < \frac{c_{13} c_{23}}{c_{14} c_{24} |b_1| |b_2| |c_1| |c_2|} \begin{bmatrix} c_{11} c_{21} \\ c_{12} c_{22} \end{bmatrix}^{1/2} \tag{2.8.25}$$

as a sufficient condition for exponential stability. Since $(c_{11} c_{21})/(c_{12} c_{22}) \le 1$, condition (2.8.25) is in general more conservative than condition (2.8.24).

2.8.26. Example. The purpose of the present simple example is to provide a case where the equilibrium is asymptotically stable but not necessarily exponentially stable and to present an example where the results of Section 2.5 involving M-matrices do not predict stability while those of Section 2.4 do predict stability, using identical Lyapunov functions for the subsystems. Specifically, consider

$$\begin{aligned} \dot{x}_1 &= -x_1{}^3 - 1.5 x_1 |x_2|^3 \\ \dot{x}_2 &= -x_2{}^5 + x_1{}^2 x_2{}^2 \end{aligned} \tag{2.8.27}$$

where $x_1 \in R$ and $x_2 \in R$. This system may be viewed as a nonlinear interconnection of isolated subsystems $(\mathscr{S}_1)$, $(\mathscr{S}_2)$,

$$\dot{x}_1 = -x_1{}^3, \tag{$\mathscr{S}_1$}$$

$$\dot{x}_2 = -x_2{}^5. \tag{$\mathscr{S}_2$}$$

Choosing $v_1(x_1) = x_1{}^2$ and $v_2(x_2) = x_2{}^2$, we have $Dv_{1(\mathscr{S}_1)}(x_1) = -2x_1{}^4$ and $Dv_{2(\mathscr{S}_2)}(x_2) = -2x_2{}^6$. Using the notation of Theorem 2.4.2 we make the identifications $\psi_{11}(r) = \psi_{12}(r) = r^2$, $\psi_{13}(r) = r^4$, $\psi_{21}(r) = \psi_{22}(r) = r^2$, and $\psi_{23}(r) = r^6$. The interconnecting structure of Eq. (2.8.27) is characterized by $g_1(x_1, x_2) = -1.5 x_1 |x_2|^3$ and $g_2(x_1, x_2) = x_1{}^2 x_2{}^2$, where the notation of

Eq. (2.3.10) has been used. We now have

$$\nabla v_1(x_1)g_1(x_1, x_2) = (2x_1)(-1.5x_1)|x_2|^3 = |x_1|^2(-3)|x_2|^3$$
$$= \psi_{13}(|x_1|)^{1/2}(-3)\psi_{23}(|x_2|)^{1/2}$$

and

$$\nabla v_2(x_2)g_2(x_1, x_2) = (2x_2)(x_1{}^2 x_2{}^2) \le |x_2|^3(2)|x_1|^2$$
$$= \psi_{23}(|x_2|)^{1/2}(2)\psi_{13}(|x_1|)^{1/2}.$$

In the notation of Theorem 2.4.2 we now have $a_{11} = a_{22} = 0$, $a_{12} = -3$, $a_{21} = 2$, $\sigma_1 = -2$, and $\sigma_2 = -2$. Choosing $\alpha_1 = \alpha_2 = 1$, matrix S of Theorem 2.4.2 assumes the form

$$S = \begin{bmatrix} -2 & -\frac{1}{2} \\ -\frac{1}{2} & -2 \end{bmatrix}.$$

Since S is negative definite it follows from Theorem 2.4.2 that the equilibrium $(x_1, x_2)^T = x = 0$ of Eq. (2.8.27) is asymptotically stable in the large. However, since the comparison functions ψ_{ij}, $i = 1, 2, j = 1, 2, 3$, are not all of the same order of magnitude, we cannot conclude that the equilibrium is exponentially stable (see Theorem 2.4.20).

Next, note that $-S$ has positive off-diagonal terms and thus the results of Section 2.5 are not directly applicable. However, using the *same* Lyapunov functions as above, let us attempt to establish stability, using a test matrix $D = [d_{ij}]$ for which $d_{ij} \le 0$, $i \ne j$. In the notation of Theorem 2.5.11 we have the estimates

$$\nabla v_1(x_1)g_1(x_1, x_2) \le \psi_{13}(|x_1|)^{1/2}(3)\psi_{23}(|x_2|)^{1/2}$$
$$\nabla v_2(x_2)g_2(x_1, x_2) \le \psi_{23}(|x_2|)^{1/2}(2)\psi_{13}(|x_1|)^{1/2}.$$

Hypotheses (i) and (ii) of Theorem 2.5.11 are now satisfied. The test matrix of hypothesis (iii) assumes the form

$$D = \begin{bmatrix} 2 & -3 \\ -2 & 2 \end{bmatrix}.$$

Since the successive principal minors of D are not both positive, Theorem 2.5.11 fails to predict asymptotic stability. Thus, using the *same* Lyapunov functions $v_1(z_1)$ and $v_2(z_2)$ we were able to predict stability for system (2.8.27) using Theorem 2.4.2, but failed to do so using Theorem 2.5.11.

2.8.28. Example. (*The Hicks Conditions in Economics.*) Let $p_i \ge 0$, $i = 1, ..., n$, denote the prices of n interrelated commodities supplied from the same or related sources which are demanded by the same or related industries and let $f_i(p_1, ..., p_n)$ denote the excess demand function of commodity i. Such a multimarket system may be represented by the equation

$$\dot{p} = f(p) \qquad (2.8.29)$$

where $p^{\mathrm{T}} = (p_1, ..., p_n)$ and where $f: R^n \to R^n$ is assumed to be continuously differentiable with respect to all of its arguments. Assume that $\bar{p} > 0$ is an isolated equilibrium price vector, so that $f(\bar{p}) = 0$. Linearizing Eq. (2.8.29) about the equilibrium $\bar{p}$, we obtain the set of equations

$$\dot{p}_i = \sum_{j=1}^{n} a_{ij}(p_j - \bar{p}_j), \qquad i = 1, ..., n, \qquad (2.8.30)$$

where $a_{ij} = [\partial f_i(p)/\partial p_j]_{p=\bar{p}}$. Here we need to restrict ourselves to "small" initial deviations from the equilibrium. As well as providing mathematical simplicity, this restriction has the advantage of assuring us that in the presence of asymptotic stability, no price will become negative on its path to equilibrium, since it is *never* far from the equilibrium which we have already assumed to be positive. Here we are invoking the *principle of stability in the first approximation* (see Hahn [1, p. 122]) which asserts that we can deduce the stability properties of the equilibrium $\bar{p}$ of Eq. (2.8.29) from Eq. (2.8.30), provided that $A = [a_{ij}]$ is not a critical matrix (i.e., A has only eigenvalues with negative real parts or at least one eigenvalue with positive real part).

Letting $x_i = p_i - \bar{p}_i$, Eq. (2.8.30) assumes the form

$$\dot{x}_i = a_{ii} x_i + \sum_{j=1, i \neq j}^{n} a_{ij} x_j, \qquad i = 1, ..., n. \qquad (2.8.31)$$

Assuming that all commodities are gross substitutes for one another, we have the conditions $a_{ij} \geq 0$, $i \neq j$. We also make the realistic assumptions $a_{ii} < 0$, $i = 1, ..., n$.

Noting that the asymptotic stability results of Sections 2.4 and 2.5 can be modified in an obvious way to yield local conditions rather than global ones, we may view system (2.8.31) as a composite system of n isolated subsystems $(\mathscr{S}_i)$,

$$\dot{x}_i = a_{ii} x_i, \qquad (\mathscr{S}_i)$$

which are interconnected by means of the relations $g_i(x) = \sum_{j=1, i \neq j}^{n} a_{ij} x_j$, $x^{\mathrm{T}} = (x_1, ..., x_n)$, where the notation of Eq. (2.3.10) is used.

For each isolated subsystem $(\mathscr{S}_i)$ we choose

$$v_i(x_i) = |x_i|.$$

Note that v_i is Lipschitzian with $L_i = 1$. Along solutions of $(\mathscr{S}_i)$ we have

$$Dv_{i(\mathscr{S}_i)}(x_i) = a_{ii}|x_i|.$$

We are now in a position to apply Theorem 2.5.15. The test matrix $D = [d_{ij}]$

of this theorem is specified by

$$d_{ij} = \begin{cases} |a_{ii}|, & i = j \\ -|a_{ij}|, & i \neq j. \end{cases}$$

It follows from Theorem 2.5.15 that the equilibrium $x = 0$ of Eq. (2.8.31) (and hence, the equilibrium $\bar{p}$ of Eq. (2.8.29)) is asymptotically stable if the successive principal minors of matrix D are all positive. To put it an equivalent way, the equilibrium price vector $\bar{p}$ of Eq. (2.8.29) is asymptotically stable if

$$(-1)^k \begin{vmatrix} a_{11} & a_{12} & \cdots & a_{1k} \\ a_{21} & a_{22} & \cdots & a_{2k} \\ \vdots & \vdots & \cdots & \vdots \\ a_{k1} & a_{k2} & \cdots & a_{kk} \end{vmatrix} > 0, \qquad k = 1, \ldots, n. \qquad (2.8.32)$$

Conditions (2.8.32) are called the *Hicks conditions* in economics (see, e.g., Quirk and Saposnik [1], Metzler [1], McKenzie [1]). Since D is an M-matrix, we can express these conditions equivalently by requiring the existence of real constants $\lambda_i > 0$, $i = 1, \ldots, l$, such that the inequalities

$$|a_{ii}| - \sum_{j=1, i \neq j}^{n} (\lambda_j / \lambda_i) |a_{ji}| > 0, \qquad i = 1, \ldots, n, \qquad (2.8.33)$$

are true (see Corollary 2.5.3).

2.8.34. Example. (*Linear Resistor–Time Varying Capacitor Circuits.*) A large class of time varying capacitor–linear resistor networks (see Sandberg [10], Mitra and So [1]) can be described by equations of the form

$$\dot{x} + [AD_1(t) + BD_2(t)] x = b(t) \qquad (2.8.35)$$

where $x \in R^n$, $A = [a_{ij}]$, and $B = [b_{ij}]$ are constant $n \times n$ matrices with $a_{ii} > 0$ and $b_{ii} > 0$, $i = 1, \ldots, n$, $D_1(t)$ and $D_2(t)$ are diagonal matrices whose diagonal elements $[D_1(t)]_{ii}$, $[D_2(t)]_{ii}$ are continuous nonnegative functions on $J = [t_0, \infty)$, and $b(t)$ is an n-vector bounded and continuous on J. Henceforth it is assumed that $[D_1(t)]_{ii} + [D_2(t)]_{ii} \geq \delta > 0$, $i = 1, \ldots, n$ for all $t \in J$.

Presently we are interested in the asymptotic stability of the equilibrium $x = 0$. For this reason we let $b(t) = 0$ for all $t \in J$ and consider the free or unforced system

$$\dot{x} + [AD_1(t) + BD_2(t)] x = 0. \qquad (2.8.36)$$

This system may be viewed as n isolated subsystems $(\mathscr{S}_i)$,

$$\dot{x}_i = -\{a_{ii}[D_1(t)]_{ii} + b_{ii}[D_2(t)]_{ii}\} x_i \triangleq m_{ii}(t) x_i, \qquad (\mathscr{S}_i)$$

which are interconnected by the relations $g_{ij}(x_j, t)$ (using the notation of Eq. (2.3.16)),

$$g_{ij}(x_j, t) = -\{a_{ij}[D_1(t)]_{jj} + b_{ij}[D_2(t)]_{jj}\} x_j \triangleq m_{ij}(t) x_j, \qquad i \neq j,$$

where $m_{ij}(t)$ is defined in the obvious way. For $(\mathcal{S}_i)$ we choose

$$v_i(x_i) = \lambda_i |x_i|$$

where $\lambda_i > 0$ is a constant. Note that v_i is Lipschitz continuous for all $x_i \in R$ with Lipschitz constant $L_i = \lambda_i$. Along solutions of $(\mathcal{S}_i)$ we have

$$Dv_{i(\mathcal{S}_i)}(x_i, t) = \lambda_i m_{ii}(t) |x_i|.$$

It now follows from Corollary 2.5.16 that the equilibrium $x = 0$ of Eq. (2.8.36) is uniformly asymptotically stable in the large if

$$F_j(t) \triangleq |m_{jj}(t)| - \sum_{i=1, i \neq j}^{n} (\lambda_i/\lambda_j) |m_{ij}(t)| \geq \gamma > 0, \qquad j = 1, \dots, n.$$
$$(2.8.37)$$

Now assume that

$$a_{jj} - \sum_{i=1, i \neq j}^{n} (\lambda_i/\lambda_j) |a_{ij}| \geq \mathscr{E} > 0, \qquad b_{jj} - \sum_{i=1, i \neq j}^{n} (\lambda_i/\lambda_j) |b_{ij}| \geq \mathscr{E} > 0,$$
$$j = 1, \dots, n.$$
$$(2.8.38)$$

Applying the definition of $m_{ij}(t)$ and the properties of $D_1(t), D_2(t)$ to (2.8.37), we obtain

$$F_j(t) \geq \left(a_{jj} - \sum_{i=1, i \neq j}^{n} (\lambda_i/\lambda_j) |a_{ij}| \right) [D_1(t)]_{jj}$$
$$+ \left(b_{jj} - \sum_{i=1, i \neq j}^{n} (\lambda_i/\lambda_j) |b_{ij}| \right) [D_2(t)]_{jj}$$
$$\geq \mathscr{E} [D_1(t) + D_2(t)]_{jj} \geq \mathscr{E}\delta > 0.$$

Therefore, the equilibrium $x = 0$ of the free system (2.8.36) is uniformly asymptotically stable in the large if inequalities (2.8.38) are true.

It is interesting to note that stability conditions (2.8.38) were obtained by Mitra and So [1] (see also Sandberg [10]) by methods which differ significantly from the present approach.

2.8.39. Example. Using Theorem 2.7.3, we now establish the following estimate for the trajectory behavior of composite system (2.8.35). Assume that for unforced system (2.8.36) there exist constants $\lambda_i > 0$, $i = 1, \dots, n$, and $\mathscr{E} > 0$, such that the inequalities (2.8.38) hold. There is no loss of generality in assuming that $\sum_{i=1}^{n} \lambda_i^2 = 1$. Also assume that for forced system (2.8.35),

$\sum_{i=1}^{n} \lambda_i |b_i(t)| \leq k$ for all $t \in J$, where $b_i(t)$ denotes the ith component of vector $b(t)$. Let $c = \mathscr{E}\delta$, where $\delta > 0$ is defined in Example 2.8.34. If $\alpha > k/c$ and $\sum_{i=1}^{n} \lambda_i |x_{i0}| \leq \alpha$ for the initial vector $x_0 = (x_{10}, \ldots, x_{n0})^T$, then for $t \geq t_0$,

$$|x(t; x_0, t_0)| \leq [\alpha - k/c] e^{-c(t-t_0)} + k/c. \tag{2.8.40}$$

To obtain this estimate, we choose again $v(x) = \sum_{i=1}^{n} \lambda_i |x_i|$. Simple computations yield

$$Dv_{(2.8.35)}(x) \leq -cv(x) + k,$$

where the constants c and k are defined above. In Theorem 2.7.3 let $G(v, t) = -cv + k$. Solving the equation

$$\dot{p} = G(p, t)$$

we obtain

$$p(t; p_0, t_0) = (p_0 - k/c) e^{-c(t-t_0)} + k/c.$$

Choosing

$$S_0(t_0) = \left\{ x = (x_1, \ldots, x_n)^T : \sum_{i=1}^{n} \lambda_i |x_i| \leq \alpha \right\}$$

and

$$S(t) = B([\alpha - k/c] e^{-c(t-t_0)} + k/c),$$

recalling that $\sum_{i=1}^{n} \lambda_i^2 = 1$ by assumption, and noting that

$$\inf_{x \in S(t)} v(x) = (\alpha - k/c) e^{-c(t-t_0)} + k/c$$

it is clear that all hypotheses of Theorem 2.7.3 are satisfied. Thus, estimate (2.8.40) follows directly from the properties of the boundary of $S(t)$, $\partial S(t)$.

It is interesting to note that estimate (2.8.40) was obtained by Mitra and So [1] (refer also to Sandberg [10]) by quite a different method.

Observe that in Example 2.8.34, the Lyapunov function $v(x) = \sum_{i=1}^{n} \lambda_i |x_i|$ and the derivative $Dv_{(2.8.36)}(x)$ can be estimated by comparison functions $\psi_1, \psi_2, \psi_3 \in KR$ which are of the same order of magnitude. In accordance with Theorem 2.2.26, we can actually deduce that the unforced system (2.8.36) is exponentially stable in the large. Indeed, letting $b(t) \equiv 0$, we obtain from inequality (2.8.40) the estimate

$$|x(t; x_0, t_0)| \leq \alpha e^{-c(t-t_0)}, \qquad t \geq t_0, \qquad \sum_{i=1}^{n} \lambda_i |x_{i0}| \leq \alpha.$$

2.8.41. Example. (*Nonlinear Transistor–Linear Resistor Networks.*) Consider nonlinear time-varying systems described by

$$\dot{x} + Af(x, t) + Bg(x, t) = b(t) \tag{2.8.42}$$

where $x \in R^n$, $A = [a_{ij}]$, and $B = [b_{ij}]$ are constant matrices with $a_{ii} > 0$ and $b_{ii} > 0$, where $f: R^n \times J \to R^n$ and $g: R^n \times J \to R^n$ are continuously differentiable in x and continuous in t, where $f(x, t) = 0$ and $g(x, t) = 0$ for all $t \in J$ if and only if $x = 0$, where $f_i(x, t) = f_i(x_i, t)$ and $g_i(x, t) = g_i(x_i, t)$, and where $b(t)$ is a bounded continuous n-vector. Henceforth it is assumed that $[f_i(x_i, t)/x_i] \geq \delta > 0$, $[g_i(x_i, t)/x_i] \geq \delta > 0$ for all $x_i \neq 0$, $t \in J$, and that $[\partial f_i(x_i, t)/\partial x_i]|_{x_i=0} \geq \eta > 0$, $[\partial g_i(x_i, t)/\partial x_i]|_{x_i=0} \geq \eta > 0$ for all $t \in J$. Presently, the asymptotic stability of the equilibrium $x = 0$ is of interest. For this reason we let $b(t) = 0$ for all $t \in J$ and consider the free system

$$\dot{x} + Af(x, t) + Bg(x, t) = 0. \tag{2.8.43}$$

Equation (2.8.42) can be used to model a great variety of physical systems. For example, a class of nonlinear transistor–linear resistor networks is described by Eq. (2.8.42), where the functions $f_i(x_i, t) \equiv f_i(x_i)$, $g_i(x_i, t) \equiv g_i(x_i)$ are assumed to be monotonically increasing in x_i (see Sandberg [10]).

The free system (2.8.43) may be viewed as n isolated subsystems $(\mathscr{S}_i)$,

$$\dot{x}_i = m_{ii}(x_i, t) x_i \tag{$\mathscr{S}_i$}$$

which are interconnected by the relations $g_{ij}(x_j, t)$ (using the notation of Eq. (2.3.16))

$$g_{ij}(x_j, t) = m_{ij}(x_j, t) x_j, \qquad i \neq j,$$

where

$$m_{kl}(x_l, t) = -\{a_{kl}[f_l(x_l, t)/x_l] + b_{kl}[g_l(x_l, t)/x_l]\}, \qquad x_l \neq 0, \tag{2.8.44}$$

and

$$m_{kl}(x_l, t) = -\{a_{kl}[\partial f_l(x_l, t)/\partial x_l] + b_{kl}[\partial g_l(x_l, t)/\partial x_l]\}, \qquad x_l = 0. \tag{2.8.45}$$

For $(\mathscr{S}_i)$ we choose

$$v_i(x_i) = \lambda_i |x_i|, \qquad \lambda_i > 0 \quad \text{a constant.}$$

Along solutions of $(\mathscr{S}_i)$ we have

$$Dv_{i(\mathscr{S}_i)}(x_i, t) = \lambda_i m_{ii}(x_i, t) |x_i|.$$

It follows from Corollary 2.5.16 that the equilibrium $x = 0$ of Eq. (2.8.43) is uniformly asymptotically stable in the large if

$$F_j(x_j, t) \triangleq |m_{jj}(x_j, t)| - \sum_{i=1, i \neq j}^{n} (\lambda_i/\lambda_j) |m_{ij}(x_j, t)| \geq \gamma > 0, \tag{2.8.46}$$
$$j = 1, \ldots, n,$$

for all $x \in R^n$ and $t \in J$. Applying the definition of $m_{ij}(x_j, t)$ and taking the properties of $f_i(x_i, t)$ and $g_i(x_i, t)$ into account, it follows similarly as in Example 2.8.34 that the equilibrium $x = 0$ of unforced system (2.8.43) is uniformly asymptotically stable in the large if there exist constants $\lambda_i > 0$, $i = 1, \ldots, n$, such that

$$a_{jj} - \sum_{i=1, i \neq j}^{n} (\lambda_i/\lambda_j) |a_{ij}| \geq \mathscr{E} > 0, \qquad b_{jj} - \sum_{i=1, i \neq j}^{n} (\lambda_i/\lambda_j) |b_{ij}| \geq \mathscr{E} > 0,$$

$$j = 1, \ldots, n.$$
$$(2.8.47)$$

Following a procedure similar to that of Example 2.8.39, it is also possible to obtain an estimate for $|x(t; x_0, t_0)|$, by invoking Theorem 2.7.3. Interestingly, the stability conditions (2.8.47) were obtained earlier by quite different methods (see Mitra and So [1], Sandberg [10]).

2.8.48. Example. Consider the composite system described by the set of equations

$$\dot{z}_i = A_i(t) z_i + \sum_{j=1}^{l} C_{ij}(t) z_j, \qquad i = 1, \ldots, l, \qquad (2.8.49)$$

where $t \in J$, $z_i \in R^{n_i}$, and $A_i(t)$ and $C_{ij}(t)$ are real continuous $n_i \times n_i$ and $n_i \times n_j$ matrices, respectively. Let $\sum_{i=1}^{l} n_i = n$, let $x^T = (z_1^T, \ldots, z_l^T)$, and let $g_i(x, t) = \sum_{j=1}^{l} C_{ij}(t) z_j$, so that

$$\dot{z}_i = A_i(t) z_i + g_i(x, t), \qquad i = 1, \ldots, l. \qquad (2.8.50)$$

This system is clearly a special case of system $(\mathscr{S}'')$ with decomposition (Σ_i'').

Now assume that $\|C_{ij}(t)\| < 1$, $i, j = 1, \ldots, l$, for all $t \in J$. Choose $S^i(t) \equiv B^i(\beta_i) = \{z_i \in R^{n_i} : |z_i| < \beta_i\}$, $S_0^i(t) \equiv B^i(\alpha_i)$, $0 < \alpha_i < \beta_i$, $i = 1, \ldots, l$, and

$$D_i(t) = \left\{ g_i(x, t) : |g_i(x, t)| < \sum_{j=1}^{l} \beta_j \text{ for all } x \in \underset{j=1}{\overset{l}{\times}} B^j(\beta_j) \text{ and } t \in J \right\}.$$

Since $|g_i(x, t)| = |\sum_{j=1}^{l} C_{ij}(t) z_j| \leq \sum_{j=1}^{l} \|C_{ij}(t)\| |z_j| < \sum_{j=1}^{l} |z_j|$ it follows that $g_i(x, t) \in D_i(t)$ whenever $x \in \times_{j=1}^{l} S^j(t) = \times_{j=1}^{l} B^j(\beta_j)$ and $t \in J$. Hence, hypothesis (i) of Theorem 2.7.13 is satisfied.

Now let $W_i(t) = [A_i^T(t) + A_i(t)]/2$ and choose $v_i(z_i) = \ln z_i^T z_i$ for all $z_i \in B^i(\beta_i) - B^i(\alpha_i)$. Using Theorem 2.7.9, it is an easy matter to show that each system described by Eq. (2.8.50) is uniformly totally stable with respect to $\{B^i(\alpha_i), B^i(\beta_i), D_i(t)\}$, $0 < \alpha_i < \beta_i$, if

$$\int_{t_1}^{t_2} \left\{ \lambda_M [W_i(t)] + (1/\alpha_i) \sum_{j=1}^{l} \beta_j \right\} dt < \ln(\beta_i/\alpha_i) \qquad (2.8.51)$$

for all $t_1 < t_2$, $t_1, t_2 \in J$.

Direct application of Theorem 2.7.13 yields now the following result. Composite system (2.8.49) is uniformly stable with respect to

$$\left\{ \bigtimes_{j=1}^{l} B^j(\alpha_j),\ \bigtimes_{j=1}^{l} B^j(\beta_j) \right\} \qquad \text{if}\quad \|C_{ij}(t)\| < 1, \qquad i,j = 1,\dots,l,$$

for all $t \in J$ and if inequality (2.8.51) holds for each i.

It is possible to generalize the preceding result. Let $\alpha_i: J \to R^+$, $\beta_i: J \to R^+$ be continuous functions and remove the restriction $\|C_{ij}(t)\| < 1$. Following a similar procedure as before, we can show that composite system (2.8.49) is uniformly stable with respect to $\{\bigtimes_{j=1}^{l} B^j(\alpha_j(t)), \bigtimes_{j=1}^{l} B^j(\beta_j(t))\}$, $0 < \delta \le \alpha_j(t) < \beta_j(t), j = 1,\dots,l, t \in J$, if for each i the inequality

$$\int_{t_1}^{t_2} \left\{ \lambda_M[W_i(t)] + (1/\alpha_i(t)) \sum_{j=1}^{l} \|C_{ij}(t)\|\,\beta_j(t) \right\} dt < \ln[\beta_i(t)/\alpha_i(t)]$$

is satisfied for all $t_1 < t_2$, $t_1, t_2 \in J$.

2.8.52. Example. (*Multiple Feedback Systems: Direct Control Case.*) We begin by considering l isolated subsystems ($\mathscr{S}_i$) described by the set of equations

$$\dot{z}_i = A_i z_i - b_i f_i(\sigma_i) \qquad\qquad (\mathscr{S}_i)$$
$$\sigma_i = d_{ii} c_i^{\mathrm{T}} z_i,$$

$i = 1,\dots,l$, where $z_i \in R^{n_i}$, A_i is a stable $n_i \times n_i$ matrix, $b_i \in R^{n_i}$, $c_i \in R^{n_i}$, $d_{ii} \in R$, $\sigma_i \in R$ and $f_i: R \to R$ is a continuous function such that $f_i(\sigma_i) = 0$ if and only if $\sigma_i = 0$ and $0 < \sigma_i f_i(\sigma_i) \le k_i \sigma_i^2$ for all $\sigma_i \ne 0$, where $k_i > 0$ is a constant. In this case we call f_i an "admissible nonlinearity." System ($\mathscr{S}_i$) is known as the direct control problem (see, e.g., Aizerman and Gantmacher [1], Lefschetz [1]).

Next we consider the composite system

$$\dot{z}_i = A_i z_i - b_i f_i(\sigma_i)$$
$$y_i = c_i^{\mathrm{T}} z_i \qquad\qquad (2.8.53)$$
$$\sigma_i = d_i^{\mathrm{T}} y = \sum_{j=1}^{l} d_{ij}\, y_j,$$

$i = 1,\dots,l$, where $y^{\mathrm{T}} = (y_1,\dots,y_l) \in R^l$ and $d_i^{\mathrm{T}} = (d_{i1},\dots,d_{il}) \in R^l$. Let $\sum_{i=1}^{l} n_i = n$, let $x^{\mathrm{T}} = (z_1^{\mathrm{T}},\dots,z_l^{\mathrm{T}}) \in R^n$, and assume that $x = 0$ is the only equilibrium of system (2.8.53).

It is important to note that in this example the interconnecting structure of composite system (2.8.53) does *not* enter additively into the system description. Given ($\mathscr{S}_i$) as above, system (2.8.53) is *not* a special case of composite system ($\mathscr{S}$) with decomposition (Σ_i). However, it is a special case of composite system ($\mathscr{S}''$) with decomposition (Σ_i'').

Let us first apply Theorem 2.5.22. We choose

$$v_i(z_i) = z_i^T P_i z_i$$

where $P_i = P_i^T$ is a positive definite matrix which we will need to determine for some particular choice of a positive definite matrix $C_i = C_i^T$ via the Lyapunov equation

$$-C_i = A_i^T P_i + P_i A_i. \tag{2.8.54}$$

We have

$$\lambda_m(P_i)|z_i|^2 \le v_i(z_i) \le \lambda_M(P_i)|z_i|^2,$$

and using the notation of Theorems 2.4.29 and 2.5.22, we have

$$\nabla v_i(z_i)^T h_i(x) \le |z_i| \left\{ -\lambda_m(C_i) + 2k_i \lambda_M(P_i)|b_i|\,|d_{ii}|\,|c_i| \right\} |z_i|$$

$$+ |z_i| \left\{ \sum_{j=1, i \ne j}^{l} 2k_i \lambda_M(P_i)|b_i|\,|d_{ij}|\,|c_j| \right\} |z_j|.$$

Hypotheses (i) and (ii) of Theorem 2.4.29 are now clearly satisfied. The matrix $D = [d_{ij}]$ of Theorem 2.5.22 is given by

$$d_{ij} = \begin{cases} \lambda_m(C_i) - 2k_i \lambda_M(P_i)|b_i|\,|d_{ii}|\,|c_i|, & i = j \\ -2k_i \lambda_M(P_i)|b_i|\,|d_{ij}|\,|c_j|, & i \ne j. \end{cases}$$

From Theorem 2.5.22 it now follows that the equilibrium $x = 0$ of composite system (2.8.53) is exponentially stable in the large for every admissible nonlinearity (i.e., system (2.8.53) is absolutely stable) if all principal minors of matrix D are positive. In order to obtain the least conservative results, we need to choose matrix C_i so that the ratio $\lambda_m(C_i)/\lambda_M(P_i)$ is maximized.

Finally it should be noted that Eq. (2.8.53) can serve as a model of a large class of practical systems endowed with multiple nonlinearities.

2.8.55. Example. We have pointed out before that in a sense the method of analysis advanced herein is more important than the individual results of the preceding sections. To demonstrate this, we reconsider system (2.8.53) in an attempt to obtain less conservative stability conditions. This time we choose for each ($\mathscr{S}_i$) a Lyapunov function of the Luré form (see, e.g., Aizerman and Gantmacher [1], Lefschetz [1]),

$$v_i(z_i) = z_i^T P_i z_i + \beta_i \int_0^{\sigma_i} f_i(\eta)\, d\eta \tag{2.8.56}$$

where $P_i = P_i^T$ is a positive definite matrix and $\beta_i > 0$ is a constant. Clearly, $v_i(z_i)$ is positive definite and radially unbounded.

For composite system (2.8.53) we choose a Lyapunov function of the form

$$v(x) = \sum_{i=1}^{l} \alpha_i v_i(z_i) \tag{2.8.57}$$

where $\alpha^T = (\alpha_1, \ldots, \alpha_l) > 0$ is a weighting vector. This function is positive definite for all $x \in R^n$ and is radially unbounded. Along solutions of Eq. (2.8.53) we have

$$Dv_{(2.8.53)}(x) = \sum_{i=1}^{l} \alpha_i z_i^T [-C_i] z_i - 2 \sum_{i=1}^{l} \alpha_i b_i^T P_i z_i f_i(\sigma_i)$$

$$+ \sum_{i=1}^{l} \alpha_i \beta_i f_i(\sigma_i) \sum_{j=1}^{l} d_{ij} c_j^T A_j z_j$$

$$- \sum_{i=1}^{l} \alpha_i \beta_i f_i(\sigma_i) \sum_{j=1}^{l} d_{ij} c_j^T b_j f_j(\sigma_j),$$

where $C_i = -(A_i^T P_i + P_i A_i)$. Adding and subtracting the nonnegative quantity $\sum_{i=1}^{l} \alpha_i(\sigma_i - (f_i(\sigma_i)/k_i)) f_i(\sigma_i)$ to $Dv_{(2.8.53)}(x)$, we obtain

$$Dv_{(2.8.53)}(x) = w^T \begin{bmatrix} C & S \\ S^T & R \end{bmatrix} w - \sum_{i=1}^{l} \alpha_i(\sigma_i - (f_i(\sigma_i)/k_i)) f_i(\sigma_i)$$

where $w^T = (z_1^T, \ldots, z_l^T, f_1(\sigma_1), \ldots, f_l(\sigma_l))$, where

$$C = \begin{bmatrix} -\alpha_1 C_1 & & 0 \\ & \ddots & \\ 0 & & -\alpha_l C_l \end{bmatrix}$$

is an $n \times n$ matrix, where $S = [s_{ij}]$ is an $n \times l$ matrix, where each s_{ij},

$$s_{ij} = \begin{cases} \frac{1}{2}A_i^T c_i d_{ii} \alpha_i \beta_i + \frac{1}{2} c_i d_{ii} \alpha_i - P_i b_i \alpha_i, & i = j \\ \frac{1}{2}A_i^T c_i d_{ji} \alpha_j \beta_j + \frac{1}{2} c_i d_{ji} \alpha_j, & i \neq j \end{cases}$$

is an n_i-vector, and $R = [r_{ij}]$ is an $l \times l$ matrix, and where

$$r_{ij} = \begin{cases} -\alpha_i \beta_i d_{ii} c_i^T b_i - (\alpha_i/k_i), & i = j \\ -\frac{1}{2}(\alpha_i \beta_i d_{ij} c_j^T b_j + \alpha_j \beta_j d_{ji} c_i^T b_i), & i \neq j. \end{cases}$$

It now follows that system (2.8.53) is absolutely stable if the symmetric $(n+l) \times (n+l)$ matrix

$$\begin{bmatrix} C & S \\ S^T & R \end{bmatrix} \tag{2.8.58}$$

is negative definite.

It is interesting to note that this result is somewhat similar to one obtained by Lefschetz [1], using quite different methods.

Under suitable conditions, frequency domain techniques can be used to construct Lyapunov functions of the form (2.8.56) for subsystem $(\mathscr{S}_i)$ (see Bose [1], Bose and Michel [1, 2]). If the pairs (A_i, b_i) and (A_i^T, c_i) are completely controllable and completely observable, respectively, (see Kalman, Ho, and Narendra [1]) then the Popov criterion (see Popov [1]) along with additional reasonable conditions guarantee the existence of a Lyapunov function for $(\mathscr{S}_i)$ of the Luré type (by the Kalman–Yacubovich Lemma) whose derivative is negative definite (see Kalman [1], Yacubovich [1]). In fact, an algorithm based on the proof of the Kalman–Yacubovich Lemma can be implemented on the computer to construct the Lyapunov function (2.8.56). However, this approach is not particularly satisfactory, for it is not explicit. We will make extensive use of graphical methods (including Popov-type conditions) in Chapter VI, where we establish results which are much easier to use and which are applicable to a larger class of problems.

2.8.59. Example. (*Multiple Feedback Systems: Indirect Control Case.*) Consider systems described by the set of equations

$$\dot{z}_i = A_i z_i - b_i f_i(\sigma_i)$$
$$y_i = c_i^T z_i - \gamma_i f_i(\sigma_i) \qquad (2.8.60)$$
$$\dot{\sigma}_i = d_i^T y = \sum_{j=1}^{l} d_{ij} y_j,$$

$i = 1, \dots, l$, where $\gamma_i > 0$ is a constant and all other symbols are defined similarly as those in Eq. (2.8.53). In addition, assume that $\lim_{|\sigma_i| \to \infty} \int_0^\sigma f_i(\eta) \, d\eta = \infty$. This system may be viewed as a nonlinear interconnection of isolated subsystems $(\mathscr{S}_i)$,

$$\dot{z}_i = A_i z_i - b_i f_i(\sigma_i)$$
$$\dot{\sigma}_i = d_{ii}[c_i^T z_i - \gamma_i f_i(\sigma_i)], \qquad (\mathscr{S}_i)$$

$i = 1, \dots, l$. For each $(\mathscr{S}_i)$ we choose a Luré type Lyapunov function

$$v_i(z_i, \sigma_i) = z_i^T P_i z_i + \beta_i \int_0^{\sigma_i} f_i(\eta) \, d\eta \qquad (2.8.61)$$

where $P_i = P_i^T$ is a positive definite matrix and $\beta_i > 0$ is a constant. For system (2.8.60) we choose a Lyapunov function of the form

$$v(x, \sigma) = \sum_{i=1}^{l} \alpha_i v_i(z_i, \sigma_i) \qquad (2.8.62)$$

where $\sigma^T = (\sigma_1, \dots, \sigma_l)$, $x^T = (z_1^T, \dots, z_l^T)$, and $\alpha^T = (\alpha_1, \dots, \alpha_l) > 0$. This function is clearly positive definite and radially unbounded. Along solutions of

Eq. (2.8.60) we have

$$Dv_{(2.8.60)}(x, \sigma) = w^{\mathrm{T}} \begin{bmatrix} \tilde{C} & \tilde{S} \\ \tilde{S}^{\mathrm{T}} & \tilde{R} \end{bmatrix} w$$

where w is defined as in Example 2.8.55, $\tilde{C}$ is of the same form as submatrix C of (2.8.58), $\tilde{S} = [\tilde{s}_{ij}]$ is given by

$$\tilde{s}_{ij} = \begin{cases} \frac{1}{2} c_i d_{ii} \alpha_i \beta_i - \alpha_i P_i b_i, & i = j \\ \frac{1}{2} c_i d_{ji} \alpha_j \beta_j, & i \neq j, \end{cases}$$

and $\tilde{R} = [\tilde{r}_{ij}]$ is specified by

$$\tilde{r}_{ij} = \tfrac{1}{2}(\alpha_i \beta_i d_{ij} \gamma_j + \alpha_j \beta_j d_{ji} \gamma_i).$$

Noting that $(x^{\mathrm{T}}, \sigma^{\mathrm{T}}) = 0$ is the only invariant subset of the set $E = \{(x^{\mathrm{T}}, \sigma^{\mathrm{T}}) \in R^{n+l} : Dv_{(2.8.60)}(x, \sigma) = 0\}$, it follows from Theorem 2.2.30 that the origin $(x^{\mathrm{T}}, \sigma^{\mathrm{T}}) = 0$ of system (2.8.60) is asymptotically stable in the large for all admissible nonlinearities if the matrix

$$\begin{bmatrix} \tilde{C} & \tilde{S} \\ \tilde{S}^{\mathrm{T}} & \tilde{R} \end{bmatrix}$$

is negative definite.

This result is somewhat similar to one obtained by Lefschetz [1], using a different approach. It is also interesting to note that Pai and Narayan [1] have used a set of equations similar to Eq. (2.8.60) and a Lyapunov function of the form (2.8.62) to analyze *multi-machine power systems*.

2.8.63. Example. (*Stabilization.*) Consider systems described by equations of the form

$$\dot{z}_i = A_i z_i + \sum_{j=1, i \neq j}^{l} C_{ij} z_j, \qquad i = 1, \dots, l, \qquad (2.8.64)$$

where $z_i \in R^{n_i}$, A_i is an $n_i \times n_i$ matrix, and C_{ij} is an $n_i \times n_j$ matrix. Suppose this system, called an *uncompensated system*, is unstable or that its degree of stability is not acceptable. Associated with (2.8.64), consider the *compensated system*

$$\dot{z}_i = A_i z_i + \sum_{j=1, i \neq j}^{l} C_{ij} z_j + B_i u_i, \qquad i = 1, \dots, l, \qquad (2.8.65)$$

where $u_i \in R^{m_i}$ and B_i is an $n_i \times m_i$ matrix. Let $B_i u_i$ be of the form

$$B_i u_i = B_i \sum_{j=1}^{l} E_{ij} z_j \triangleq \sum_{j=1}^{l} F_{ij} z_j,$$

where E_{ij} is an $m_i \times n_j$ matrix. System (2.8.65) may be viewed as a linear interconnection of l isolated subsystems $(\mathscr{S}_i)$,

$$\dot{z}_i = A_i z_i + B_i E_{ii} z_i = A_i z_i + F_{ii} z_i, \qquad (\mathscr{S}_i)$$

$i = 1, ..., l$. If (A_i, B_i) is controllable, we can choose the local feedback matrix E_{ii} in such a fashion that $(\mathscr{S}_i)$ has a desired degree of exponential stability. To determine this property, choose

$$v_i(z_i) = z_i^T P_i z_i, \qquad P_i = P_i^T$$

where P_i is a positive definite matrix which needs to be determined for some choice of a positive definite matrix C_i, via the Lyapunov equation

$$-C_i = (A_i + F_{ii})^T P_i + P_i (A_i + F_{ii}).$$

This yields the estimates $\lambda_m(P)|z_i|^2 \leq v_i(z_i) \leq \lambda_M(P_i)|z_i|^2$, $|\nabla v_i(z_i)| \leq 2\lambda_M(P_i)|z_i|$, and $Dv_{i(\mathscr{S}_i)}(z_i) \leq -\lambda_m(C_i)|z_i|^2$. In accordance with Corollary 2.5.21, the compensated system (2.8.65) will be exponentially stable in the large if the principal minors of the test matrix $D = [d_{ij}]$ are all positive, where

$$d_{ij} = \begin{cases} \lambda_m(C_i), & i = j \\ -2\lambda_M(P_i)\|C_{ij} + F_{ij}\|, & i \neq j. \end{cases}$$

This is the case if for example

$$\lambda_m(C_i)/[2\lambda_M(P_i)] - \sum_{j=1, i \neq j}^{l} \|C_{ij} + F_{ij}\| > 0, \qquad i = 1, ..., l. \quad (2.8.66)$$

Thus, for given matrices B_i, $i = 1, ..., l$, we need to choose feedback matrices E_{ij}, $i, j = 1, ..., l$, and matrices C_i, $i = 1, ..., l$, in such a fashion that the ratios $\lambda_m(C_i)/\lambda_M(P_i)$, $i = 1, ..., l$ are maximized and such that $\|C_{ij} + F_{ij}\| \leq \|C_{ij}\|$, $i, j = 1, ..., l$, $i \neq j$. Finally, we can invoke Corollary 2.5.23 to obtain an indication of the degree of stability of the compensated system, by computing $\mu = \min_k \mathrm{Re}[\lambda_k(D)]$, $k = 1, ..., l$.

2.9 Notes and References

For existence and uniqueness results of solutions for ordinary differential equations refer to Coddington and Levinson [1]. Standard references on the Lyapunov theory include Hahn [2], Yoshizawa [1], Krasovskii [1], and LaSalle and Lefschetz [1]. Our exposition is heavily influenced by Hahn [2], where an extensive discussion of comparison functions of class K is given (Hahn [2, pp. 95–97]). For a good overview of comparison theorems and the comparison principle, refer to Lakshmikantham and Leela [1, 2], Szarski [1], and Walter [1]. The original results in this area include Müller [1] and Kamke [1]. For subsequent work, refer to Wazewski [1], Grimmer [1] and LaSalle [1]. Standard references on absolute stability are the monographs by Aizerman and Gantmacher [1] and Lefschetz [1]. Nice sources on M-matrices include the papers by Ostrowski [1] and Fiedler and Ptak [1] and the books by Bellman [3] and Gantmacher [1, 2]. See also the report by Araki [2].

Sections 2.3 and 2.4 are based on references by Porter and Michel [1, 2], Michel and Porter [3], Michel [5, 7], Bose and Michel [1, 2], and Bose [1]. For related results refer to Thompson [1, 2] and Thompson and Koenig [1]. Section 2.5 is an adaptation and expansion of results reported by Michel and Porter [3], Araki and Kondo [1], Michel [5-7, 9], Araki [1, 4], Rasmussen and Michel [2, 4], and Rasmussen [1].

The introduction of vector Lyapunov functions by Bellman [2] gave rise to the qualitative analysis of large scale systems, using the approach advanced herein. Bailey [1, 2] was the first to apply the comparison principle to vector Lyapunov functions in analyzing nonlinear composite systems with linear interconnecting structure. Subsequent extensions include the work of Matrosov [1, 2] whose approach we follow in Section 2.6. Refer also to Piontkovskii and Rutkovskaya [1].

Additional related references on Lyapunov stability of composite systems include Weissenberger [1], Matzer [1], Grujić and Siljak [2], Tokumaru, Adachi, and Amemiya [2], and Athans, Sandell, and Varaiya [1].

Section 2.7 is based on results by Michel [2, 4, 9]. Related work is contained in Michel [1, 3], Michel and Heinen [1-4], and Michel and Porter [1, 2]. For results on practical stability and finite time stability, refer to LaSalle and Lefschetz [1], Weiss and Infante [1], and Michel and Porter [4].

The source of Examples 2.8.1, 2.8.9, and 2.8.22 is Piontkovskii and Rutkovskaya [1]. Our treatment of these examples follows Michel [5, 7]. For a good qualitative treatment of economic systems refer to Quirk and Saposnik [1]. The sources of Examples 2.8.34, 2.8.39, and 2.8.41 are Sandberg [10], Mitra and So [1], and the related work by Rosenbrock [1, 2]. In our approach to these examples we follow Michel [9]. Multiple feedback systems similar to those of Examples 2.8.52 and 2.8.59 are treated by Lefschetz [1] .The present results, which are not entirely identical to those of Lefschetz [1], are based on references by Bose [1] and Bose and Michel [1, 2]. Sources concerned with systems containing many nonlinearities are numerous. For some of these refer to Narendra and Neuman [1], Narendra and Taylor [1], McClamroch and Ianculescu [1], and Blight and McClamroch [1].

CHAPTER III

Discrete Time Systems
and Sampled Data Systems

In the present chapter we continue the stability analysis of large scale systems described on finite-dimensional spaces. In particular, we show how the methods of Chapter II need to be modified for the case of discrete time systems described by ordinary difference equations and also for the case of sampled data systems. Although we develop only selected results in this chapter, it will become apparent that most of the results and comments of Chapter II can in principle be modified to accommodate corresponding systems described by ordinary difference equations. However, it will also become apparent that these modifications are not completely obvious in every case.

Difference equations are important in their own right in applications. They are also frequently used to represent sampled data systems at discrete points in time. However, a complete description of sampled data systems, at all points in time, provides us with the first important class of hybrid systems. Later, in Chapter V, we will be able to analyze hybrid systems (systems described by a mixture of equations) defined on infinite-dimensional spaces.

This chapter consists of five parts. In the first section we briefly introduce the types of difference equations considered, while in the second section we characterize the classes of composite systems which we will treat. In the third section we state and prove selected results for stability, asymptotic stability, exponential stability, and instability of interconnected systems described by ordinary difference equations. In the fourth section we apply these results to some specific examples. In the fifth section we consider the stability analysis of a class of composite sampled data systems, valid for all points in time. We conclude our presentation with a brief discussion of pertinent literature in the last section.

As in Chapter II, the objective in this chapter will be to analyze large scale systems in terms of lower order subsystems and in terms of interconnecting structure.

3.1   Systems Described by Difference Equations

We consider discrete time systems described by difference equations of the form

$$x(\tau+1) = g[x(\tau), \tau] \tag{3.1.1}$$

where $\tau \in I \triangleq \{t_0 + k\}$, $t_0 \geq 0$, $k = 0, 1, 2, \ldots$, $x \in R^n$, and $g: R^n \times I \to R^n$. Note that for every $x_0 \in R^n$ and for every $t_0 \geq 0$, Eq. (3.1.1) has a unique solution $x(\tau; x_0, t_0)$ which is defined for all $\tau \in I$, with $x(t_0; x_0, t_0) = x_0$. We assume that $g(x, \tau) = x$ for all $\tau \in I$ if and only if $x = 0$. Thus, Eq. (3.1.1) admits the trivial solution $x = 0$ which is in fact the only equilibrium point.

We can characterize the equilibrium of Eq. (3.1.1) as being uniformly stable, uniformly asymptotically stable, exponentially stable, unstable, completely unstable, etc., and we can characterize the solutions of Eq. (3.1.1) as being uniformly bounded, uniformly ultimately bounded, etc., by rephrasing Definitions 2.2.1–2.2.11 *verbatim*, save that $t \in J$ is replaced by $\tau \in I$.

The Lyapunov results for system (3.1.1) involve the existence of mappings $v: R^n \times I \to R$ and the first difference $Dv(x, \tau)$ along solutions of Eq. (3.1.1) expressed by

$$Dv_{(3.1.1)}(x, \tau) = v[g(x, \tau), \tau+1] - v(x, \tau).$$

Such functions are characterized as being positive definite, negative definite, radially unbounded, decrescent, positive semidefinite, and negative semidefinite by modifying Definitions 2.2.16–2.2.20, respectively, in an obvious way, replacing $t \in J$ by $\tau \in I$.

Lyapunov theorems for stability, uniform stability, uniform asymptotic stability, exponential stability, uniform asymptotic stability in the large, exponential stability in the large, instability and complete instability of the equilibrium of Eq. (3.1.1) as well as Lyapunov-type theorems for uniform boundedness and uniform ultimate boundedness of the solutions of Eq. (3.1.1) have the form of Theorems 2.2.21–2.2.28, with $t \in J$ replaced by $\tau \in I$.

3.2 Large Scale Systems

We consider discrete time systems described by equations of the form

$$(\Sigma_i) \qquad z_i(\tau+1) = f_i[z_i(\tau), \tau] + g_i[z_1(\tau), ..., z_l(\tau), \tau], \qquad (3.2.1)$$

$i = 1, ..., l$, where $\tau \in I$, $z_i \in R^{n_i}$, $f_i: R^{n_i} \times I \to R^{n_i}$, and $g_i: R^{n_1} \times \cdots \times R^{n_l} \times I \to R^{n_i}$. Letting $\sum_{i=1}^l n_i = n$, $x^T = (z_1^T, ..., z_l^T)$, $f(x, \tau)^T = [f_1(z_1, \tau)^T, ..., f_l(z_l, \tau)^T]$, and

$$g(x, \tau)^T = [g_1(z_1, ..., z_l, \tau)^T, ..., g_l(z_1, ..., z_l, \tau)^T] \triangleq [g_1(x, \tau)^T, ..., g_l(x, \tau)^T],$$

we can represent Eq. (3.2.1) equivalently as

$$(\mathscr{S}) \qquad x(\tau+1) = f[x(\tau), \tau] + g[x(\tau), \tau] \triangleq h[x(\tau), \tau]. \qquad (3.2.2)$$

We call system $(\mathscr{S})$, which is of the same form as Eq. (3.1.1), a **composite system** or an **interconnected system with decomposition** (Σ_i). System $(\mathscr{S})$ may be viewed as a nonlinear and time varying interconnection of l isolated subsystems $(\mathscr{S}_i)$ described by equations of the form

$$(\mathscr{S}_i) \qquad z_i(\tau+1) = f_i[z_i(\tau), \tau]. \qquad (3.2.3)$$

We denote the unique solutions of $(\mathscr{S})$ and $(\mathscr{S}_i)$ by $x(\tau; x_0, t_0)$ and $z_i(\tau; z_{i_0}, t_0)$, respectively. Furthermore, we assume that $x = 0$ and $z_i = 0$ are the only equilibrium points of $(\mathscr{S})$ and $(\mathscr{S}_i)$, respectively.

The interconnecting structure enters additatively into the description of system $(\mathscr{S})$. As such, this system is a special case of composite systems described by equations of the form

$$(\Sigma_i'') \qquad z_i(\tau+1) = f_i[z_i(\tau), g_i(z_1(\tau), ..., z_l(\tau), \tau), \tau]$$
$$= f_i[z_i(\tau), g_i(x(\tau), \tau), \tau] \triangleq h_i[x(\tau), \tau], \qquad (3.2.4)$$

$i = 1, ..., l$, where $z_i \in R^{n_i}$, $\tau \in I$, $\sum_{i=1}^l n_i = n$, $x^T = (z_1^T, ..., z_l^T) \in R^n$, $g_i: R^n \times I \to R^{r_i}$, $f_i: R^{n_i} \times R^{r_i} \times I \to R^{n_i}$, and $h_i: R^n \times I \to R^{n_i}$. Letting $h(x, \tau)^T = [h_1(x, \tau)^T, ..., h_l(x, \tau)^T]$, Eq. (3.2.4) can be rewritten as

$$(\mathscr{S}'') \qquad x(\tau+1) = h[x(\tau), \tau]. \qquad (3.2.5)$$

Given $t_0 \geq 0$ and $x_0 \in R^n$, we let $x(\tau; x_0, t_0)$ denote the unique solution of Eq. (3.2.5) for $\tau \in I$ such that $x(t_0; x_0, t_0) = x_0$. Furthermore, we assume that Eq. (3.2.5) possesses only one equilibrium, $x = 0$.

3.3 Stability and Instability of Large Scale Systems

We now state and prove selected qualitative results for composite systems $(\mathscr{S})$ and $(\mathscr{S}'')$.

3.3.1. Definition. Isolated subsystem $(\mathscr{S}_i)$ possesses **Property A** if there exists a function $v_i : R^{n_i} \times I \to R$, $\psi_{i1}, \psi_{i2} \in KR$, $\psi_{i3} \in K$, and a constant $\sigma_i \in R$ such that

$$\psi_{i1}(|z_i|) \le v_i(z_i, \tau) \le \psi_{i2}(|z_i|), \qquad Dv_{i(\mathscr{S}_i)}(z_i, \tau) \le \sigma_i \psi_{i3}(|z_i|)$$

for all $z_i \in R^{n_i}$ and $\tau \in I$.

If $\sigma_i < 0$, then the equilibrium of $(\mathscr{S}_i)$ is uniformly asymptotically stable in the large. If $\sigma_i = 0$, the equilibrium is uniformly stable and if $\sigma_i > 0$, the equilibrium of $(\mathscr{S}_i)$ may be unstable.

3.3.2. Theorem. The equilibrium $x = 0$ of composite system $(\mathscr{S})$ with decomposition (Σ_i) is **uniformly asymptotically stable in the large** if the following conditions are satisfied.

  (i)   Each isolated subsystem $(\mathscr{S}_i)$ possesses Property A;
  (ii) given v_i of hypothesis (i), there is a constant $L_i > 0$ such that

$$|v_i(z_i', \tau) - v_i(z_i'', \tau)| \le L_i |z_i' - z_i''|$$

for all $z_i', z_i'' \in R^{n_i}$ and $\tau \in I$;
  (iii) given ψ_{i3} of hypothesis (i), there exist constants $a_{ij} \ge 0$, $i, j = 1, ..., l$, such that

$$|g_i(z_1, ..., z_l, \tau)| \le \sum_{j=1}^{l} a_{ij} \psi_{j3}(|z_j|), \qquad i = 1, ..., l,$$

for all $z_i \in R^{n_i}$, $i = 1, ..., l$, and $\tau \in I$; and
  (iv) given σ_i of hypothesis (i), the successive principal minors of the test matrix $D = [d_{ij}]$ are all positive, where

$$d_{ij} = \begin{cases} -(\sigma_i + L_i a_{ii}), & i = j \\ -L_i a_{ij}, & i \ne j. \end{cases}$$

Proof. Given v_i of hypothesis (i), let $\alpha^{T} = (\alpha_1, ..., \alpha_l) > 0$ be an arbitrary constant vector and choose as a Lyapunov function for $(\mathscr{S})$,

$$v(x, \tau) = \sum_{i=1}^{l} \alpha_i v_i(z_i, \tau). \tag{3.3.3}$$

In view of hypothesis (i), v is positive definite, decrescent, and radially unbounded. Along solutions of $(\mathscr{S})$ we have, taking hypotheses (i)–(iv) into account

$$Dv_{(\mathscr{S})}(x,\tau) = \sum_{i=1}^{l} \alpha_i v_i[f_i(z_i,\tau)+g_i(x,\tau),\ \tau+1] - \sum_{i=1}^{l} \alpha_i v_i(z_i,\tau)$$

$$= \sum_{i=1}^{l} \alpha_i\{v_i[f_i(z_i,\tau)+g_i(x,\tau),\ \tau+1] - v_i[f_i(z_i,\tau),\ \tau+1]\}$$

$$+ \sum_{i=1}^{l} \alpha_i\{v_i[f_i(z_i,\tau),\ \tau+1] - v_i(z_i,\tau)\}$$

$$\leq \sum_{i=1}^{l} \alpha_i Dv_{i(\mathscr{S}_i)}(z_i,\tau) + \sum_{i=1}^{l} \alpha_i L_i|g_i(x,\tau)|$$

$$\leq \sum_{i=1}^{l} \alpha_i\left\{\sigma_i\psi_{i3}(|z_i|) + L_i\sum_{j=1}^{l} a_{ij}\psi_{j3}(|z_j|)\right\}$$

$$= -\alpha^{T}Dw \triangleq -y^{T}w$$

where $w^{T} = [\psi_{13}(|z_1|),...,\psi_{13}(|z_l|)]$, $\alpha^{T}D = y^{T}$, and matrix D is defined in hypothesis (iv). Since D is an M-matrix, it follows from Theorem 2.5.2 that $D^{-1} \geq 0$ exists. Thus, $\alpha = (D^{-1})^{T}y$. Using a similar argument as in the proof of Theorem 2.5.15, we can choose $y > 0$ so that $\alpha > 0$. Thus,

$$Dv_{(\mathscr{S})}(x,\tau) \leq -y^{T}w < 0, \qquad x \neq 0,$$

i.e., $Dv_{(\mathscr{S})}(x,\tau)$ is negative definite for all $x \in R^n$ and $\tau \in I$. Therefore, the equilibrium of $(\mathscr{S})$ is uniformly asymptotically stable in the large. ∎

Next, we prove an instability result. Recall the notation $B_i(r_i) = \{z_i \in R^{n_i} : |z_i| < r_i\}$.

3.3.4. Definition. For $(\mathscr{S}_i)$ let there exist a function $v_i: B_i(r_i) \times I \to R$, $\psi_{i1}, \psi_{i2}, \psi_{i3} \in K$, and constants $\delta_{i1}, \delta_{i2}, \sigma_i \in R$ such that

$$\delta_{i1}\psi_{i1}(|z_i|) \leq v_i(z_i,\tau) \leq \delta_{i2}\psi_{i2}(|z_i|), \qquad Dv_{i(\mathscr{S}_i)}(z_i,\tau) \leq \sigma_i\psi_{i3}(|z_i|)$$

for all $z_i \in B_i(r_i)$ and $\tau \in I$. If $\delta_{i1} = \delta_{i2} = -1$ we say that $(\mathscr{S}_i)$ possesses **Property C**. If $\delta_{i1} = \delta_{i2} = 1$ we say that $(\mathscr{S}_i)$ has **Property A″**.

If $(\mathscr{S}_i)$ has Property C with $\sigma_i < 0$, then the equilibrium $z_i = 0$ is completely unstable. If $(\mathscr{S}_i)$ has Property A″ and $\sigma_i < 0$, the equilibrium is uniformly asymptotically stable.

3.3.5. Theorem. Let $N \neq \varnothing$, $N \subset L = \{1,...,l\}$. Assume that for composite system $(\mathscr{S})$ with decomposition (Σ_i) the following conditions are true.

(i) If $i \in N$, then isolated subsystem $(\mathscr{S}_i)$ has Property C and if $i \notin N$, $i \in L$, then $(\mathscr{S}_i)$ has Property A″;

(ii)   given v_i of hypothesis (i), there is a constant $L_i > 0$ such that

$$|v_i(z_i', \tau) - v_i(z_i'', \tau)| \leq L_i |z_i' - z_i''|$$

for all $z_i', z_i'' \in B_i(r_i)$ and $\tau \in I$;

(iii)   given ψ_{i3} of hypothesis (i), there exist constants $a_{ij} \geq 0$, $i, j \in L$, such that

$$|g_i(z_1, \ldots, z_l, \tau)| \leq \sum_{j=1}^{l} a_{ij} \psi_{j3}(|z_j|), \qquad i = 1, \ldots, l,$$

for all $z_i \in B_i(r_i)$, $i \in L$, and $\tau \in I$; and

(iv)   given σ_i of hypothesis (i), the successive principal minors of the test matrix $D = [d_{ij}]$ are all positive, where

$$d_{ij} = \begin{cases} -(\sigma_i + L_i a_{ii}), & i = j \\ -L_i a_{ij}, & i \neq j. \end{cases}$$

If $N \neq L$, the equilibrium $x = 0$ of system $(\mathscr{S})$ is **unstable**. If $N = L$, the equilibrium of $(\mathscr{S})$ is **completely unstable**.

Proof. For system $(\mathscr{S})$ we choose the Lyapunov function given in Eq. (3.3.3). Invoking hypotheses (i)–(iv), it is clear that $Dv_{(\mathscr{S})}(x, \tau)$ is negative definite for all $z_i \in B_i(r_i)$, $i \in L$, and $\tau \in I$. Now consider the set

$$D = \{(x, \tau) \in R^n \times I : z_i \in B_i(r), \; r < \min r_i, \; i \in N, \text{ and } z_i = 0, \; i \notin N, \text{ and } \tau \in I\}.$$

For $(x, \tau) \in D$ we have $-\sum_{i \in N} \alpha_i \psi_{i1}(|z_i|) \leq v(x, \tau) \leq -\sum_{i \in N} \alpha_i \psi_{i2}(|z_i|)$. It follows that in every neighborhood of the origin $x = 0$ there is at least one point $x' \neq 0$ for which $v(x', \tau) < 0$ for all $\tau \in I$. Also, $v(x, \tau)$ is bounded from below on D. Thus, if $N \neq L$, it follows from Chetaev's theorem (Theorem 2.2.28), modified for systems described by difference equations, that the equilibrium of $(\mathscr{S})$ is unstable. If $N = L$, then v is negative definite and the equilibrium is completely unstable. ∎

Because of hypothesis (ii), Theorem 3.3.2 can be satisfied only for Lyapunov functions of the norm type. It is possible to obtain results using higher order Lyapunov functions. However, as will be seen, in this case the results are not quite as simple as corresponding ones for systems described by ordinary differential equations.

3.3.6. Theorem. The equilibrium $x = 0$ of composite system $(\mathscr{S})$ with decomposition (Σ_i) is **uniformly asymptotically stable in the large** if the following conditions are satisfied.

(i)   Each isolated subsystem $(\mathscr{S}_i)$ possesses Property A;

(ii)   given v_i and ψ_{i3} of hypothesis (i), there exist constants $\delta_i, \beta_{ij} \in R$,

$i, j = 1, ..., l$, such that

$$v_i[f_i(z_i, \tau) + g_i(z_1, ..., z_l, \tau), \tau + 1] - v_i[f_i(z_i, \tau), \tau + 1]$$

$$\leq 2\delta_i [\psi_{i3}(|z_i|)]^{1/2} \sum_{j=1}^{l} \beta_{ij} [\psi_{j3}(|z_j|)]^{1/2}$$

$$+ \left\{ \sum_{j=1}^{l} \beta_{ij} [\psi_{j3}(|z_j|)]^{1/2} \right\}^2, \qquad i = 1, ..., l,$$

for all $z_i \in R^{n_i}$, $z_j \in R^{n_j}$, $i, j = 1, ..., l$, and $\tau \in I$;

(iii) given σ_i of hypothesis (i), there exists a vector $\alpha^T = (\alpha_1, ..., \alpha_l) > 0$ such that the matrix $B = [b_{ij}]$ is positive definite, where

$$b_{ij} = \begin{cases} \alpha_i c_i^2 - \sum_{k=1}^{l} \alpha_k d_{ki}^2, & i = j \\ -\sum_{k=1}^{l} \alpha_k d_{ki} d_{kj}, & i \neq j \end{cases} \qquad (3.3.7)$$

and

$$c_i = (-\sigma_i + \delta_i^2)^{1/2} \qquad (3.3.8)$$

and

$$d_{ij} = \begin{cases} \delta_i + \beta_{ii}, & i = j \\ \beta_{ij}, & i \neq j. \end{cases} \qquad (3.3.9)$$

Proof. For $(\mathscr{S})$ we choose the Lyapunov function (3.3.3) which is clearly positive definite, radially unbounded, and decrescent. Along solutions of $(\mathscr{S})$ we have, using hypotheses (i)–(iii),

$$Dv_{(\mathscr{S})}(x, \tau) = \sum_{i=1}^{l} \alpha_i \{v_i[f_i(z_i, \tau) + g_i(x, \tau), \tau + 1] - v_i(z_i, \tau)\}$$

$$= \sum_{i=1}^{l} \alpha_i \{v_i[f_i(z_i, \tau), \tau + 1] - v_i(z_i, \tau)\}$$

$$+ \sum_{i=1}^{l} \alpha_i \{v_i[f_i(z_i, \tau) + g_i(x, \tau), \tau + 1] - v_i[f_i(z_i, \tau), \tau + 1]\}$$

$$\leq \sum_{i=1}^{l} \alpha_i \left\{ (\sigma_i - \delta_i^2) \psi_{i3}(|z_i|) \right.$$

$$\left. + \left[\delta_i \psi_{i3}(|z_i|)^{1/2} + \sum_{j=1}^{l} \beta_{ij} \psi_{j3}(|z_j|)^{1/2} \right]^2 \right\}$$

$$= -\sum_{i=1}^{l} \sum_{j=1}^{l} b_{ij} \psi_{i3}(|z_i|)^{1/2} \psi_{j3}(|z_j|)^{1/2}.$$

Since $B = [b_{ij}]$ is positive definite, it follows that $Dv_{(\mathscr{S})}(x, \tau)$ is negative definite for all $x \in R^n$ and $\tau \in I$. This completes the proof. ∎

In order to satisfy hypothesis (iii) of Theorem 2.3.6, we need $-\sigma_i + \delta_i^2 > 0$ and $\alpha_i c_i^2 - \sum_{k=1}^{l} \alpha_k d_{ki}^2 > 0$. Note that if in this theorem all $\psi_{ij} \in KR$ and are of the same order of magnitude, then we can conclude that the equilibrium $x = 0$ of $(\mathscr{S})$ is **exponentially stable in the large**. Also, if matrix B is only positive semidefinite, then the equilibrium is **uniformly stable**.

Now let $C = \mathrm{diag}[c_1, \ldots, c_l]$, $A = \mathrm{diag}[\alpha_1, \ldots, \alpha_l]$, $D = [d_{ij}]$, and $B = [b_{ij}]$, with elements as specified in (3.3.7)–(3.3.9). Then clearly

$$B = CAC - D^{\mathrm{T}}AD. \tag{3.3.10}$$

We now prove the following simpler result involving M-matrices.

3.3.11. Corollary. Assume that hypotheses (i) and (ii) of Theorem 3.3.6 are true with $-\sigma_i + \delta_i^2 > 0$, $\delta_i + \beta_{ii} \geq 0$, $i = 1, \ldots, l$, and $\beta_{ij} \geq 0$, $i, j = 1, \ldots, l$, $i \neq j$. The equilibrium $x = 0$ of system $(\mathscr{S})$ is **uniformly asymptotically stable in the large** if all successive principal minors of the test matrix $C - D$ are positive.

Proof. By the above assumptions, $D \geq 0$, $C > 0$, and $C - D$ is an M-matrix. From Corollary 2.5.8 it follows that there exists $A = \mathrm{diag}[\alpha_1, \ldots, \alpha_l] > 0$ such that the matrix (3.3.10) is positive definite. Therefore all assumptions of Theorem 3.3.6 are satisfied, which completes the proof. ∎

Combining the ideas in the proofs of Theorems 3.3.5 and 3.3.6 and Corollary 3.3.11, we can prove the following instability results.

3.3.12. Theorem. Assume that hypothesis (i) of Theorem 3.3.5 is true. Assume that hypotheses (ii) and (iii) of Theorem 3.3.6 are true for $z_i \in B_i(r_i)$, $i = 1, \ldots, l$, and $\tau \in I$. If $N \neq L$, the equilibrium of $(\mathscr{S})$ is **unstable** and if $N = L$, the equilibrium of $(\mathscr{S})$ is **completely unstable**.

3.3.13. Corollary. Assume that hypothesis (i) of Theorem 3.3.5 is true. Assume that hypotheses (ii) and (iii) of Theorem 3.3.6 are true for $z_i \in B_i(r_i)$, $i = 1, \ldots, l$, and $\tau \in I$, with $-\sigma_i + \delta_i^2 > 0$, $\delta_i + \beta_{ii} \geq 0$, $i = 1, \ldots, l$, and $\beta_{ij} \geq 0$, $i, j = 1, \ldots, l$, $i \neq j$. Assume that all successive principal minors of the test matrix $C - D$ are positive. If $N \neq L$, the equilibrium of $(\mathscr{S})$ is **unstable** and if $N = L$, the equilibrium of $(\mathscr{S})$ is **completely unstable**.

In the remaining results we consider composite system $(\mathscr{S}'')$. These results are of course also applicable to system $(\mathscr{S})$.

3.3.14. Theorem. The equilibrium $x = 0$ of composite system $(\mathscr{S}'')$ with decomposition (Σ_i'') is **uniformly asymptotically stable in the large** if the following conditions are satisfied.

  (i)   There exist functions $v_i \colon R^n \times I \to R$, $i = 1, \ldots, l$, and $\psi_{i1}, \psi_{i2} \in KR$, $i = 1, \ldots, l$, such that

$$\psi_{i1}(|z_i|) \leq v_i(x, \tau) \leq \psi_{i2}(|x|)$$

for all $z_i \in R^{n_i}$, $x \in R^n$, and $\tau \in I$;

(ii) there exist $\psi_{i4} \in K$, $i = 1, ..., l$, constants $b_{ij} \in R$, $i, j = 1, ..., l$, and constants $a_{ij} \in R$, $i, j = 1, ..., l$, $i \neq j$, such that

$$v_i[h_i(x, \tau), \tau+1] - v_i(x, \tau) \leq \sum_{j=1}^{l} b_{ij}\,\psi_{j4}(|z_j|)$$

$$+ \sum_{j=1, i \neq j}^{l} a_{ij}\,\psi_{i4}(|z_i|)^{1/2}\,\psi_{j4}(|z_j|)^{1/2}$$

for all $z_i \in R^{n_i}$, $x \in R^n$, $\tau \in I$, $i = 1, ..., l$; and

(iii) there exists a vector $\alpha^T = (\alpha_1, ..., \alpha_l) > 0$ such that the test matrix $S = [s_{ij}]$ is negative definite, where

$$s_{ij} = \begin{cases} \sum_{k=1}^{l} \alpha_k b_{ki}, & i = j \\ (\alpha_i a_{ij} + \alpha_j a_{ji})/2, & i \neq j. \end{cases}$$

Proof. Given v_i in hypothesis (i) and α in hypothesis (iii) we choose

$$v(x, \tau) = \sum_{i=1}^{l} \alpha_i v_i(x, \tau). \qquad (3.3.15)$$

This function is clearly positive definite, radially unbounded, and decrescent. In view of hypotheses (i)–(iii) we have along solutions of $(\mathscr{S}'')$,

$$Dv_{(\mathscr{S}'')}(x, \tau) = \sum_{i=1}^{l} \alpha_i\{v_i[h_i(x, \tau), \tau+1] - v_i(x, \tau)\}$$

$$\leq \sum_{i=1}^{l} \alpha_i\left\{\sum_{j=1}^{l} b_{ij}\,\psi_{j4}(|z_j|) + \sum_{j=1, i \neq j}^{l} a_{ij}\,\psi_{i4}(|z_i|)^{1/2}\,\psi_{j4}(|z_j|)^{1/2}\right\}$$

$$= u^T R u = u^T((R+R^T)/2)u = u^T S u \leq \lambda_M(S)|u|^2$$

where $u^T = [\psi_{14}(|z_1|)^{1/2}, ..., \psi_{l4}(|z_l|)^{1/2}]$, matrix $R = [r_{ij}]$ is specified by

$$r_{ij} = \begin{cases} \sum_{k=1}^{l} \alpha_k b_{ki}, & i = j \\ \alpha_i a_{ij}, & i \neq j \end{cases}$$

and $S = [s_{ij}]$ is given in hypothesis (iii). Since S is negative definite we have $\lambda_M(S) < 0$. Therefore $Dv_{(\mathscr{S}'')}(x, \tau)$ is negative definite for all $x \in R^n$ and $\tau \in I$, which completes the proof. ∎

Note that if $\psi_{ij} \in KR$ and are of the same order of magnitude then the equilibrium $x = 0$ of system $(\mathscr{S}'')$ is **exponentially stable in the large**. Also, if in the above theorem matrix S is only negative semidefinite, then the equilibrium of $(\mathscr{S}'')$ is **uniformly stable**.

3.3.16. Theorem. Let $L = \{1, \ldots, l\} \supset N \neq \varnothing$. Assume that for composite system $(\mathscr{S}'')$ with decomposition (Σ_i'') the following conditions hold.

(i) There exist functions $v_i \colon \bigtimes_{i=1}^{l} B_i(r_i) \times I \to R$, for some $r_i > 0$, $i = 1, \ldots, l$, and $\psi_{i1}, \psi_{i2} \in K$ such that

$$-\psi_{i1}(|x|) \leq v_i(x, \tau) \leq -\psi_{i2}(|z_i|), \qquad i \in N$$

$$\psi_{i1}(|z_i|) \leq v_i(x, \tau) \leq \psi_{i2}(|x|), \qquad i \notin N, \quad i \in L$$

for all $z_i \in B_i(r_i)$, $i \in L$, $x \in \bigtimes_{i=1}^{l} B_i(r_i)$, and $\tau \in I$;

(ii) there exist $\psi_{i4} \in K$, $i \in L$, constants $b_{ij} \in R$, $i, j \in L$, and constants $a_{ij} \in R$, $i, j \in L$, $i \neq j$, such that

$$v_i[h_i(x, \tau), \tau+1] - v_i(x, \tau) \leq \sum_{j=1}^{l} b_{ij} \psi_{j4}(|z_j|)$$

$$+ \sum_{j=1, \, i \neq j}^{l} a_{ij} \psi_{i4}(|z_i|)^{1/2} \psi_{j4}(|z_j|)^{1/2}$$

for all $z_i \in B_i(r_i)$, $i \in L$, $x \in \bigtimes_{i=1}^{l} B_i(r_i)$, and $\tau \in I$;

(iii) there exists a vector $\alpha^{\mathrm{T}} = (\alpha_1, \ldots, \alpha_l) > 0$ such that the matrix $S = [s_{ij}]$ is negative definite, where

$$s_{ij} = \begin{cases} \sum_{k=1}^{l} \alpha_k b_{ki}, & i = j \\ (\alpha_i a_{ij} + \alpha_j a_{ji})/2, & i \neq j. \end{cases}$$

If $N \neq L$, the equilibrium $x = 0$ of system $(\mathscr{S}'')$ is **unstable**. If $N = L$, the equilibrium of $(\mathscr{S}'')$ is **completely unstable**.

Proof. For $(\mathscr{S}'')$ we choose the Lyapunov function (3.3.15). In view of (ii) and (iii), $Dv_{(\mathscr{S}'')}(x, \tau)$ is negative definite for all $x \in R^n$ and $\tau \in I$. From (i) it follows that in every neighborhood of $x = 0$ there exists a point $x' \neq 0$ such that $v(x', \tau) < 0$ for all $\tau \in I$. Also, v is bounded on $B(r) \times I$ where $r < \min r_i$. The conclusion of the theorem follows from Theorem 2.2.28 (modified for difference equations). ∎

We conclude this section by noting that it is also possible to establish results for boundedness and ultimate boundedness of solutions for systems $(\mathscr{S})$ and $(\mathscr{S}'')$ described by Eqs. (3.2.2) and (3.2.5), respectively. In addition, it is also possible to formulate results for discrete time systems which are analogous to those of Sections 2.6 and 2.7.

3.4 Examples

We now apply the preceding results to some specific examples.

3.4.1. Example. Consider systems described by the set of equations

$$z_1(\tau+1) = A_1 z_1(\tau) + b_1 f_1(\eta_1(\tau))$$
$$\eta_1(\tau) = c_1{}^{\mathrm{T}} z_2(\tau)$$
$$z_2(\tau+1) = A_2 z_2(\tau) + b_2 f_2(\eta_2(\tau)) \qquad (3.4.2)$$
$$\eta_2(\tau) = c_2{}^{\mathrm{T}} z_1(\tau)$$

where $\tau \in I$, $z_i \in R^{n_i}$, A_i is an $n_i \times n_i$ matrix such that $\|A_i\| = (\lambda_M(A_i{}^{\mathrm{T}} A_i))^{1/2} < 1$, $i = 1, 2$, $c_1 \in R^{n_2}$, $c_2 \in R^{n_1}$, $f_i: R \to R$, $f_i(\eta_i) = 0$ if and only if $\eta_i = 0$, $0 < \eta_i f_i(\eta_i) < k_i \eta_i{}^2$ for all $\eta_i \neq 0$, and $k_i > 0$ is a constant, $i = 1, 2$. System (3.4.2) may be viewed as a nonlinear time-invariant interconnection of two isolated subsystems $(\mathscr{S}_1)$, $(\mathscr{S}_2)$,

$$z_i(\tau+1) = A_i z_i(\tau), \qquad (\mathscr{S}_i)$$

$i = 1, 2$, interconnected by the relations $g_1(x) = b_1 f_1(\eta_1)$ and $g_2(x) = b_2 f_2(\eta_2)$. System (3.4.2) is clearly a special case of system (3.2.1).

For $(\mathscr{S}_i)$ choose $v_i(z_i) = |z_i|$ which is Lipschitz continuous with $L_i = 1$, $i = 1, 2$. Since $Dv_{i(\mathscr{S}_i)}(z_i) \leq (\|A_i\| - 1)|z_i|$, hypotheses (i) and (ii) of Theorem 3.3.2 are satisfied. Hypothesis (iii) is also satisfied with $a_{11} = a_{22} = 0$, $a_{12} = k_1 |b_1| |c_1|$, and $a_{21} = k_2 |b_2| |c_2|$. Matrix D in hypothesis (iv) is given by

$$D = \begin{bmatrix} 1 - \|A_1\| & -k_1 |b_1| \cdot |c_1| \\ -k_2 |b_2| \cdot |c_2| & 1 - \|A_2\| \end{bmatrix}.$$

It follows from Theorem 3.3.2 that the equilibrium $x^{\mathrm{T}} = (z_1{}^{\mathrm{T}}, z_2{}^{\mathrm{T}}) = 0$ of system (3.4.2) is uniformly asymptotically stable in the large if all successive principal minors of D are positive, i.e., if

$$k_1 k_2 < \frac{(1 - \|A_1\|)(1 - \|A_2\|)}{|b_1| \cdot |b_2| \cdot |c_1| \cdot |c_2|}. \qquad (3.4.3)$$

3.4.4. Example. We reconsider system (3.4.2). This time we assume that there exists a constant $a_1 > 1$ such that $|A_1 z_1| \geq a_1 |z_1|$ for all $z_1 \in R^{n_1}$ and that $\|A_2\| < 1$. For $(\mathscr{S}_1)$ choose $v_1(z_1) = -|z_1|$ and for $(\mathscr{S}_2)$ choose $v_2(z_2) = |z_2|$. Then v_i is Lipschitz continuous with $L_i = 1$, $i = 1, 2$. Since $Dv_{1(\mathscr{S}_1)}(z_1) \leq (1 - a_1)|z_1|$ and $Dv_{2(\mathscr{S}_2)}(z_2) \leq (\|A_2\| - 1)|z_2|$, hypotheses (i) and (ii) of Theorem 3.3.5 are satisfied. Hypothesis (iii) of this theorem is also true with $a_{11} = a_{22} = 0$, $a_{12} = k_1 |b_1| |c_1|$, and $a_{21} = k_2 |b_2| |c_2|$. Matrix D of hypothesis

(iv) is given by

$$D = \begin{bmatrix} a_1 - 1 & -k_1|b_1||c_1| \\ -k_2|b_2||c_2| & 1 - \|A_2\| \end{bmatrix}.$$

It follows from Theorem 3.3.5 that the equilibrium of system (3.4.2) is unstable if

$$k_1 k_2 < \frac{(a_1 - 1)(1 - \|A_2\|)}{|b_1| \cdot |b_2| \cdot |c_1| \cdot |c_2|}.$$

3.4.5. Example. Let us reconsider Example 3.4.1, where we assume that $\|A_i\| < 1$, $i = 1, 2$. For $(\mathscr{S}_i)$ we choose $v_i(z_i) = |z_i|^2$, $i = 1, 2$. In this case hypotheses (i) and (ii) of Theorem 3.3.6 are satisfied for $i = 1, 2$, with $\sigma_i = \|A_i\|^2 - 1$, $\delta_i = \|A_i\|$, $\beta_{ii} = 0$, $\beta_{12} = k_1|b_1||c_1|$, $\beta_{21} = k_2|b_2||c_2|$, and $\psi_{i3}(|z_i|) = |z_i|^2$. Using the notation of Eqs. (3.3.7)–(3.3.10), we have

$$C = \begin{bmatrix} 1 & 0 \\ 0 & 1 \end{bmatrix}, \quad D = \begin{bmatrix} \|A_1\| & -k_1 \cdot |b_1| \cdot |c_1| \\ -k_2 \cdot |b_2| \cdot |c_2| & \|A_2\| \end{bmatrix}$$

so that

$$C - D = \begin{bmatrix} 1 - \|A_1\| & -k_1 \cdot |b_1| \cdot |c_1| \\ -k_2 \cdot |b_2| \cdot |c_2| & 1 - \|A_2\| \end{bmatrix}.$$

It follows from Corollary 3.3.11 that the equilibrium of system (3.4.2) is uniformly asymptotically stable in the large if all principal minors of $C - D$ are positive. This is the case if and only if inequality (3.4.3) is true. Thus, given our choices of Lyapunov functions, Theorem 3.3.2 and Corollary 3.3.11 yield the same stability condition for system (3.4.2).

3.4.6. Example. We consider two subsystems $(\mathscr{S}_i)$, $i = 1, 2$, described by

$$z_i(\tau + 1) = A_i z_i(\tau) + b_i f_i(\sigma_i(\tau)) \tag{$\mathscr{S}_i$}$$

$$\sigma_i(\tau) = c_{ii}^{\mathrm{T}} z_i(\tau)$$

where $z_i \in R^{n_i}$, A_i is an $n_i \times n_i$ matrix, $b_i \in R^{n_i}$, $c_{ii} \in R^{n_i}$, $f_i: R \to R$, $f_i(\sigma_i) = 0$ if and only if $\sigma_i = 0$, $0 < \sigma_i f_i(\sigma_i) \le k_i \sigma_i^2$ for all $\sigma_i \ne 0$, and $k_i > 0$ is a constant for $i = 1, 2$.

We interconnect $(\mathscr{S}_1)$ and $(\mathscr{S}_2)$ to form a system described by

$$z_1(\tau + 1) = A_1 z_1(\tau) + b_1 f_1(\sigma_1(\tau))$$

$$\sigma_1(\tau) = c_{11}^{\mathrm{T}} z_1(\tau) + c_{12}^{\mathrm{T}} z_2(\tau)$$

$$z_2(\tau + 1) = A_2 z_2(\tau) + b_2 f_2(\sigma_2(\tau)) \tag{3.4.7}$$

$$\sigma_2(\tau) = c_{21}^{\mathrm{T}} z_1(\tau) + c_{22}^{\mathrm{T}} z_2(\tau)$$

where $c_{12} \in R^{n_2}$ and $c_{21} \in R^{n_1}$. Note that in this case the interconnecting structure does not enter additatively into the system description. Thus, with $(\mathscr{S}_1)$ and $(\mathscr{S}_2)$ specified as above, system (3.4.7) may be viewed as a special case of system $(\mathscr{S}'')$ with decomposition (Σ_i'').

If we choose $v_i(z_i) = |z_i|^2$, then hypothesis (i) of Theorem 3.3.14 is satisfied. Now

$$
\begin{aligned}
Dv_1(z_1) &= |A_1 z_1 + b_1 f_1(\sigma_1)|^2 - |z_1|^2 \\
&\leq \{\|A_1\|^2 - 1 + 2k_1\|A_1\| \cdot |b_1| \cdot |c_{11}| + k_1{}^2 |b_1|^2 \cdot |c_{11}|^2\}|z_1|^2 \\
&\quad + \{2k_1\|A_1\| \cdot |b_1| \cdot |c_{12}| + 2k_1{}^2 |b_1|^2 \cdot |c_{11}| \cdot |c_{12}|\}|z_1| \cdot |z_2| \\
&\quad + \{k_1{}^2 |b_1|^2 \cdot |c_{12}|^2\}|z_2|^2
\end{aligned}
$$

with a similar inequality being also true for $Dv_2(z_2)$. Let $\psi_{i4}(|z_i|) = |z_i|^2$, $i = 1, 2$. Then hypothesis (ii) of Theorem 3.3.14 is satisfied with

$$
\begin{aligned}
b_{11} &= \{\|A_1\| + k_1 |b_1| \cdot |c_{11}|\}^2 - 1, \qquad b_{12} = k_1{}^2 |b_1|^2 \cdot |c_{12}|^2 \\
b_{22} &= \{\|A_2\| + k_2 |b_2| \cdot |c_{22}|\}^2 - 1, \qquad b_{21} = k_2{}^2 |b_2|^2 \cdot |c_{21}|^2 \\
a_{12} &= 2k_1\|A_1\| \cdot |b_1| \cdot |c_{12}| + 2k_1{}^2 |b_1|^2 \cdot |c_{11}| \cdot |c_{12}| \\
a_{21} &= 2k_2\|A_2\| \cdot |b_2| \cdot |c_{21}| + 2k_2{}^2 |b_2|^2 \cdot |c_{22}| \cdot |c_{21}|.
\end{aligned}
$$

Choosing $\alpha_1 = \alpha_2 = 1$, matrix $S = [s_{ij}]$ of Theorem 3.3.14 is specified by

$$
\begin{aligned}
s_{11} &= \{\|A_1\| + k_1 |b_1| \cdot |c_{11}|\}^2 - 1 + k_2{}^2 |b_2|^2 \cdot |c_{21}|^2 \\
s_{22} &= \{\|A_2\| + k_2 |b_2| \cdot |c_{22}|\}^2 - 1 + k_1{}^2 |b_1|^2 \cdot |c_{12}|^2 \\
s_{12} &= s_{21} = k_1\|A_1\| \cdot |b_1| \cdot |c_{12}| + k_1{}^2 |b_1|^2 \cdot |c_{11}| \cdot |c_{12}| \\
&\quad + k_2\|A_2\| \cdot |b_2| \cdot |c_{21}| + k_2{}^2 |b_2|^2 \cdot |c_{22}| \cdot |c_{21}|.
\end{aligned}
$$

It follows from Theorem 3.3.14 that the equilibrium $x^T = (z_1{}^T, z_2{}^T) = 0$ of system (3.4.7) is uniformly asymptotically stable in the large if matrix S is negative definite, i.e., if

$$
\{\|A_1\| + k_1 |b_1| \cdot |c_{11}|\}^2 + k_2{}^2 |b_2|^2 \cdot |c_{21}|^2 < 1
$$
$$
\{\|A_2\| + k_2 |b_2| \cdot |c_{22}|\}^2 + k_1{}^2 |b_1|^2 \cdot |c_{12}|^2 < 1
$$

and

$$
\begin{aligned}
&\{[\|A_1\| + k_1 |b_1| \cdot |c_{11}|]^2 - 1 + k_2{}^2 |b_2|^2 \cdot |c_{21}|^2\} \\
&\times \{[\|A_2\| + k_2 |b_2| \cdot |c_{22}|]^2 - 1 + k_1{}^2 |b_1|^2 \cdot |c_{12}|^2\} \\
&> \{k_1\|A_1\| \cdot |b_1| \cdot |c_{12}| + k_1{}^2 |b_1|^2 \cdot |c_{11}| \cdot |c_{12}| + k_2\|A_2\| \cdot |b_2| \cdot |c_{21}| \\
&\quad + k_2{}^2 |b_2|^2 \cdot |c_{22}| \cdot |c_{21}|\}^2.
\end{aligned}
$$

3.5 Sampled Data Systems

We consider systems described by equations of the form

$$\dot{z}_i(t) = f_i(z_i(t), t) + \sum_{j=1, i \neq j}^{l} C_{ij} z_j(kT), \qquad kT \leq t < (k+1)T, \qquad (3.5.1)$$
$$k = 0, 1, 2, ...,$$

$i = 1, ..., l$, where $z_i \in R^{n_i}$, $f_i: R^{n_i} \times [0, \infty) \to R^{n_i}$, $f_i(z_i, t) = 0$ for all $t \in [0, \infty)$ if and only if $z_i = 0$, C_{ij} is an $n_i \times n_j$ matrix, $T > 0$ is a constant, called the **sampling time**, and $\dot{z}_i(t)$ denotes the right-hand derivative of $z(t)$ with respect to t. This system may be viewed as a linear interconnection of l isolated subsystems $(\mathscr{S}_i)$, described by

$$(\mathscr{S}_i) \qquad\qquad\qquad \dot{z}_i = f_i(z_i, t). \qquad\qquad\qquad (3.5.2)$$

Letting $x^T = (z_1^T, ..., z_l^T)$, $f(x, t)^T = [f_1(z_1, t)^T, ..., f_l(z_l, t)^T]$, $\sum_{i=1}^{l} n_i = n$, and

$$C = \begin{bmatrix} 0 & C_{12} & \cdots & C_{1l} \\ C_{21} & 0 & \cdots & C_{2l} \\ \vdots & \vdots & \cdots & \vdots \\ C_{l1} & C_{l2} & \cdots & 0 \end{bmatrix},$$

system (3.5.1) can be represented by the equation

$$(\mathscr{S}) \qquad \begin{aligned} \dot{x}(t) &= f[x(t), t] + Cx(kT) \triangleq F[x(t), x(kT), t], \\ & kT \leq t < (k+1)T, \qquad k = 0, 1, 2, ..., \end{aligned} \qquad (3.5.3)$$

where $x \in R^n$, $f: R^n \times [0, \infty) \to R^n$, C is an $n \times n$ matrix, and $f(x, t) = 0$ for all $t \in [0, \infty)$ if and only if $x = 0$. Henceforth we assume that Eq. (3.5.2) possesses for every $z_i(0) = z_{i_0}$ a unique solution $z_i(t; z_i(0), 0)$ for all $t \geq 0$ and that Eq. (3.5.3) has for every $x_0 = x(0)$ a unique solution $x(t; x(0), 0)$ for all $t \geq 0$.

It should be noted that the terms

$$u_i(t) = u_i(kT) \triangleq \sum_{j=1, i \neq j}^{l} C_{ij} z_j(kT) \qquad (3.5.4)$$

are forcing functions which assume constant values over each interval $[kT, (k+1)T)$, $k = 0, 1, 2, ...$, and which change their values for each such interval according to $z_j(t)$, $j = 1, ..., l$, sampled at the discrete points in time kT, $k = 0, 1, 2, ...$

3.5.5. Lemma. Assume that for each isolated subsystem $(\mathcal{S}_i)$ described by Eq. (3.5.2) there exists a continuous function $v_i: R^{n_i} \times [0, \infty) \to R$ and three positive constants c_{i1}, c_{i2}, L_i such that

(i) $|z_i| \le v_i(z_i, t) \le c_{i1}|z_i|$

(ii) $D_+ v_{i(\mathcal{S}_i)}(z_i, t) = \lim\inf_{h \to 0^+} (1/h)\{v_i[z_i + h \cdot f_i(z_i, t), t+h] - v_i(z_i, t)\}$

$$\ge -c_{i2}|z_i|$$

(iii) $|v_i(z_i', t) - v_i(z_i'', t)| \le L_i|z_i' - z_i''|$

for all $z_i, z_i', z_i'' \in R^{n_i}$ and $t \in [0, \infty)$. Then the solution $z_i(t; z_i(kT), kT)$ (with $z_i(kT; z_i(kT), kT) = z_i(kT)$) of the forced system

$$\dot{z}_i(t) = f_i(z_i(t), t) + \sum_{j=1, i \ne j}^{l} C_{ij} z_j(kT), \qquad kT \le t < (k+1)T,$$

$$(3.5.6)$$

satisfies for all $t \in [kT, (k+1)T]$ the estimate

$$|z_i(t; z_i(kT), kT)|$$

$$\ge (1/c_{i1})[e^{c_{i2}(kT)}][e^{-c_{i2}t}]|z_i(kT)| - \{(1/c_{i1})(L_i/c_{i2})[1 - e^{c_{i2}(kT)}e^{-c_{i2}t}]\}$$

$$\times \left\{ \sum_{j=1, i \ne j}^{l} \|C_{ij}\| \cdot |z_j(kT)| \right\}$$

$$\ge (1/c_{i1})[e^{-c_{i2}T}]|z_i(kT)| - \{(1/c_{i1})(L_i/c_{i2})[1 - e^{-c_{i2}T}]\}$$

$$\times \left\{ \sum_{j=1, i \ne j}^{l} \|C_{ij}\| \cdot |z_j(kT)| \right\}.$$

Proof. Let $t, t+h \in [kT, (k+1)T)$. Along solutions of Eq. (3.5.6) we have

$$v_i[z_i(t+h; z_i, t), t+h] - v_i(z_i, t)$$

$$= v_i\left[z_i + h \cdot f_i(z_i, t) + h \cdot \sum_{j=1, i \ne j}^{l} C_{ij} z_j(kT) + o(h), t+h \right] - v_i(z_i, t)$$

$$= \left\{ v_i\left[z_i + h \cdot f_i(z_i, t) + h \cdot \sum_{j=1, i \ne j}^{l} C_{ij} z_j(kT) + o(h), t+h \right] \right.$$

$$\left. - v_i[z_i + h \cdot f_i(z_i, t) + o(h), t+h] \right\}$$

$$+ \{v_i[z_i + h \cdot f_i(z_i, t) + o(h), t+h] - v_i(z_i, t)\}$$

$$\ge -h \cdot L_i \left| \sum_{j=1, i \ne j}^{l} C_{ij} z_j(kT) \right| + v_i[z_i + h \cdot f_i(z_i, t) + o(h), t+h]$$

$$- v_i(z_i, t)$$

where use of hypothesis (iii) has been made. Invoking hypothesis (ii) we have

$$D_+ v_{i(3.5.6)}(z_i, t) \geq D_+ v_{i(\mathcal{S}_i)}(z_i, t) - L_i \left| \sum_{j=1, i \neq j}^{l} C_{ij} z_j(kT) \right|$$

$$\geq -c_{i2}|z_i| - L_i \left| \sum_{j=1, i \neq j}^{l} C_{ij} z_j(kT) \right|$$

for all $t \in [kT, (k+1)T)$. Let

$$k_{i1} = L_i \left| \sum_{j=1, i \neq j}^{l} C_{ij} z_j(kT) \right| \leq L_i \sum_{j=1, i \neq j}^{l} \|C_{ij}\| |z_j(kT)|$$

and let $u_i(t) = v_i[z_i(t; z_i(kT), kT), t]$. Then

$$D_+ u_i \geq -c_{i2} u_i - k_{i1} \tag{3.5.7}$$

and along solutions of (3.5.7) we have for all $t \in [kT, (k+1)T)$,

$$u_i(t) \geq e^{-c_{i2}(t-kT)} u_i(kT) - k_{i1} \int_{kT}^{t} e^{-c_{i2}(t-\tau)} d\tau$$

$$= e^{c_{i2}(kT)} e^{-c_{i2}t} u_i(kT) - (k_{i1}/c_{i2})(e^{-c_{i2}t} e^{c_{i2}\tau}) \Big|_{kT}^{t}$$

$$= e^{c_{i2}(kT)} e^{-c_{i2}t} u_i(kT) - (k_{i1}/c_{i2})[1 - e^{c_{i2}(kT)} e^{-c_{i2}t}].$$

Using hypothesis (i) we have

$$c_{i1}|z_i(t; z(kT), kT)| \geq e^{c_{i2}(kT)} c^{-c_{i2}t} |z_i(kT)|$$

$$- (k_{i1}/c_{i2})[1 - e^{c_{i2}(kT)} e^{-c_{i2}t}]$$

or

$$|z_i(t; z(kT), kT)| \geq (1/c_{i1})[e^{c_{i2}(kT)}][e^{-c_{i2}t}]|z_i(kT)|$$

$$- \{(1/c_{i1})(L_i/c_{i2})[1 - e^{c_{i2}(kT)} e^{-c_{i2}t}]\}$$

$$\times \left\{ \sum_{j=1, i \neq j}^{l} \|C_{ij}\| |z_j(kT)| \right\}$$

$$\geq (1/c_{i1})[e^{-c_{i2}T}]|z_i(kT)| - \{(1/c_{i1})(L_i/c_{i2})[1 - e^{-c_{i2}T}]\}$$

$$\times \left\{ \sum_{j=1, i \neq j}^{l} \|C_{ij}\| |z_j(kT)| \right\}$$

for all $t \in [kT, (k+1)T)$, which concludes the proof. ∎

3.5.8. Definition. Subsystem $(\mathcal{S}_i)$ described by Eq. (3.5.2) is said to possess **Property $\tilde{A}$** if there exists a continuous function $v_i: R^{n_i} \times [0, \infty) \to R$ which satisfies hypotheses (i)–(iii) of Lemma 3.5.5.

We are now in a position to prove the following result.

3.5.9. Theorem. The equilibrium $x = 0$ of composite system $(\mathscr{S})$ with decomposition (3.5.1) is **uniformly asymptotically stable in the large** if the following conditions are satisfied.

(i)   Each isolated subsystem $(\mathscr{S}_i)$ possesses Property $\overset{\approx}{A}$;

(ii)   given v_i of hypothesis (i), there exists a constant $c_{i3} > 0$ such that

$$Dv_{i(\mathscr{S}_i)}(z_i, t) = \lim_{h \to 0^+} \sup(1/h)\{v_i[z_i((t+h); z_i, t), t+h] - v_i(z_i, t)\}$$

$$\leq -c_{i3}|z_i|$$

for all $z_i \in R^{n_i}$ and $t \in [0, \infty)$; and

(iii)   all successive principal minors of the test matrix $S = [s_{ij}]$ are positive, where

$$s_{ij} = \begin{cases} (c_{i3}/c_{i1})\, e^{-c_{i2}T}, & i = j \\ -\{[c_{i3}\, L_i/(c_{i1}\, c_{i2})]\,[1 - e^{-c_{i2}T}] + L_i\}\, \|C_{ij}\|, & i \neq j. \end{cases}$$

Proof. Given v_i of hypothesis (i), let $\alpha^T = (\alpha_1, \ldots, \alpha_l) > 0$ be an arbitrary vector and choose as a Lyapunov function for $(\mathscr{S})$,

$$v(x, t) = \sum_{i=1}^{l} \alpha_i v_i(z_i, t).$$

Using hypothesis (i) we obtain

$$\sum_{i=1}^{l} \alpha_i|z_i| \leq v(x, t) \leq \sum_{i=1}^{l} \alpha_i c_{i1}|z_i|$$

for all $z_i \in R^{n_i}$, $i = 1, \ldots, l$, $x \in R^n$ and $t \in [0, \infty)$. Thus, v is positive definite, decrescent, and radially unbounded. Let $t, t+h \in [kT, (k+1)T]$. In view of hypotheses (i) and (ii) we have

$$Dv_{(\mathscr{S})}(x, t) = \lim_{h \to 0^+} \sup(1/h)\left\{\sum_{i=1}^{l} \alpha_i v_i[z_i(t+h; z_i, t), t+h] - \alpha_i v_i(z_i, t)\right\}$$

$$\leq \sum_{i=1}^{l} \alpha_i Dv_{i(\mathscr{S}_i)}(z_i, t) + \sum_{i=1}^{l} \alpha_i L_i\left\{\sum_{j=1, i \neq j}^{l} \|C_{ij}\| \cdot |z_j(kT)|\right\}$$

$$\leq \sum_{i=1}^{l} (-\alpha_i c_{i3})|z_i| + \sum_{i=1}^{l} (\alpha_i L_i)\left\{\sum_{j=1, i \neq j}^{l} \|C_{ij}\| \, |z_j(kT)|\right\}.$$

Utilizing the estimate for $|z_i(t; z(kT), kT)|$ obtained in Lemma 3.5.5, we have for all $t \in [kT, (k+1)T)$, $k = 0, 1, 2, \ldots,$

$$Dv_{(\mathscr{S})}(x, t) \leq \sum_{i=1}^{l} \alpha_i\{(-c_{i3}/c_{i1})[e^{-c_{i2}T}]|z_i(kT)|\}$$

$$+ \sum_{i=1}^{l} \alpha_i\{[c_{i3}\, L_i/(c_{i1}\, c_{i2})][1 - e^{-c_{i2}T}] + L_i\} \sum_{j=1, i \neq j}^{l} \|C_{ij}\| \cdot |z_j(kT)|$$

$$= (-\alpha^T S)\, w(kT) = (-y^T)\, w(kT),$$

where $w(kT)^{\mathrm{T}} = (|z_1(kT)|, ..., |z_l(kT)|)$ and S is the matrix defined in hypothesis (iii). Noting that $w(kT) = 0$ if and only if $x = 0$, and noting that S is an M-matrix, we conclude that $Dv_{(\mathscr{S})}(x, t)$ is negative definite for all $x \in R^n$ and $t \in [0, \infty)$, using the same argument as in the proof of Theorem 2.5.15. Therefore, the equilibrium $x = 0$ of composite system $(\mathscr{S})$ is uniformly asymptotically stable in the large. ∎

It is possible to establish stability results for other classes of sampled data systems by the present approach. Using appropriate modifications, one can also obtain instability and boundedness conditions for sampled data systems by the present method.

3.6   Notes and References

For a treatment of the Lyapunov theory of systems described by ordinary difference equations, refer to Hahn [1, 2] and Kalman and Bertram [2]. Sampled data systems are treated in numerous engineering texts (see, e.g., Kuo [1]).

Theorems 3.3.2 and 3.3.5 are based on results reported by Michel [5, 7] while Theorems 3.3.14 and 3.3.16 are new. Theorems 3.3.6 and 3.3.12 and Corollaries 3.3.11 and 3.3.13 constitute an adaptation and expansion of results by Araki [4]. Some of the examples of Section 3.4 are given in Michel [5, 7]. Stability results for composite sampled data systems, such as those given in Section 3.5, are from Michel [5, 7]. For related stability results dealing with discrete time interconnected systems, refer to Grujić and Siljak [1, 3].

CHAPTER IV

Systems Described by
Stochastic Differential Equations

Disturbances are important considerations in the qualitative analysis of large scale systems. For this reason, we now turn our attention to systems described by a large class of stochastic differential equations. Although our primary objective in this chapter is to establish conditions for asymptotic stability with probability one, exponential stability with probability one, and exponential stability in the quadratic mean of systems described by Ito differential equations, the present results can readily be modified to be applicable to other types of stochastic differential equations (and stochastic difference equations) and also, to stochastic stability types not explicitly considered herein. As in the preceding chapters, our objective is to analyze large scale systems in terms of lower order (and simpler) subsystems and in terms of the system interconnecting structure.

This chapter consists of six parts. In the first section necessary nomenclature is established. In the second section, which consists of background material, the stochastic differential equations of interest are introduced and selected stability concepts and results are presented. Composite systems described by stochastic differential equations are formulated in the third section.

Scalar and vector Lyapunov functions are used in the fourth and fifth section, respectively, to establish stability results for the interconnected systems considered. These results are applied to specific examples in the sixth section. The chapter is concluded with a brief discussion of the pertinent literature in the last section.

4.1 Notation

In the present chapter we find it convenient to use the following additional nomenclature. Let $\nabla_u = \partial/\partial u$ and $\nabla_{uv} = \partial^2/(\partial u\, \partial v)$, where u and v may be scalars or vectors. For example, if $x \in R^n$ and $v: R^n \to R$, then $\nabla_x v$ denotes the gradient vector of v and $\nabla_{xx} v$ represents the matrix defined by the elements $\partial^2 v/(\partial x_i\, \partial x_j)$, $i, j = 1, \ldots, n$.

The trace of a square matrix A is denoted by $\operatorname{tr} A$. As before, we define for an arbitrary matrix D the norm $\|D\|$ by $\|D\| = [\lambda_M(D^T D)]^{1/2}$ and we define the norm $\|D\|_m$ by $\|D\|_m = [\operatorname{tr}(D^T D)]^{1/2}$.

Let $T = R^+ = [0, \infty)$ and let $(\Omega, \mathscr{A}, P)$ denote a probability space with probability measure P defined on the σ-algebra $\mathscr{A}$ of ω-sets ($\omega \in \Omega$) in the sample space Ω. Any $\mathscr{A}$-measurable function on Ω is called a random variable. A sequence of random variables indexed by $t \in T$, $\{x_t(\omega) \in R^n,\ t \in T\}$, is called a (continuous time parameter) stochastic process. Henceforth the ω dependence is suppressed and we always assume that $x_0(\omega) \triangleq x$ is known. Throughout this chapter we consider only Markov processes and we sometimes simply write x_t in place of $\{x_t, t \in T\}$ for such processes. For $A \in \mathscr{A}$, $P(A)$ denotes the probability of event A and $P(A \mid B)$ denotes the conditional probability of A under the condition $B \in \mathscr{A}$. Let E denote the expectation operator and let $\{x_t, t \in T\}$ be a Markov process. Then $E_{x,s} x_t$ denotes the expected value of x_t at $t \in T$ if it is known that $x_s = x$. If $s = 0$, the notation $E_x x_t$ is used.

Subsequently, we will have occasion to use the quasimonoticity condition introduced earlier in Chapter II. We will find it convenient to use the following convention.

4.1.1. Definition. A function $b: R^l \times T \to R$ is said to belong to **class K^i** (i.e., $b \in K^i$) if for two vectors $u, v \in R^l$, such that $0 \le u_j < v_j$ for all $i \ne j$ and such that $0 \le u_i = v_i$, we have $b(u,t) < b(v,t)$ for all $t \in T$.

4.2 Systems Described by Stochastic Differential Equations

We consider systems which may be represented by equations of the form

$$dx_t = f(x_t, t)\, dt + \sigma(x_t, t)\, d\xi_t \tag{4.2.1}$$

where $t \in T$, $x_t \in R^n$, $f: R^n \times T \to R^n$, $\sigma: R^n \times T \to R^{n \cdot m}$, and $\{\xi_t, t \in T\}$ is an independent increment Markov process. Subsequently we consider two specific types of stochastic processes for Eq. (4.2.1): (a) normalized m-dimensional Wiener processes with independent components, denoted by $\{\xi_t, t \in T\} \triangleq \{z_t, t \in T\}$, and (b) normalized m-dimensional Poisson step processes with independent components, denoted by $\{\xi_t, t \in T\} \triangleq \{q_t, t \in T\}$.

For an appropriate interpretation of Eq. (4.2.1) as an integral equation, refer to Arnold [1], Wong [1], and Kushner [1]. Henceforth we assume that the initial condition $x_0 = x$ is known and that the trivial solution $\{x_t = 0, t \in T\}$ is the only equilibrium of Eq. (4.2.1). The following is an existence and uniqueness result for Eq. (4.2.1).

4.2.2. Theorem. In Eq. (4.2.1) let $f(\cdot)$ and $\sigma(\cdot)$ be measurable functions in (x, t) and let $f(\cdot)$ and $\sigma(\cdot)$ satisfy the conditions

$$|f(x, t)|^2 \le k(1 + |x|^2)$$

$$\|\sigma(x, t)\|_m^2 \le k(1 + |x|^2)$$

$$|f(x + u, t) - f(x, t)| \le k|u|$$

$$\|\sigma(x + u, t) - \sigma(x, t)\|_m \le k|u|$$

for all $x \in R^n$, $u \in R^n$, $t \in T$, and some constant $k > 0$. Then there exists a separable and measurable Markov process $\{x_t, t \in T\}$ satisfying Eq. (4.2.1) which is unique and sample continuous with probability one.

In the following definitions, the qualitative behavior of the trivial solution of Eq. (4.2.1) is characterized.

4.2.3. Definition. The trivial solution of Eq. (4.2.1) is **stable with probability one (stable w.p.1)** if for any $\rho > 0$ and any $\mathscr{E} > 0$ there is a $\delta = \delta(\rho, \mathscr{E}) > 0$ such that, if $|x_0| < \delta$, then

$$P\left\{\sup_{t \in T} |x_t| \ge \mathscr{E} \mid x_0 = x\right\} \le \rho.$$

If the trivial solution is not stable w.p.1, then it is said to be **unstable w.p.1**. The trivial solution of Eq. (4.2.1) is **asymptotically stable with probability one (AS w.p.1)** if it is stable w.p.1 and if $\lim_{t \to \infty} x_t = 0$ with probability one, for all x_0 in some neighborhood U of the origin. If $U = R^n$, the trivial solution is said to be **asymptotically stable in the large w.p.1 (ASL w.p.1)**.

4.2.4. Definition. The trivial solution of Eq. (4.2.1) is **exponentially stable in the large with probability one (ESL w.p.1)** if the trivial solution is stable w.p.1 and for all $r > 0$ and for all $\tau \in T$ and $\mathscr{E} > 0$, there exist positive constants $M = M(r)$ and β such that if $|x_0| < r$, then

$$P\left\{\sup_{t \ge \tau} |x_t| \ge \mathscr{E} \mid x_0 = x\right\} \le Me^{-\beta\tau}.$$

The above definitions pertain to the sample functions x_t on T, which are of particular interest in applications, because the observable behavior of a system is generally exhibited via sample functions. However, moments are also frequently of interest.

4.2.5. Definition. The trivial solution of Eq. (4.2.1) is **exponentially stable in the large in the quadratic mean (ESL q.m.)** if, given $x_0 = x$

(i)   for every $\mathscr{E} > 0$ there is a $\delta = \delta(\mathscr{E}) > 0$ such that for any $x \in R^n$, $|x| < \delta$ implies

$$\sup_{t \in T} E_x |x_t|^2 \leq \mathscr{E},$$

and

(ii)   given $r > 0$, there exist positive constants $M(r)$ and β such that when $|x| < r$, then

$$E_x |x_t|^2 \leq M(r) e^{-\beta t}, \qquad t \in T.$$

Stability results for Eq. (4.2.1) involve the existence of Lyapunov type functions $v: R^n \times T \to R$. Henceforth we assume that $v(x, t)$ possesses continuous first and second order partial derivatives in x and continuous first partial derivatives in t. Stability results for Eq. (4.2.1) also require that $v(x, t)$ be in the **domain of the weak infinitesimal generator** A for Eq. (4.2.1), given by

$$Av(x, t) = \lim_{\delta \to 0^+} [E_{x,t} v(x_{t+\delta}, t+\delta) - v(x, t)]/\delta. \qquad (4.2.6)$$

Note that A is a linear operator.

If in Eq. (4.2.1) $\{\xi_t = z_t, \ t \in T\}$ is a normalized Wiener process with independent components, then

$$Av(x, t) \triangleq \mathscr{L}v(x, t) = \nabla_t v(x, t) + [\nabla_x v(x, t)]^T f(x, t)$$
$$+ \tfrac{1}{2} \mathrm{tr}[\sigma(x, t)^T \nabla_{xx} v(x, t) \sigma(x, t)]. \qquad (4.2.7)$$

If in Eq. (4.2.1) $\{\xi_t = q_t, \ t \in T\}$ is a normalized Poisson step process with independent components q_i (which experience a jump in any interval of length Δt with probability $p_i \Delta t + o(\Delta t)$) and with zero mean jump distribution $P_i(\cdot)$, then

$$Av(x, t) \triangleq \mathscr{D}v(x, t) = \nabla_t v(x, t) + [\nabla_x v(x, t)]^T f(x, t)$$
$$+ \sum_{i=1}^m \int_{q_i} [v(x + \sigma_i(x, t) q_i, t) - v(x, t)] p_i P_i(dq_i)$$
$$(4.2.8)$$

where $\sigma_i(x, t)$ denotes the ith column of $\sigma(x, t)$, i.e.,

$$\sigma(x, t) = [\sigma_1(x, t), \ldots, \sigma_m(x, t)].$$

If for the autonomous version of Eq. (4.2.1) we consider $v(x) = x^T P x$, where P is a positive definite matrix, then straightforward computation yields $\mathscr{L}v(x) = \mathscr{D}v(x) = 2f(x)^T P x + \mathrm{tr}[\sigma(x)^T P \sigma(x)]$.

In the following results, $Av(x, t)$ denotes either $\mathscr{L}v(x, t)$ or $\mathscr{D}v(x, t)$.

4.2.9. Theorem. Suppose there exists a Lyapunov function $v: R^n \times T \to R$ in the domain of the infinitesimal generator A and that there exist $\psi_1, \psi_2 \in KR$ and $\psi_3 \in K$ such that

(i) $\psi_1(|x|) \le v(x, t) \le \psi_2(|x|)$ and
(ii) $Av(x, t) \le -\psi_3(|x|)$

for all $x \in R^n$ and $t \in T$. Then the trivial solution of Eq. (4.2.1) is **ASL w.p.1.**

4.2.10. Theorem. If in Theorem 4.2.9 $\psi_1(r) = c_1 r^2$, $\psi_2(r) = c_2 r^2$, and $\psi_3(r) = c_3 r^2$ where c_1, c_2, and c_3 are positive constants, then the trivial solution of Eq. (4.2.1) is **ESL w.p.1** and **ESL q.m.**

4.3 Composite Systems

We consider composite or interconnected systems which may be described by equations of the form

$$(\Sigma_i) \qquad dw_t^i = f_i(w_t^i, t)\, dt + g_i(w_t^1, \ldots, w_t^l, t)\, dt + \sum_{j=1}^l \sigma_{ij}(w_t^j, t)\, d\xi_t^j, \qquad (4.3.1)$$

$$i = 1, \ldots, l,$$

where $w_t^i \in R^{n_i}$, $f_i: R^{n_i} \times T \to R^{n_i}$, $\sigma_{ij}: R^{n_j} \times T \to R^{n_i \cdot m_j}$, $g_i: R^{n_1} \times \cdots \times R^{n_l} \times T \to R^{n_i}$, $\xi_t^i \in R^{m_i}$, and $\{\xi_t^i, t \in T\}$ is an independent increment Markov process.

Letting $\sum_{j=1}^l n_j = n$, $\sum_{j=1}^l m_j = m$, $x^T = [(w^1)^T, \ldots, (w^l)^T] \in R^n$, $f(x, t)^T = [f_1(w^1, t)^T, \ldots, f_l(w^l, t)^T]$, $g(x, t)^T = [g_1(w^1, \ldots, w^l, t)^T, \ldots, g_l(w^1, \ldots, w^l, t)^T] \triangleq [g_1(x, t)^T, \ldots, g_l(x, t)^T]$, $\sigma(x, t) = [\sigma_{ij}(w^j, t)]$, and $\xi^T = [(\xi^1)^T, \ldots, (\xi^l)^T]$, we can represent Eq. (4.3.1) equivalently as

$$(\mathscr{S}) \qquad dx_t = f(x_t, t)\, dt + \sigma(x_t, t)\, d\xi_t + g(x_t, t)\, dt$$

$$\triangleq F(x_t, t)\, dt + \sigma(x_t, t)\, d\xi_t \qquad (4.3.2)$$

where $f: R^n \times T \to R^n$, $\sigma: R^n \times T \to R^{n \cdot m}$, $g: R^n \times T \to R^n$, and $\xi_t \in R^m$.

We call system (4.3.2), which is of the same form as Eq. (4.2.1), a **composite** or **interconnected system** $(\mathscr{S})$ **with decomposition** (Σ_i). It may be viewed as a nonlinear and time varying interconnection (under disturbances) of l **isolated subsystems** $(\mathscr{S}_i)$ described by equations of the form

$$(\mathscr{S}_i) \qquad dw_t^i = f_i(w_t^i, t)\, dt + \sigma_{ii}(w_t^i, t)\, d\xi_t^i. \qquad (4.3.3)$$

We assume that the Markov process $\{\xi_t, t \in T\}$ in Eq. (4.3.2) has independent components. Furthermore, we assume once and for all that systems $(\mathscr{S})$ and $(\mathscr{S}_i)$ satisfy conditions of the type given in Theorem 4.2.2. We denote the unique solutions of $(\mathscr{S})$ and $(\mathscr{S}_i)$ by $\{x_t, t \in T\}$ (or simply x_t) and $\{w_t^i, t \in T\}$ (or simply w_t^i), respectively, with $x_0 \triangleq x$ and $w_0^i \triangleq w^i$ known. Also, we assume that the origin is the only equilibrium position of $(\mathscr{S})$ and $(\mathscr{S}_i)$.

In the following definition we let A_i denote the infinitesimal generator (defined by Eq. (4.2.6)) for subsystem $(\mathscr{S}_i)$.

4.3.4. Definition. Isolated subsystem $(\mathscr{S}_i)$ is said to possess **Property A** if there exists a Lyapunov function v_i in the domain of the infinitesimal generator A_i and if there exist $\psi_{i1}, \psi_{i2} \in KR$, $\psi_{i3} \in K$, and a constant $\sigma_i \in R$, such that

(i) $\psi_{i1}(|w^i|) \le v_i(w^i, t) \le \psi_{i2}(|w^i|)$ and
(ii) $A_i v_i(w^i, t) \le \sigma_i \psi_{i3}(|w^i|)$

for all $w^i \in R^{n_i}$ and for all $t \in T$.

4.3.5. Definition. If in Definition 4.3.4 $\psi_{i1}(r) = c_{i1} r^2$, $\psi_{i2}(r) = c_{i2} r^2$, and $\psi_{i3}(r) = r^2$, where c_{i1} and c_{i2} are positive constants, then isolated subsystem $(\mathscr{S}_i)$ is said to possess **Property B**.

If $\sigma_i < 0$, then the conditions in Definitions 4.3.4 and 4.3.5 correspond to the hypotheses of stability Theorems 4.2.9 and 4.2.10, respectively. As such, σ_i plays a similar role in the present case as it did in the preceding chapters, i.e., it is a measure of the degree of stability for $(\mathscr{S}_i)$.

Subsequently, we require certain identities involving the infinitesimal generator of Lyapunov functions v_i for system (Σ_i) described by Eq. (4.3.1), or equivalently, for system $(\mathscr{S})$ described by Eq. (4.3.2). In the following remarks, $A_i = \mathscr{L}_i$ denotes the operation $\mathscr{L}$ specified by Eq. (4.2.7) for the ith isolated subsystem $(\mathscr{S}_i)$. Similarly, $A_i = \mathscr{D}_i$ denotes the operation $\mathscr{D}$ specified by Eq. (4.2.8) for $(\mathscr{S}_i)$.

4.3.6. Remark. For system (Σ_i), or equivalently, for system $(\mathscr{S})$, let $\{\xi_t^i = z_t^i, t \in T\}$, $i = 1, \ldots, l$, be normalized Wiener processes and consider a Lyapunov function $v_i: R^{n_i} \times T \to R$. Let δ_{ij} denote the Kronecker delta symbol. Then

$$\mathscr{L}v_i(x, t)$$
$$= F(x, t)^{\mathrm{T}} \nabla_x v_i(w^i, t) + \tfrac{1}{2} \operatorname{tr}[\sigma(x, t)^{\mathrm{T}} \nabla_{xx} v_i(w^i, t) \sigma(x, t)] + \nabla_t v_i(w^i, t)$$
$$= \sum_{j=1}^{l} [f_j(w^j, t) + g_j(x, t)]^{\mathrm{T}} \nabla_{w^j} v_i(w^i, t)$$
$$+ \tfrac{1}{2} \sum_{j, k, m = 1}^{l} \operatorname{tr}[\sigma_{kj}(w^j, t)^{\mathrm{T}} \nabla_{w^k w^m} v_i(w^i, t) \sigma_{mj}(w^j, t)] + \nabla_t v_i(w^i, t)$$

$$= \sum_{j=1}^{l} [f_j(w^j, t) + g_j(x, t)]^T \delta_{ij} \nabla_{w^i} v_i(w^i, t) + \nabla_t v_i(w^i, t)$$

$$+ \tfrac{1}{2} \sum_{j,k,m=1}^{l} \text{tr}[\sigma_{kj}(w^j, t)^T \delta_{ki} \delta_{mi} \nabla_{w^i w^i} v_i(w^i, t) \sigma_{mj}(w^j, t)]$$

$$= [f_i(w^i, t) + g_i(x, t)]^T \nabla_{w^i} v_i(w^i, t) + \nabla_t v_i(w^i, t)$$

$$+ \tfrac{1}{2} \sum_{j=1}^{l} \text{tr}[\sigma_{ij}(w^j, t)^T \nabla_{w^i w^i} v_i(w^i, t) \sigma_{ij}(w^j, t)]$$

$$= \mathcal{L}_i v_i(w^i, t) + g_i(x, t)^T \nabla_{w^i} v_i(w^i, t)$$

$$+ \tfrac{1}{2} \sum_{j=1,\, i \neq j}^{l} \text{tr}[\sigma_{ij}(w^j, t)^T \nabla_{w^i w^i} v_i(w^i, t) \sigma_{ij}(w^j, t)],$$

i.e., we have

$$\mathcal{L} v_i(x, t) = \mathcal{L}_i v_i(w^i, t) + g_i(x, t)^T \nabla_{w^i} v_i(w^i, t)$$

$$+ \tfrac{1}{2} \sum_{j=1,\, i \neq j}^{l} \text{tr}[\sigma_{ij}(w^j, t)^T \nabla_{w^i w^i} v_i(w^i, t) \sigma_{ij}(w^j, t)].$$

(4.3.7)

If in particular, $\sigma_{ij}(w^j, t) \equiv 0$ for all $i \neq j$, then Eq. (4.3.7) reduces to

$$\mathcal{L} v_i(x, t) = \mathcal{L}_i v_i(w^i, t) + g_i(x, t)^T \nabla_{w^i} v_i(w^i, t).$$ (4.3.8)

4.3.9. Remark. For system (Σ_i), or equivalently, for system ($\mathscr{S}$), let $\{\xi_t^i = q_t^i,\ t \in T\}$, $i = 1, \ldots, l$, be Poisson processes, assume that $\sigma_{ij}(w^j, t) \equiv 0$ for all $i \neq j$, and consider a Lyapunov function $v_i: R^{n_i} \times T \to R$. Straightforward computation yields

$$\mathcal{D} v_i(x, t) = \mathcal{D}_i v_i(w^i, t) + g_i(x, t)^T \nabla_{w^i} v_i(w^i, t).$$ (4.3.10)

4.3.11. Remark. From Eqs. (4.3.8) and (4.3.10) it now follows that if $\xi_t^i = z_t^i$ or $\xi_t^i = q_t^i$, and if $\sigma_{ij}(w^j, t) \equiv 0$ for all $i \neq j$, then

$$A v_i(x, t) = A_i v_i(w^i, t) + g_i(x, t)^T \nabla_{w^i} v_i(w^i, t)$$ (4.3.12)

with $A_i = \mathcal{L}_i$ when $\xi_t^i = z_t^i$ and $A_i = \mathcal{D}_i$ when $\xi_t^i = q_t^i$.

4.4 Analysis by Scalar Lyapunov Functions

We are now in a position to state and prove several stability results for composite system ($\mathscr{S}$) with decomposition (Σ_i). In the present section we employ scalar Lyapunov functions consisting of a weighted sum of Lyapunov functions for the isolated subsystems ($\mathscr{S}_i$). In our first result we consider systems with disturbances confined to the subsystem structure.

4.4.1. Theorem. Assume that composite system ($\mathscr{S}$) satisfies the following conditions.

   (i) $\{\xi_t, t \in T\}$ is either a Wiener process or a Poisson process and there are no stochastic disturbances in the interconnecting structure, i.e., $\sigma_{ij}(w^j, t) \equiv 0$, $i \neq j$;

   (ii)   each isolated subsystem ($\mathscr{S}_i$) possesses Property A;

   (iii)   given the Lyapunov functions v_i and comparison functions ψ_{i3}, $i = 1, ..., l$, of hypothesis (ii), there exist real constants b_{ij}, $i, j = 1, ..., l$, such that

$$g_i(x, t)^T \nabla_{w^i} v_i(w^i, t) \leq [\psi_{i3}(|w^i|)]^{1/2} \sum_{j=1}^{l} b_{ij} [\psi_{j3}(|w^j|)]^{1/2}$$

for all $x \in R^n$ and $t \in T$; and

   (iv)   there exist positive constants α_i, $i = 1, ..., l$, such that the test matrix $S = [s_{ij}]$, $i, j = 1, ..., l$, defined by

$$s_{ij} = \begin{cases} \alpha_i(\sigma_i + b_{ii}), & i = j \\ \frac{1}{2}(\alpha_i b_{ij} + \alpha_j b_{ji}), & i \neq j \end{cases}$$

is negative definite, where σ_i is given in hypothesis (ii).
   Then the trivial solution of ($\mathscr{S}$) is **ASL w.p.1.**

   Proof. Given the Lyapunov functions v_i of hypothesis (ii) and the vector $\alpha^T = (\alpha_1, ..., \alpha_l) > 0$ of hypothesis (iv), choose for system ($\mathscr{S}$) the Lyapunov function

$$v(x, t) = \sum_{i=1}^{l} \alpha_i v_i(w^i, t).$$

Since each subsystem ($\mathscr{S}_i$) possesses Property A, we have

$$\sum_{i=1}^{l} \alpha_i \psi_{i1}(|w^i|) \leq v(x, t) \leq \sum_{i=1}^{l} \alpha_i \psi_{i2}(|w^i|)$$

for all $x \in R^n$ and $t \in T$. Thus, $v(x, t)$ is positive definite, decrescent, and radially unbounded, and there exist $\psi_1, \psi_2 \in KR$ such that

$$\psi_1(|x|) \leq v(x, t) \leq \psi_2(|x|)$$

for all $x \in R^n$ and $t \in T$.
   From hypothesis (i) we have $\sigma_{ij}(w^j, t) \equiv 0$ for all $i \neq j$, and from Remark 4.3.11 we have

$$Av_i(x, t) = A_i v_i(w^i, t) + g_i(x, t)^T \nabla_{w^i} v_i(w^i, t).$$

From the linearity of A and from hypotheses (ii) and (iii) it now follows that

$$Av(x,t) = \sum_{i=1}^{l} \alpha_i Av_i(x,t)$$

$$= \sum_{i=1}^{l} \alpha_i [A_i v_i(w^i, t) + g_i(x,t)^T \nabla_{w^i} v_i(w^i, t)]$$

$$\leq \sum_{i=1}^{l} \alpha_i \left\{ \sigma_i \psi_{i3}(|w^i|) + [\psi_{i3}(|w^i|)]^{1/2} \sum_{j=1}^{l} b_{ij} [\psi_{j3}(|w^j|)]^{1/2} \right\}$$

$$= u^T S u,$$

where $u^T = ([\psi_{13}(|w^1|)]^{1/2}, \dots, [\psi_{l3}(|w^l|)]^{1/2})$, and S is the test matrix given in hypothesis (iv). Since S is by assumption negative definite, it follows that $\lambda_M(S) < 0$ and

$$Av(x,t) \leq \lambda_M(S) \sum_{i=1}^{l} \psi_{i3}(|w^i|)$$

for all $x \in R^n$ and $t \in T$. Therefore, $Av(x,t)$ is negative definite and there exists a function $\psi_3 \in K$ such that

$$Av(x,t) \leq -\psi_3(|x|)$$

for all $x \in R^n$ and $t \in T$. Thus, all conditions of Theorem 4.2.9 are satisfied for system $(\mathscr{S})$, which proves the theorem. ∎

Note that Theorem 4.4.1 is quite similar in form to Theorem 2.4.2. In fact, the test matrices of these two results are identical. Therefore, the observations made in Chapter II for results such as Theorem 2.4.2 are also presently applicable. Furthermore, if $s_{ij} \geq 0$ for all $i \neq j$, then we can use the theory of M-matrices to eliminate the weighting vector α and obtain a result which is similar in form to Theorem 2.5.11.

Next, we consider systems with disturbances in the interconnecting structure and in the subsystem structure.

4.4.2. Theorem. Assume that composite system $(\mathscr{S})$ satisfies the following conditions.

(i) $\{\xi_t, t \in T\}$ is a Wiener process and in general the interconnecting structure disturbances are nonzero, i.e., $\sigma_{ij}(w^j, t) \not\equiv 0$, $i, j = 1, \dots, l$;

(ii) each isolated subsystem $(\mathscr{S}_i)$ possesses Property A;

(iii) given the Lyapunov functions v_i and comparison functions ψ_{i3}, $i = 1, \dots, l$, of hypothesis (ii), there exist real constants b_{ij}, $i, j = 1, \dots, l$, such that

$$g_i(x,t)^T \nabla_{w^i} v_i(w^i, t) \leq [\psi_{i3}(|w^i|)]^{1/2} \sum_{j=1}^{l} b_{ij} [\psi_{j3}(|w^j|)]^{1/2}$$

for all $x \in R^n$ and $t \in T$;

(iv)   for each v_i, $i = 1, ..., l$, there is a positive constant e_i such that

$$(u^i)^T \nabla_{w^i w^i} v_i(w^i, t) u^i \leq e_i |u^i|^2$$

holds uniformly in w^i and $t \in T$ for all $u^i \in R^{n_i}$;

(v)   for each $\sigma_{ij}(w^j, t)$, $i, j = 1, ..., l$, $i \neq j$, there exists a constant $d_{ij} \geq 0$ such that

$$\|\sigma_{ij}(w^j, t)\|_m^2 \leq d_{ij} \psi_{j3}(|w^j|)$$

for all $w^j \in R^{n_j}$, $j = 1, ..., l$; and

(vi)   there exist positive constants α_i, $i = 1, ..., l$, such that the test matrix $S = [s_{ij}]$, $i, j = 1, ..., l$, defined by

$$s_{ij} = \begin{cases} \alpha_i(\sigma_i + b_{ii}) + \tfrac{1}{2} \sum\limits_{k=1, k \neq i}^{l} \alpha_k e_k d_{ki}, & i = j \\ \tfrac{1}{2}[\alpha_i b_{ij} + \alpha_j b_{ji}], & i \neq j \end{cases}$$

is negative definite, where σ_i is given in hypothesis (ii).

Then the trivial solution of composite system $(\mathscr{S})$ is **ASL w.p.1**.

Proof. As in the proof of Theorem 4.4.1 we choose for system $(\mathscr{S})$ the Lyapunov function

$$v(x, t) = \sum_{i=1}^{l} \alpha_i v_i(z_i, t).$$

Since each subsystem $(\mathscr{S}_i)$ possesses Property A, it follows again that $v(x, t)$ is positive definite, decrescent, and radially unbounded for all $x \in R^n$ and $t \in T$. Thus, there exist $\psi_1, \psi_2 \in KR$ such that

$$\psi_1(|x|) \leq v(x, t) \leq \psi_2(|x|)$$

for all $x \in R^n$ and $t \in T$.

Using the linearity of the operator $\mathscr{L}$, using hypotheses (ii)–(vi), and invoking Eq. (4.3.7), we obtain for system $(\mathscr{S})$,

$$\mathscr{L}v(x, t) = \sum_{i=1}^{l} \alpha_i \mathscr{L}v_i(x, t)$$

$$= \sum_{i=1}^{l} \alpha_i \Big\{ \mathscr{L}_i v_i(w^i, t) + g_i(x, t)^T \nabla_{w^i} v_i(w^i, t)$$

$$+ \tfrac{1}{2} \sum_{j=1, i \neq j}^{l} \text{tr}[\sigma_{ij}(w^j, t)^T \nabla_{w^i w^i} v_i(w^i, t) \sigma_{ij}(w^j, t)] \Big\}$$

$$\leq \sum_{i=1}^{l} \alpha_i \Big\{ \sigma_i \psi_{i3}(|w^i|) + [\psi_{i3}(|w^i|)]^{1/2} \sum_{j=1}^{l} b_{ij} [\psi_{j3}(|w^j|)]^{1/2}$$

$$+ \tfrac{1}{2} \sum_{j=1, i \neq j}^{l} e_i \|\sigma_{ij}(w^j, t)\|_m^2 \Big\}$$

$$\leq \sum_{i=1}^{l} \alpha_i(\sigma_i + b_{ii}) \{[\psi_{i3}(|w^i|)]^{1/2}\}^2$$

$$+ \sum_{i,j=1; i \neq j}^{l} \tfrac{1}{2}(\alpha_i b_{ij} + \alpha_j b_{ji}) [\psi_{i3}(|w^i|)]^{1/2} [\psi_{j3}(|w^j|)]^{1/2}$$

$$+ \sum_{i,j=1; i \neq j}^{l} \tfrac{1}{2}\alpha_i e_i d_{ij} \{[\psi_{j3}(|w^j|)]^{1/2}\}^2 = u^T S u$$

where $u^T = ([\psi_{13}(|w^1|)]^{1/2}, ..., [\psi_{13}(|w^l|)]^{1/2})$ and S is the test matrix given in hypothesis (vi). Since S is negative definite, it follows that $\lambda_M(S) < 0$ and $\mathscr{L}v(x, t) \leq u^T S u$ is negative definite. Thus, there exists a function $\psi_3 \in K$ such that

$$\mathscr{L}v(x,t) \leq -\psi_3(|x|)$$

for all $x \in R^n$ and $t \in T$. This completes the proof. ∎

4.4.3. Theorem. If in Theorems 4.4.1 and 4.4.2 each isolated subsystem $(\mathscr{S}_i)$ possesses Property B, then the trivial solution of composite system $(\mathscr{S})$ is **ESL w.p.1** and **ESL q.m.**

Proof. Since each subsystem $(\mathscr{S}_i)$ possesses Property B it is easily shown that the comparison functions ψ_1, ψ_2, and ψ_3 in the proofs of Theorems 4.4.1 and 4.4.2 can be chosen as

$$\psi_1(r) = \min_i (\alpha_i c_{i1}) r^2, \qquad \psi_2(r) = \max_i (\alpha_i c_{i2}) r^2, \qquad \psi_3(r) = \lambda_M(S) r^2.$$

The conclusion of the theorem follows now from Theorem 4.2.10. ∎

In our next result we consider the *autonomous version* of composite system $(\mathscr{S})$ with decomposition (Σ_i). We refer to the autonomous versions of $(\mathscr{S})$, (Σ_i), and $(\mathscr{S}_i)$ as $(\mathscr{S}')$, (Σ_i'), and $(\mathscr{S}_i')$, respectively. For each isolated subsystem $(\mathscr{S}_i')$ we choose a quadratic Lyapunov function

$$v_i(w^i) = (w^i)^T P_i w^i \tag{4.4.4}$$

where $w^i \in R^{n_i}$ and $P_i = P_i^T$ is a positive definite $(n_i \times n_i)$ matrix.

4.4.5. Theorem. Assume that for composite system $(\mathscr{S}')$ the following conditions hold.

(i) $\{\xi_t, t \in T\}$ is either a Wiener process or a Poisson process and in general the interconnecting structure disturbances are nonzero, i.e., $\sigma_{ij}(w^j) \not\equiv 0$, $i, j = 1, ..., l$;

(ii) each isolated subsystem $(\mathscr{S}_i')$ possesses Property B with $v_i(w_i)$ specified by Eq. (4.4.4);

(iii)   given the matrices P_i of hypothesis (ii), there exist real constants b_{ij}, $i, j = 1, \dots, l$, such that

$$g_i(x)^{\mathrm{T}} P_i w^i \le \tfrac{1}{2} |w^i| \left| \sum_{j=1}^{l} b_{ij} \right| w^j|$$

for all $x \in R^n$;

(iv)   for each $\sigma_{ij}(w^j)$, $i, j = 1, \dots, l$, $i \ne j$, there exists a constant $d_{ij} \ge 0$ such that

$$\| \sigma_{ij}(w^j) \|_m^2 \le d_{ij} |w^j|^2$$

for all $w^j \in R^{n_j}$, $j = 1, \dots, l$; and

(v)   there exist positive constants α_i, $i = 1, \dots, l$, such that the test matrix $S = [s_{ij}]$, $i, j = 1, \dots, l$, defined by

$$s_{ij} = \begin{cases} \alpha_i (\sigma_i + b_{ii}) + \displaystyle\sum_{k=1, \, k \ne i}^{l} \alpha_k \lambda_M(P_k) \, d_{ki}, & i \ne j \\[2ex] \tfrac{1}{2}(\alpha_i b_{ij} + \alpha_j b_{ji}), & i \ne j \end{cases}$$

is negative definite.

Then the trivial solution of composite system $(\mathscr{S}')$ is **ESL w.p.1** and **ESL q.m.**

Proof.   The proof follows directly from Theorems 4.4.2 and 4.4.3 and from the fact that $\mathscr{L} = \mathscr{D}$, given the quadratic functions v_i.   ∎

Note that in Theorems 4.4.2, 4.4.3, and 4.4.5, the interconnection disturbance terms d_{ki}, $k, i = 1, \dots, l$, $k \ne i$, which express the magnitude of the disturbances, occur in the diagonal of the test matrix S. Their effect on the negative definiteness of S is to make more restrictive the conditions on the remaining parameters of this matrix. Thus, the disturbance terms have in general a degrading influence on the stability properties of composite system $(\mathscr{S})$. It is also interesting to note that the test matrices of Theorems 4.4.2, 4.4.3, and 4.4.5 have a similar structure as the test matrix of Theorem 3.3.14.

Note that the stabilization procedure indicated in Chapter II is also applicable in the present case.

The presence of additional terms in the diagonal of the test matrices S in Theorems 4.4.2, 4.4.3, and 4.4.5 prevents us from eliminating the weighting vector α by the methods of Chapter II. However, as can be seen from our next result, it is still possible to use M-matrix results to establish stability conditions for system $(\mathscr{S})$ which involve nonsymmetric test matrices without the presence of weighting factors.

4.4.6.   Theorem.   Assume that for composite system $(\mathscr{S})$ the following conditions hold.

(i) $\{\xi_t, t \in T\}$ is a Wiener process and in general the interconnecting disturbances are nonzero, i.e., $\sigma_{ij}(w^j, t) \not\equiv 0$, $i, j = 1, ..., l$;

(ii) each isolated subsystem $(\mathscr{S}_i)$ possesses Property A;

(iii) given the Lyapunov functions v_i and the comparison functions ψ_{i3}, $i = 1, ..., l$, of hypothesis (ii), there exist real constants a_{ij}, $i, j = 1, ..., l$, such that

(a) $a_{ij} \geq 0$ for all $i \neq j$ and

(b) $\nabla_{w^i} v_i(w^i, t)^{\mathrm{T}} g_i(x, t) \leq \sum_{j=1}^{l} a_{ij} \psi_{j3}(|w^j|)$ for all $x \in R^n$ and $t \in T$;

(iv) for each v_i, $i = 1, ..., l$, there is a positive constant e_i such that

$$(u^i)^{\mathrm{T}} \nabla_{w^i w^i} v_i(w^i, t) u^i \leq e_i |u^i|^2$$

holds uniformly in w^i and $t \in T$ for all $u^i \in R^{n_i}$;

(v) for each $\sigma_{ij}(w^j, t)$, $i, j = 1, ..., l$, $i \neq j$, there exists a constant $d_{ij} \geq 0$ such that

$$\|\sigma_{ij}(w^j, t)\|_m^2 \leq d_{ij} \psi_{j3}(|w^j|)$$

for all $w^j \in R^{n_j}$, $j = 1, ..., l$, and $t \in T$; and

(vi) the test matrix $S = [s_{ij}]$ defined by

$$s_{ij} = \begin{cases} -(\sigma_i + a_{ii}), & i = j \\ -a_{ij} - \tfrac{1}{2} e_i d_{ij}, & i \neq j \end{cases}$$

has positive successive principal minors.

Then the trivial solution of composite system $(\mathscr{S})$ is **ASL w.p.1.**

Proof. Given the Lyapunov functions v_i of hypothesis (ii), let $\alpha^{\mathrm{T}} = (\alpha_1, ..., \alpha_l) > 0$ be an arbitrary constant vector and choose for system $(\mathscr{S})$ the Lyapunov function

$$v(x, t) = \sum_{i=1}^{l} \alpha_i v_i(w^i, t).$$

Since each subsystem $(\mathscr{S}_i)$ possesses Property A it follows that $v(x, t)$ is positive definite, decrescent, and radially unbounded for all $x \in R^n$ and $t \in T$.

As in the proof of Theorem 4.4.2, we have

$$\mathscr{L}v(x, t) = \sum_{i=1}^{l} \alpha_i \left\{ \mathscr{L}_i v_i(w^i, t) + g_i(x, t)^{\mathrm{T}} \nabla_{w^i} v_i(w^i, t) \right.$$
$$\left. + \frac{1}{2} \sum_{j=1, i \neq j}^{l} \operatorname{tr}\left[\sigma_{ij}(w^j, t)^{\mathrm{T}} \nabla_{w^i w^i} v_i(w^i, t) \sigma_{ij}(w^j, t) \right] \right\}.$$

Usage of hypotheses (iii)–(v) yields

$$\mathscr{L}v(x,t) \le \sum_{i=1}^{l} \alpha_i \left\{ \sigma_i \psi_{i3}(|w^i|) + \sum_{j=1}^{l} a_{ij} \psi_{j3}(|w^j|) \right.$$

$$\left. + \frac{1}{2} \sum_{j=1,\, i \ne j}^{l} e_i \|\sigma_{ij}(w^j,t)\|_m^2 \right\}$$

$$\le \sum_{i=1}^{l} \alpha_i (\sigma_i + a_{ii}) \psi_{i3}(|w^i|) + \sum_{i,j=1;\, i \ne j}^{l} \alpha_i a_{ij} \psi_{j3}(|w^j|)$$

$$+ \sum_{i,j=1;\, i \ne j}^{l} \tfrac{1}{2} \alpha_i e_i d_{ij} \psi_{j3}(|w^j|).$$

Letting $u^{\mathrm{T}} = (\psi_{13}(|w^1|), \ldots, \psi_{13}(|w^l|))$, defining $S = [s_{ij}]$ as in hypothesis (vi), defining $y^{\mathrm{T}} = \alpha^{\mathrm{T}} S$, and using the identical argument as in the proof of Theorem 2.5.15, we can readily show that

$$\mathscr{L}v(x,t) \le -y^{\mathrm{T}}u < 0, \qquad x \ne 0,$$

i.e., $\mathscr{L}v(x,t)$ is negative definite for all $x \in R^n$ and $t \in T$. This completes the proof. ∎

There are several differences between the test matrices of Theorems 4.4.2, 4.4.3, 4.4.5, and the test matrix of Theorem 4.4.6. In the case of the former, S is symmetric, there are no sign restrictions on the off-diagonal terms, S involves the weighting vector α, and the disturbance terms d_{ij} enter into the diagonal terms s_{ii}. In the case of the latter, S is in general not symmetric, the off-diagonal terms have to be nonpositive, S does not involve any weighting factors, and the disturbance terms enter into the off-diagonal terms s_{ij}, $i \ne j$. On the other hand, in *all* results of this section, the disturbances for the isolated subsystems $(\mathscr{S}_i)$ are reflected in the diagonal elements of the test matrix S.

Finally, it is important to note that if $\sigma_{ij}(w^j,t) \equiv 0$ for all $i, j = 1, \ldots, l$, then $(\mathscr{S})$ becomes an ordinary differential equation. In that case, all results of the present section reduce to corresponding results established for deterministic systems described by ordinary differential equations which we treated in Chapter II.

4.5   Analysis by Vector Lyapunov Functions

In the present section we concern ourselves once more with composite system $(\mathscr{S})$ with decomposition (Σ_i). To simplify matters, we consider only the case when $\{\xi_t, t \in T\}$ is a Wiener process. In contrast to the approach of

Section 4.4, where we utilized scalar Lyapunov functions, the results of the present section depend on the existence of vector Lyapunov functions constructed from scalar Lyapunov functions for the isolated subsystems $(\mathscr{S}_i)$. Henceforth we assume that all such functions are in the domain of their respective infinitesimal generator $\mathscr{L}_i$.

4.5.1. Definition. Isolated subsystem $(\mathscr{S}_i)$ is said to possess **Property E** if there exists a scalar function $v_i: R^{n_i} \times T \to R$ in the domain of $\mathscr{L}_i$, a function $\psi_i \in KR$, and a continuous scalar function $a_i: R \times T \to R$ such that

$$v_i(w^i, t) \geq \psi_i(|w^i|), \qquad v_i(0, t) \equiv 0,$$

and

$$\mathscr{L}_i v_i(w^i, t) \leq a_i(v_i(w^i, t), t)$$

for all $w^i \in R^{n_i}$ and $t \in T$ and if $|v_i|$ is decrescent.

If isolated subsystem $(\mathscr{S}_i)$ possesses Property E and if there exists a function $\varphi_i \in K$ such that $a_i(v_i(w^i, t), t) \leq -\varphi_i(|w^i|)$ for all $w^i \in R^{n_i}$ and $t \in T$, then the trivial solution of $(\mathscr{S}_i)$ is ASL w.p.1. If $(\mathscr{S}_i)$ possesses Property E and if for some $c_i > 0$, $a_i(v_i, t) \leq -c_i v_i$, then the trivial solution of $(\mathscr{S}_i)$ is ESL w.p.1 (see Kushner [1] and Arnold [1]).

In our first result, which is a comparison theorem, we use functions of class K^i defined in Section 4.1. This result involves vector Lyapunov functions of the form $v(x, t)^T = (v_1(w^1, t), \ldots, v_l(w^l, t))$.

4.5.2. Theorem. Assume that for composite system $(\mathscr{S})$ the following conditions hold.

(i)   Each isolated subsystem $(\mathscr{S}_i)$ possesses Property E;

(ii)   for each $i = 1, \ldots, l$, there exists a function $b_i: R^l \times T \to R$, $b_i \in K^i$, such that

$$\nabla_{w^i} v_i(w^i, t)^T g_i(x, t) + \frac{1}{2} \sum_{j=1, i \neq j}^{l} \mathrm{tr}[\sigma_{ij}(w^j, t)^T \nabla_{w^i w^i} v_i(w^i, t) \sigma_{ij}(w^j, t)]$$

$$< b_i(v(x, t), t) \tag{4.5.3}$$

where $v(x, t)^T = (v_1(w^1, t), \ldots, v_l(w^l, t))$;

(iii)   let $a(\cdot)^T = (a_1(\cdot), \ldots, a_l(\cdot))$ and $b(\cdot)^T = (b_1(\cdot), \ldots, b_l(\cdot))$ where $a_i(\cdot)$ and $b_i(\cdot)$, $i = 1, \ldots, l$, are defined in hypotheses (i) and (ii), and assume that the vector function $[a(\cdot) + b(\cdot)]$ satisfies the inequalities

$$|a(v, t) + b(v, t)|^2 \leq k(1 + |v|^2)$$

and

$$|a(v, t) + b(v, t) - a(w, t) - b(w, t)| \leq k|v - w|$$

for all $v, w \in R^l$, $v \geq 0$, $w \geq 0$, for all $t \in T$, and some $k > 0$;

(iv)   there exists a function $p: R^l \times T \to R$ such that

$$p(y,t) \geq \sup_{v(x,t) \leq y} \sum_{i,j=1}^{l} |\nabla_{w^i} v_i(w^i,t)^\mathrm{T} \sigma_{ij}(w^j,t)|^2 \qquad (4.5.4)$$

and

$$p(y,t) \leq k(1+|y|^2). \qquad (4.5.5)$$

Then $v(x_t,t) \leq y_t$ with probability one for all $t \in T$, where y_t is a solution of the vector stochastic differential equation

$$dy_t = [a(y_t,t)+b(y_t,t)]\,dt + Y_t\,dz_t \qquad (4.5.6)$$

with $y_0 \geq v(x_0,0)$ and with Y_t a matrix-valued random process such that

$$\|Y_t\|_m^2 \leq p(y_t,t). \qquad (4.5.7)$$

Proof. Choose $v(x,t)^\mathrm{T} = (v_1(w^1,t), ..., v_l(w^l,t))$ as a vector Lyapunov function for composite system $(\mathscr{S})$ with $v_i(w^i,t)$, $i=1,...,l$, given in hypothesis (i). By Ito's lemma (see Wong [1]) we have

$$dv(x_t,t)^\mathrm{T} = (dv_1(w^1,t), ..., dv_l(w^l,t))$$

where

$$\begin{aligned}
dv_i(w^i_t,t) &= \mathscr{L}v_i(w^i_t,t)\,dt + \nabla_x v_i(w^i_t,t)^\mathrm{T}\sigma(x_t,t)\,dz_t\\
&= \{\nabla_t v_i(w^i_t,t)+\nabla_x v_i(w^i_t,t)^\mathrm{T}[f(x_t,t)+g(x_t,t)]\\
&\quad +\tfrac{1}{2}\mathrm{tr}[\sigma(x_t,t)^\mathrm{T}\nabla_{xx}v_i(w^i_t,t)\sigma(x_t,t)]\}\,dt\\
&\quad + \nabla_x v_i(w^i_t,t)^\mathrm{T}\sigma(x_t,t)\,dz_t\\
&= \Big\{\nabla_t v_i(w^i_t,t)+\nabla_{w^i}v_i(w^i_t,t)^\mathrm{T}[f_i(w^i_t,t)+g_i(x_t,t)]\\
&\quad +\frac{1}{2}\sum_{j=1}^{l}\mathrm{tr}[\sigma_{ij}(w^j_t,t)^\mathrm{T}\nabla_{w^iw^i}v_i(w^i_t,t)\sigma_{ij}(w^j_t,t)]\Big\}\,dt\\
&\quad + \nabla_{w^i}v_i(w^i_t,t)^\mathrm{T}\sum_{j=1}^{l}\sigma_{ij}(w^j_t,t)\,dz_t^j\\
&= \Big\{\mathscr{L}_i v_i(w^i_t,t)+\nabla_{w^i}v_i(w^i_t,t)^\mathrm{T}g_i(x_t,t)\\
&\quad +\frac{1}{2}\sum_{j=1,i\neq j}^{l}\mathrm{tr}[\sigma_{ij}(w^j_t,t)^\mathrm{T}\nabla_{w^iw^i}v_i(w^i_t,t)\sigma_{ij}(w^j_t,t)]\Big\}\,dt\\
&\quad + \sum_{j=1}^{l}\nabla_{w^i}v_i(w^i_t,t)^\mathrm{T}\sigma_{ij}(w^j_t,t)\,dz_t^j\\
&< [a_i(v_i(w^i_t,t),t)+b_i(v(x_t,t),t)]\,dt + \sum_{j=1}^{l}y_t^{ij}dz_t^j
\end{aligned}$$

where

$$y_t^{ij} \triangleq \nabla_{w^i}v_i(w^i_t,t)^\mathrm{T}\sigma_{ij}(w^j_t,t).$$

Letting $a(\cdot)^{\mathrm{T}} = (a_1(\cdot), \dots, a_l(\cdot))$, $b(\cdot)^{\mathrm{T}} = (b_1(\cdot), \dots, b_l(\cdot))$, and $Y_t = [y_t^{ij}]$, the above inequality assumes the form

$$dv(x_t, t) < [a(v(x_t, t), t) + b(v(x_t, t), t)]\, dt + Y_t\, dz_t.$$

Given y_t, a solution of Eq. (4.5.6), it follows that

$$dv_i(w_t^i, t) - dy_t^i < \{[a_i(v_i(w_t^i, t), t) - a_i(y_t^i, t)] + [b_i(v(x_t, t), t) - b_i(y_t, t)]\}\, dt.$$

Now since by assumption composite system ($\mathscr{S}$) satisfies the conditions of Theorem 4.2.2 and since $v(\cdot)$ is continuous, it follows that $v(x_t, t)$ is sample continuous with probability one. Likewise, it follows from hypotheses (iii) and (iv) and from Theorem 4.2.2 that y_t is sample continuous with probability one. A sufficient condition therefore to guarantee that $v(x_t, t) \le y_t$ with probability one for all $t \in T$, given $v(x_0, 0) \le y_0$, is that $dv_i(w_t^i, t) - dy_t^i < 0$ whenever $v_i(w_t^i, t) = y_t^i$. But since $b_i \in K^i$, then this condition is satisfied and therefore $v(x_t, t) \le y_t$ with probability one for all $t \in T$.

Equation (4.5.7) follows immediately since

$$\|Y_t\|_m^2 = \sum_{i, j=1}^{l} |y_t^{ij}|^2 = \sum_{i, j=1}^{l} |\nabla_{w^i} v_i(w_t^i, t)^{\mathrm{T}} \sigma_{ij}(w_t^j, t)|^2$$

$$\le \sup_{v(x, t) < y_t} \sum_{i, j=1}^{l} |\nabla_{w^i} v_i(w^i, t)^{\mathrm{T}} \sigma_{ij}(w^j, t)|^2 \le p(y_t, t).$$

This completes the proof. ∎

4.5.8. Remark. Hypothesis (i) and the conclusion of Theorem 4.5.2 imply that $|y_t| \ge |v(x_t, t)| \ge [\sum_{i=1}^{l} \psi_i(|w_t^i|)^2]^{1/2} \ge \psi(|x_t|)$, where the existence of a function $\psi \in KR$ follows from the continuity of the ψ_i, $i = 1, \dots, l$, and the norm $|\cdot|$ (refer to Hahn [2, pp. 97–99]). Therefore, since $|x_t| \le \psi^{-1}(|y_t|)$, and since $\psi^{-1} \in KR$, the stability of composite system ($\mathscr{S}$) is implied by the stability of **comparison system** (4.5.6).

Inequality (4.5.7) is useful in determining the stability of Eq. (4.5.6) and this fact is used to obtain the following result.

4.5.9. Theorem. Assume that for comparison system (4.5.6) the following conditions hold.

(i)   The hypotheses of Theorem 4.5.2 are true; and

(ii) there exists a continuous function $u: R^l \times T \to R$ in the domain of $\mathscr{L}_y$ (the $\mathscr{L}$ operator with respect to (4.5.6)), a function $h: R^l \times T \to R$ and two functions $\psi_1 \in KR$ and $\psi_2 \in K$ such that for all $t \in T$, $q \in R^l$, and $y \ge 0$,

$$q^{\mathrm{T}} [\nabla_{yy} u(y, t)] q \le h(y, t)|q|^2, \tag{4.5.10}$$

$$u(y, t) \ge \psi_1(|y|), \qquad u(0, t) \equiv 0, \tag{4.5.11}$$

and such that $|u|$ is decrescent and

$$\nabla_t u(y,t) + \nabla_y u(y,t)^{\mathsf{T}}[a(y,t)+b(y,t)] + \tfrac{1}{2}h(y,t)p(y,t) \leq -\psi_2(|y|).$$
(4.5.12)

Then comparison system (4.5.6), and hence composite system $(\mathscr{S})$, is **ASL w.p.1.**

Proof. With $u(y,t)$ as given in the theorem, u is positive definite, decrescent, and radially unbounded for all $y \geq 0$, and

$$\begin{aligned}
\mathscr{L}_y u(y,t) &= \nabla_t u(y,t) + \nabla_y u(y,t)^{\mathsf{T}}[a(y,t)+b(y,t)] + \tfrac{1}{2}\operatorname{tr}[y_t^{\mathsf{T}}\nabla_{yy}u(y,t)y_t] \\
&\leq \nabla_t u(y,t) + \nabla_y u(y,t)^{\mathsf{T}}[a(y,t)+b(y,t)] + \tfrac{1}{2}h(y,t)\|Y_t\|_m^2 \\
&\leq \nabla_t u(y,t) + \nabla_y u(y,t)^{\mathsf{T}}[a(y,t)+b(y,t)] + \tfrac{1}{2}h(y,t)p(y,t) \\
&\leq -\psi_2(|y|).
\end{aligned}$$

Thus, $\mathscr{L}_y u(y,t)$ is negative definite for all $y \geq 0$.

Since $y_t \geq v(x_t,t) \geq 0$, as a consequence of hypothesis (i), it follows that stability conditions given for $y \geq 0$ imply the stability of $y = 0$ for all solutions such that $y_0 \geq 0$. Therefore the above conditions on u and $\mathscr{L}_y u$ imply that Eq. (4.5.6) with the specified initial condition is ASL w.p.1 (refer to Kushner [1] and Arnold [1]) and hence as a consequence of Remark 4.5.8, the equilibrium of composite system $(\mathscr{S})$ is also ASL w.p.1. This follows because

$$P\left(\sup_{0 \leq t < \infty} |x_t| \geq \mathscr{E}\right) \leq P\left(\sup_{0 \leq t < \infty} |y_t| \geq \psi(\mathscr{E})\right)$$

for some $\psi \in K$. The remaining relationships are shown similarly, completing the proof. ∎

4.5.13. Corollary. If in Theorem 4.5.9, $\psi_2(|y|)$ is replaced by $cu(y,t)$ for some constant $c > 0$, then comparison system (4.5.6), and hence composite system $(\mathscr{S})$, is **ESL w.p.1.**

Proof. The corollary strengthens the condition on $\mathscr{L}_y u(y,t)$ to

$$\mathscr{L}_y u(y,t) \leq -cu(y,t), \qquad c > 0$$

resulting in ESLw.p.1 for system (4.5.6) and system $(\mathscr{S})$. ∎

Since the conditions on $u(\cdot)$ in the above results need to be satisfied only for $y \geq 0$, it is possible to choose u in a particularly simple form and thereby simplify the hypotheses. The following result takes advantage of this fact.

4.5.14. Theorem. Assume that for comparison system (4.5.6) the following conditions hold.

(i) The hypotheses of Theorem 4.5.2 are satisfied;

(ii)   there exists an $\alpha > 0$, $\alpha \in R^l$, and a function $\psi_2 \in K$ such that for all $y \geq 0$,

$$\alpha^T[a(y,t)+b(y,t)] \leq -\psi_2(|y|). \tag{4.5.15}$$

Then comparison system (4.5.6), and hence composite system $(\mathscr{S})$, is **ASL w.p.1**.

4.5.16. Corollary. If in Theorem 4.5.14 $\psi_2(|y|)$ is replaced by $c\alpha^T y$ for some constant $c > 0$ then the conclusion of Corollary 4.5.13 holds.

The proofs of Theorem 4.5.14 and Corollary 4.5.16 are identical to the proofs of Theorem 4.5.9 and Corollary 4.5.13, respectively, with the exception that $u(\cdot)$ is required to be of the form $u(y,t) = \alpha^T y$ for some $\alpha > 0$. This yields

$$\mathscr{L}_y u(y,t) = \alpha^T[a(y,t)+b(y,t)]$$

from which the respective conclusions follow.

4.5.17. Remark. When the functions $a(\cdot)$ and $b(\cdot)$ in Theorem 4.5.2 assume special forms, then inequality (4.5.15) may be expressed in terms of conditions on some test matrix S. For example, if there is a $w^T = (\varphi_1(|y^1|), ..., \varphi_l(|y^l|))$ for some $\varphi_i \in K$, $i = 1, ..., l$, and a constant matrix S such that

$$\alpha^T[a(y,t)+b(y,t)] \leq -\alpha^T Sw \tag{4.5.18}$$

then condition (4.5.15) may be replaced with the condition that S have positive successive principal minors and $s_{ij} \geq 0$, $i \neq j$.

Alternatively, if there is a $w^T = (\varphi_1(|y^1|), ..., \varphi_l(|y^l|))$, $\varphi_i \in K$, $i = 1, ..., l$, and constants a_{ij}, $i, j = 1, ..., l$, such that

$$a_i(y^i,t) + b_i(y,t) \leq \varphi_i(|y^i|) \sum_{j=1}^{l} a_{ij} \varphi_j(|y^j|) \tag{4.5.19}$$

and if we let $S = [s_{ij}]$, $s_{ij} = \alpha_i a_{ij}$, $i, j = 1, ..., l$, then we have

$$\alpha^T[a(y,t)+b(y,t)] \leq w^T Sw. \tag{4.5.20}$$

Condition (4.5.15) may then be replaced with the condition that $S+S^T$ be negative definite. The dependence of S on α in this case is useful since it allows the manipulation of S to improve results.

Clearly, the stability conditions involving inequalities (4.5.18) and (4.5.20) are very similar to corresponding stability conditions obtained in Section 4.4. However, the results of the present section and those of Section 4.4 do not appear to be completely equivalent.

4.6 Examples

To demonstrate applications of the preceding results, we consider some specific examples.

4.6.1. Example. Consider the *indirect control problem*

$$dw_t^1 = Aw_t^1 dt + bf(w_t^2)\, dt + \sigma_{11}(w_t^1)\, dz_t^1 + \sigma_{12}(w_t^2)\, dz_t^2$$
$$dw_t^2 = [-\rho w_t^2 - rf(w_t^2)]\, dt + a^T w_t^1 dt + \sigma_{21}(w_t^1)\, dz_t^1 + \sigma_{22}(w_t^2)\, dz_t^2$$
$$(4.6.2)$$

where $w_t^1 \in R^{n_1}$, A is a stable $n_1 \times n_1$ matrix, $b \in R^{n_1}$, $r > 0$ is a constant, $a \in R^{n_1}$, $\{z_t^i, t \in T\}$ is an m_i-dimensional normalized Wiener process with independent components, $\sigma_{ij}: R^{n_j} \to R^{n_i \cdot m_j}$, and the function $f: R \to R$ is assumed to have the following properties: (i) $f(w^2)$ is continuous for all $w^2 \in R$, (ii) $f(w^2) = 0$ if and only if $w^2 = 0$, and (iii) $0 < w^2 f(w^2) < k|w^2|^2$ for all $w^2 \in R$ where $k > 0$ is a constant. Assume that for each $\sigma_{ij}(w^j)$ there is a constant $d_{ij} > 0$ such that $\|\sigma_{ij}(w^j)\|_m^2 \le d_{ij}|w^j|^2$.

System (4.6.2), which is clearly a special case of composite system $(\mathcal{S})$, may be viewed as a nonlinear interconnection under disturbances of two isolated subsystems

$$dw_t^1 = Aw_t^1 dt + \sigma_{11}(w_t^1)\, dz_t^1 \qquad (\mathcal{S}_1)$$
$$dw_t^2 = [-\rho w_t^2 - rf(w_t^2)]\, dt + \sigma_{22}(w_t^2)\, dz_t^2 \qquad (\mathcal{S}_2)$$

with interconnecting structure under disturbances specified by

$$g_1(x_t)\, dt + \sigma_{12}(w_t^2)\, dz_t^2 \triangleq bf(w_t^2)\, dt + \sigma_{12}(w_t^2)\, dz_t^2$$
$$g_2(x_t)\, dt + \sigma_{21}(w_t^1)\, dz_t^1 \triangleq a^T w_t^1 dt + \sigma_{21}(w_t^1)\, dz_t^1,$$

where the notation of Eqs. (4.3.1) and (4.3.2) has been used.

Since A is stable, there exists a positive definite matrix P such that the matrix $A^T P + PA = -Q$ is negative definite. Choosing $v_1(w^1) = (w^1)^T P w^1$, we have for $(\mathcal{S}_1)$, using the notation of Theorem 4.4.2, $e_1 = 2\lambda_M(P)$,

$$\lambda_m(P)|w^1|^2 \le v_1(w^1) \le \lambda_M(P)|w^1|^2,$$

and

$$\mathcal{L}v_1(w^1) = (w^1)^T [A^T P + PA] w^1 + \text{tr}[\sigma_{11}(w^1)^T P \sigma_{11}(w^1)]$$
$$\le [-\lambda_m(Q) + \lambda_M(P) d_{11}]|w^1|^2.$$

Choosing $v_2(w^2) = |w^2|^2$, we have for $(\mathcal{S}_2)$, $e_2 = 2$, and

$$\mathcal{L}v_2(w^2) = -2\rho|w^2|^2 - 2rw^2 f(w^2) + [\sigma_{22}(w^2)]^2 \le (-2\rho + d_{22})|w^2|^2.$$

Thus, isolated subsystems $(\mathcal{S}_1)$ and $(\mathcal{S}_2)$ both possess Property B, with $\psi_{11}(r) = \lambda_m(P)r^2$, $\psi_{12}(r) = \lambda_M(P)r^2$, $\psi_{13}(r) = r^2$, $\sigma_1 = -\lambda_m(Q) + \lambda_M(P)d_{11}$, $\psi_{21}(r) = \psi_{22}(r) = \psi_{23}(r) = r^2$, and $\sigma_2 = -2\rho + d_{22}$.

For the interconnections we have

$$\nabla_{w^1} v_1(w^1)^T g_1(x) = 2(w^1)^T P b f(w^2) \le 2\lambda_M(P) k|b||w^1||w^2|$$

and

$$\nabla_{w^2} v_2(w^2)^{\mathsf{T}} g_2(x) = 2w^2 a^{\mathsf{T}} w^1 \le 2|a|\,|w^1|\,|w^2|.$$

In the notation of Theorem 4.4.2 we now have $b_{11} = b_{22} = 0$, $b_{12} = 2k|b|\lambda_M(P)$, and $b_{21} = 2|a|$.

Choosing $\alpha_1 = 1/\lambda_M(P)$ and $\alpha_2 = 1$, matrix S of either Theorem 4.4.2, 4.4.3, or 4.4.5 assumes the form

$$S = \begin{bmatrix} (-\lambda_m(Q)/\lambda_M(P)) + d_{11} + d_{21} & k|b| + |a| \\ k|b| + |a| & -2\rho + d_{22} + d_{12} \end{bmatrix}.$$

This matrix is negative definite if and only if

$$\rho > \tfrac{1}{2}(d_{12}+d_{22}) \tag{4.6.3}$$

and

$$k < (1/|b|)\{[(\lambda_m(Q)/\lambda_M(P))-d_{11}-d_{21}]^{1/2}[2\rho-d_{12}-d_{22}]^{1/2} - |a|\}. \tag{4.6.4}$$

Therefore, if inequalities (4.6.3) and (4.6.4) are true, then composite system (4.6.2) is ESL w.p.1 and ESL q.m.

If in Eq. (4.6.2) the Wiener processes $\{z_t^i, t \in T\}$, $i = 1,2$, are replaced by Poisson processes $\{q_t^i, t \in T\}$, $i = 1,2$, an identical analysis as the one given above yields again, in view of Theorem 4.4.5, the inequalities (4.6.3) and (4.6.4) as conditions for ESL w.p.1 and ESL q.m.

4.6.5. Example. To illustrate the applicability of the results of Section 4.5, we reconsider the indirect control problem of Example 4.6.1. Choosing the same Lyapunov functions v_1 and v_2 for $(\mathscr{S}_1)$ and $(\mathscr{S}_2)$ as before, we have

$$\mathscr{L}_1 v_1(w^1) \le [-\lambda_m(Q) + \lambda_M(P)d_{11}]|w^1|^2 \le [-(\lambda_m(Q)/\lambda_M(P)) + d_{11}]v_1(w^1)$$

and

$$\mathscr{L}_2 v_2(w^2) \le (-2\rho + d_{22})|w^2|^2 = (-2\rho + d_{22})v_2(w^2).$$

Thus, each isolated subsystem $(\mathscr{S}_i)$ possesses Property E, and hence, hypothesis (i) of Theorem 4.5.2 is satisfied. Also,

$$\nabla_{w^1} v_1(w^1)^{\mathsf{T}} g_1(x) + \tfrac{1}{2}\,\mathrm{tr}[\sigma_{12}(w^2)^{\mathsf{T}} \nabla_{w^1 w^1}(w^1)\sigma_{12}(w^2)]$$
$$= 2(w^1)^{\mathsf{T}} Pbf(w^2) + \tfrac{1}{2}\,\mathrm{tr}[2\sigma_{12}(w^2)^{\mathsf{T}} P\sigma_{12}(w^2)]$$
$$\le 2\lambda_M(P)|b|k|w^1|\,|w^2| + \lambda_M(P)d_{12}|w^2|^2$$
$$\le \lambda_M(P)|b|k(|w^1|^2 + |w^2|^2) + \lambda_M(P)d_{12}|w^2|^2$$
$$\le [\lambda_M(P)|b|k/\lambda_m(P)]v_1(w^1) + \lambda_M(P)(|b|k + d_{12})v_2(w^2). \tag{4.6.6}$$

Furthermore

$$\nabla_{w^2} v_2(w^2)^{\mathrm{T}} g_2(x) + \tfrac{1}{2} \operatorname{tr}[\sigma_{21}(w^1)^{\mathrm{T}} \nabla_{w^2 w^2} v_2(w^2) \sigma_{21}(w^1)]$$

$$= 2w^2 a^{\mathrm{T}} w^1 + \tfrac{1}{2} \operatorname{tr}[2\sigma_{21}(w^1)^{\mathrm{T}} \sigma_{21}(w^1)] \le 2|a| |w^2| |w^1| + d_{21} |w^1|^2$$

$$\le |a|(|w^1|^2 + |w^2|^2) + d_{21} |w^1|^2$$

$$\le [(|a| + d_{21})/\lambda_m(P)] v_1(w^1) + |a| v_2(w^2). \tag{4.6.7}$$

Since each coefficient in inequalities (4.6.6) and (4.6.7) is positive, hypothesis (ii) of Theorem 4.5.2 is satisfied. In addition, the linearity of the bounds in (4.6.6) and (4.6.7) ensures that hypothesis (iii) of Theorem 4.5.2 holds as well. Finally we have

$$\sup_{v<y} [|2(w^1)^{\mathrm{T}} P \sigma_{11}(w^1)|^2 + |2(w^1)^{\mathrm{T}} P \sigma_{12}(w^2)|^2$$

$$+ |2w^2 \sigma_{21}(w^1)|^2 + |2w^2 \sigma_{22}(w^2)|^2]$$

$$\le 4 \sup_{v<y} \{[\lambda_M(P)]^2 d_{11} |w^1|^4 + [\lambda_M(P)]^2 d_{12} |w^1|^2 |w^2|^2$$

$$+ d_{21} |w^1|^2 |w^2|^2 + d_{22} |w^2|^4\}$$

$$\le 4 \sup_{v<y} \left\{ \frac{[\lambda_M(P)]^2 d_{11}}{[\lambda_m(P)]^2} [v_1(w^1)]^2 + \frac{[\lambda_M(P)]^2 d_{12}}{\lambda_m(P)} v_1(w^1) v_2(w^2) \right.$$

$$\left. + \frac{d_{21}}{\lambda_m(P)} v_1(w^1) v_2(w^2) + d_{22} [v_2(w^2)]^2 \right\}$$

$$\le 4 \left\{ \frac{[\lambda_M(P)]^2 d_{11}}{[\lambda_m(P)]^2} [y^1]^2 + \frac{[\lambda_M(P)]^2 d_{12} + d_{21}}{\lambda_m(P)} [y^1 y^2] + d_{22} [y^2]^2 \right\}$$

$$\triangleq p(y),$$

and thus, hypothesis (iv) of Theorem 4.5.2 is also satisfied.

Letting $\alpha^{\mathrm{T}} = (1/\lambda_M(P), 1)$ in Theorem 4.5.14, we have

$$\alpha^{\mathrm{T}}[a(y) + b(y)]$$

$$= \{(\lambda_m(P)/\lambda_M(P))[-(\lambda_m(Q)/\lambda_M(P)) + d_{11}] + |b| k + |a| + d_{21}\}$$

$$\times (y^1/\lambda_m(P)) + [-2\rho + d_{22} + k|b| + d_{12} + |a|] y^2 \tag{4.6.8}$$

which satisfies inequality (4.5.15) for $y \ge 0$ if and only if

$$k < \frac{1}{|b|} \left[\min \left\{ \frac{\lambda_m(P)}{\lambda_M(P)} \left(\frac{\lambda_m(Q)}{\lambda_M(P)} - d_{11} \right) - d_{21}, (2\rho - d_{22} - d_{12}) \right\} - |a| \right]. \tag{4.6.9}$$

It now follows from Theorem 4.5.14 that system (4.6.2) is ASL w.p.1 if inequality (4.6.9) is true. In fact, since (4.6.8) is a linear function of y, application of Corollary 4.5.16 yields ESL w.p.1 for system (4.6.2) when condition (4.6.9) is satisfied.

Although stability conditions (4.6.3)–(4.6.4) and (4.6.9) are similar, note that these conditions are not equivalent.

4.6.10. Example. Consider the system

$$\dot{x}_t = [A(x_t) + N(t)] x_t, \qquad x_0 = x, \qquad (4.6.11)$$

where $t \in T$, $x_t \in R^l$, $A(x)$ is an $l \times l$ array of continuous bounded scalar functions $a_{ij}(x)$, $i, j = 1, \ldots, l$, and $N(t)$ is a random matrix of independent wide-band, zero mean, Gaussian random processes $\bar{\sigma}_{ij} n_{ij}(t)$, $i, j = 1, \ldots, l$. System (4.6.11), which may be considered as a nonlinear system with random parameters, can be used to represent a very large class of physical problems. For example, the transistor circuit model considered in Example 2.8.41 is a special case of the deterministic version of Eq. (4.6.11).

Letting $N(t)$ approach a white noise matrix, Eq. (4.6.11) may be transformed into an equivalent Ito differential equation as follows. Let $\sigma(x)$ be the $l \times l^2$ array of $1 \times l$ submatrices $\sigma_{ij}(x^j)$, where

$$\sigma_{ij}(x^j) = (0, \ldots, 0, \bar{\sigma}_{ij} x^j, 0, \ldots, 0), \qquad i, j = 1, \ldots, l \qquad (4.6.12)$$

(the nonzero term occurring in the ith position), and let $v(t)$ be the $l^2 \times 1$ vector given by

$$v(t)^{\mathrm{T}} = [n_{11}(t), n_{21}(t), \ldots, n_{l1}(t), n_{12}(t), \ldots, n_{ll}(t)].$$

Then Eq. (4.6.11) can be rewritten equivalently as

$$\dot{x}_t = A(x_t) x_t + \sigma(x_t) v(t). \qquad (4.6.13)$$

Following rules of transformation (see Wong [1, p. 162]), Eq. (4.6.13) can be replaced by the Ito differential equation

$$dx_t^i = \left[\sum_{j=1}^l a_{ij}(x_t) x_t^j + \tfrac{1}{2} \bar{\sigma}_{ii}^2 x_t^i \right] dt + \sum_{j=1}^l \sigma_{ij}(x_t^j) dz_t^j, \qquad (4.6.14)$$

$i = 1, \ldots, l$, where $z^j \in R^l$, $j = 1, \ldots, l$ ($\{z_t^j, t \in T\}$ is a normalized Wiener process).

In what follows we use the notation of Theorem 4.4.2. From Eq. (4.6.12) we obtain

$$\| \sigma_{ij}(x^j) \|_m^2 = \bar{\sigma}_{ij}^2 |x^j|^2, \qquad i, j = 1, \ldots, l,$$

and thus we have $d_{ij} = \bar{\sigma}_{ij}^2$, provided that $\psi_{j3}(r) = r^2$.

Choosing the isolated subsystems $(\mathscr{S}_i)$ as

$$dx_t^i = \tfrac{1}{2} \bar{\sigma}_{ii}^2 x_t^i dt + \sigma_{ii}(x_t^i) dz_t^i \qquad (\mathscr{S}_i)$$

and choosing for $(\mathcal{S}_i)$ the Lyapunov function $v_i(x^i) = \frac{1}{2}|x^i|^2$, we have (in the notation of Theorem 4.4.2) $e_i = 1$ and

$$\mathcal{L}_i v_i(x^i) = \frac{1}{2}\bar{\sigma}_{ii}^2 |x^i|^2 + \frac{1}{2} \operatorname{tr}[\sigma_{ii}(x^i)^{\mathrm{T}}\sigma_{ii}(x^i)] \leq \bar{\sigma}_{ii}^2 |x^i|^2,$$

$i = 1, \ldots, l$. Therefore, each subsystem $(\mathcal{S}_i)$ possesses Property B with $\sigma_i = \bar{\sigma}_{ii}^2$. In addition, since

$$g_i(x)^{\mathrm{T}} \nabla_{x^i} v_i(x^i) = \sum_{j=1}^{l} a_{ij}(x) x^j x^i$$

$$\leq \sup_{R^l} a_{ii}(x)|x^i|^2 + |x^i| \sum_{j=1, i\neq j}^{l} \sup_{R^l} |a_{ij}(x)| \, |x^j|$$

we have (in the notation of Theorem 4.4.2)

$$b_{ii} = \sup_{R^l} a_{ii}(x), \qquad b_{ij} = \sup_{R^l} |a_{ij}(x)|, \qquad i \neq j,$$

$i, j = 1, \ldots, l$. The test matrix $S = [s_{ij}]$ of Theorem 4.4.2 (and Theorem 4.4.3) is now given by

$$s_{ij} = \begin{cases} \alpha_i \left[\sup_{R^l} a_{ii}(x) + \bar{\sigma}_{ii}^2 \right] + \frac{1}{2} \sum_{k=1, k\neq j}^{l} \alpha_k \bar{\sigma}_{ki}^2, & i = j \\ \frac{1}{2}\left[\alpha_i \sup_{R^l} |a_{ij}(x)| + \alpha_j \sup_{R^l} |a_{ji}(x)| \right], & i \neq j. \end{cases}$$

Thus, if matrix S is negative definite for some choice of $\alpha^{\mathrm{T}} = (\alpha_1, \ldots, \alpha_l) > 0$, then the trivial solution of Eq. (4.6.11) is ESL w.p.1 and ESL q.m. by Theorem 4.4.3.

If in particular $\bar{\sigma}_{ij} = 0$, $i \neq j$, $i, j = 1, \ldots, l$ (i.e., only the diagonal elements of $A(x)$ are subject to disturbances), then Theorem 4.4.1 may be applied. In fact, since $\sup_{R^l} |a_{ij}(x)| \geq 0$, the M-matrix results of Chapter II are applicable as well. In analogy with Theorem 2.5.11, we obtain from Theorem 4.4.1 the stability condition

$$\sup_{R^l} a_{ii}(x) < -\bar{\sigma}_{ii}^2 - \sum_{j=1, i\neq j}^{l} (\lambda_j/\lambda_i) \sup_{R^l} |a_{ij}(x)| \leq 0,$$

$i = 1, \ldots, l$, for some constants $\lambda_i > 0$, $i = 1, \ldots, l$.

In the present example, as well as in the preceding ones, the weak coupling conditions and the degradation of stability due to noise are clearly apparent.

4.7   Notes and References

A standard reference on probability theory and stochastic processes is the text by Doob [1]. For an exposition of the Ito calculus, refer to the book by Wong [1]. The primary sources for Section 4.2, dealing with the stability of stochastic differential equations are the books by Kushner [1] and Arnold [1]. Additional references on stability of stochastic differential equations (and stochastic difference equations) include Kushner [2], Kats and Krasovskii [1], Bertram and Sarachik [1], and Kozin [1].

The results of Section 4.4 are based on work by Michel [8, 10, 11] and Michel and Rasmussen [1–3]. Section 4.5 is based on papers by Rasmussen and Michel [1, 3]. The examples of Section 4.6 were considered by Michel and Rasmussen [3], Rasmussen and Michel [3], and Rasmussen [1].

Using the approach presented in this chapter, it is also possible to establish stability results for discrete parameter stochastic systems (see Michel [10]) and for composite systems endowed with independent increment Markov processes other than Wiener processes (see Michel and Rasmussen [1, 3]). In addition, it is possible to establish results similar to the present ones for other stability types not considered herein.

CHAPTER V

Infinite-Dimensional Systems

In the present chapter we extend the method of analysis advanced thus far to deterministic large scale dynamical systems described on Hilbert and Banach spaces. The motivation here is primarily systems represented by partial differential equations, although the method of analysis also applies to functional differential equations and includes many of the earlier results for ordinary differential equations. In addition, the results of this chapter can be used to analyze hybrid dynamical systems (i.e., systems described by a mixture of equations) in a systematic fashion.

There are numerous difficulties that may be encountered when working in a setting of infinite-dimensional abstract spaces. For this reason we first present pertinent background material from the theory of linear and nonlinear semigroups. Then we present several important examples of semigroups. In particular, we discuss at some length how nonlinear partial differential equations can be used to construct semigroups in Hilbert space. Finally, we address ourselves to the subject on hand, that of analyzing large scale systems described on Hilbert and Banach spaces.

A brief overview of the present chapter is as follows. In the first section we establish the notation required throughout this chapter. The second section contains selected pertinent results from the theory of linear semigroups while the third section deals with required notions from the theory of nonlinear semigroups. In the fourth section we present stability definitions for semigroups, Lyapunov theorems, and selected special results for determining stability of linear systems. (Some extensions of the Lyapunov theory, which constitute background material, are also considered in the ninth section.) The fifth section contains several examples of how one constructs linear and nonlinear semigroups. In the sixth section we develop the main stability, instability, and boundedness results for large scale systems (which as before, we also call composite or interconnected systems) defined on infinite-dimensional spaces. To demonstrate the usefulness of the method of analysis advanced and to illustrate applications of individual results, we consider specific examples in the seventh section. These include an example of a *hybrid dynamical system* which may be viewed as an interconnection of two subsystems, one of which is described by an ordinary differential equation while the second subsystem is represented by a partial differential equation. Also, in this section we consider the *point kinetics model of a coupled core nuclear reactor*. In the eighth section we present special stability results for interconnected dynamical systems described by functional differential equations while in the ninth section we apply comparison theorems to vector Lyapunov functions in the stability analysis of composite systems. We conclude the chapter in the tenth section with a discussion of the literature cited.

5.1 Notation

As in the preceding chapters $R = (-\infty, \infty)$, $R^+ = [0, \infty)$, and R^n denotes n-dimensional Euclidean space with norm $|\cdot|$. Banach spaces are denoted by X or Z, with appropriate subscripts if necessary, and norms on Banach spaces are denoted by $\|\cdot\|$, with appropriate subscripts as required. Also, Hilbert spaces are denoted by X, Z, or H, with inner product $\langle \cdot, \cdot \rangle$. In this case the norm of $x \in X$ is given by $\|x\| = \langle x, x \rangle^{1/2}$.

Let A be a linear operator defined on a domain $D(A) \subset X$ with range in Z, i.e., $A: D(A) \to Z$, and assume that $D(A)$ is a dense linear subspace of X. We call A **closed** if its graph $\mathrm{Gr}(A) = \{(x, Ax) \in X \times Z : x \in D(A)\}$ is a closed subset of $X \times Z$. Also, we call A **bounded** if it maps each bounded set in X into a bounded subset of Z, or equivalently, if it is continuous. Subsequently, $I: X \to X$ will always denote the identity transformation. Given a closed linear operator $A: D(A) \to X$, $D(A) \subset X$, the **resolvent set** $\rho(A)$ consists of all points λ in the complex plane such that the linear transformation $(A - \lambda I)$ has a bounded inverse $R(\lambda, A) \triangleq (A - \lambda I)^{-1}: X \to X$. The complement of $\rho(A)$ is called the **spectral set** or simply the **spectrum** and is denoted by $\sigma(A)$.

Given a bounded linear mapping $A: D(A) \to Z$, $D(A) \subset X$, its norm is defined by

$$\|A\| = \sup\{\|Ax\|: \|x\| = 1\}.$$

Finally, $L_p(G, U)$, $1 \le p \le \infty$, denotes the usual Lebesgue space of all Lebesgue measurable functions with domain G and range U. The norm in $L_p(G, U)$ will be denoted by $\|\cdot\|_p$ (or $\|\cdot\|_{L_p}$ if more explicit notation is needed). When the range U does not need emphasis, we utilize the notation $L_p(G)$ in place of $L_p(G, U)$. If in particular $G = R^+$ and $U = R^m$, we find it convenient to simply write $L_p^m = L_p(R^+, R^m)$, and when $m = 1$, we simply write $L_p = L_p(R^+, R)$.

If $f: R \to R^m$, $g: R \to R^m$, and $p = 2$, we note that L_2^m is a Hilbert space with inner product

$$\langle f, g \rangle = \int_0^\infty f(t)^{\mathrm{T}} g(t)\, dt$$

and norm $\|f\|_{L_2} = \langle f, f \rangle^{1/2}$. On the other hand, if $p = \infty$, and $f \in L_\infty^m$, then f is an essentially bounded function with norm

$$\|f\|_\infty = \|f\|_{L_\infty} = \operatorname*{ess\,sup}_{t \ge 0} |f(t)|.$$

5.2 C_0-Semigroups

Consider a process whose evolution in time t can be described by a linear differential equation

$$\dot{x}(t) = Ax(t), \qquad x(0) = x_0 \in D(A) \tag{L}$$

for $t \in R^+$. Here $A: D(A) \to X$ is assumed to be a linear operator with domain $D(A)$ dense in X. Moreover, A is always assumed to be closed or else to have an extension $\bar{A}$ which is closed. By a **strong solution** $x(t)$ of (L) we mean a function $x: R^+ \to D(A)$ such that $\dot{x}(t)$ exists and is continuous on $R^+ \to X$ and such that (L) is true. The **abstract initial value problem** (L) is said to be **well posed** if for each $x_0 \in D(A)$ there is one and only one strong solution $x(t, x_0)$ of (L) defined on $0 \le t < \infty$ and if in addition $x(t, x_0)$ *depends continuously* on (t, x_0) in the sense that given any $N > 0$ there is an $M > 0$ such that $\|x(t, x_0)\| \le M$ when $0 \le t \le N$ and $\|x_0\| \le N$.

If (L) is well posed, there is an operator T defined by $T(t)x_0 = x(t, x_0)$ which is (for each fixed t) a bounded linear mapping on $D(A)$ to X. We call $T(t)x_0 = x(t, x_0)$ a **trajectory** of (L) for x_0. Since $T(t)$ is bounded, it has a continuous extension from $D(A)$ to the larger domain X. The trajectories

$x(t, x_0) = T(t) x_0$, for $x_0 \in X$ but $x_0 \notin D(A)$, are called **generalized solutions** of (L). The resulting family of operators $\{T(t): t \in R^+\}$ is called a C_0-**semigroup** or a **linear dynamical system** on X.

5.2.1. Definition. A family $T(t)$ of bounded linear operators on X to X is said to be a C_0-**semigroup** if

$$T(0) x = x, \qquad T(t+s) x = T(t) T(s) x$$

for all t and $s \in R^+$ and if $T(t) x$ is continuous in (t, x) on $R^+ \times X$.

Every C_0-semigroup is generated by some abstract differential equation of the form (L).

5.2.2. Definition. Given any C_0-semigroup $T(t)$, its **infinitesimal generator** is the operator defined by

$$Ax = \lim_{t \to 0^+} t^{-1} [T(t) x - x]$$

where $D(A)$ consists of all $x \in X$ for which this limit exists.

We have the following theorem.

5.2.3. Theorem. If (L) is well posed and generates a C_0-semigroup $T(t)$, then the infinitesimal generator of T is $\bar{A}$ (or A if it is already closed). Conversely, if $T(t)$ is a C_0-semigroup and A is its infinitesimal generator, then $D(A)$ is dense in X, A is closed, and for each $x_0 \in D(A)$ the function $T(t) x_0$ is a strong solution of (L).

The next result, called the Hille–Yoshida–Phillips theorem, provides necessary and sufficient conditions for a given linear operator A to be the infinitesimal generator of some C_0-semigroup.

5.2.4. Theorem. A linear operator A is the infinitesimal generator of a C_0-semigroup $T(t)$ if and only if $D(A)$ is dense in X, A is closed, and there exist two real numbers $M > 0$ and w such that whenever $\lambda > w$ one has $\lambda \in \rho(A)$ and

$$\| R(\lambda, A)^n \| \leq M (\lambda - w)^{-n}$$

for $n = 1, 2, 3, \ldots$. In this case $\| T(t) \| \leq M e^{wt}$.

The following result will also prove useful.

5.2.5. Theorem. If $A: X \to X$ generates a C_0-semigroup and if $B: X \to X$ is a bounded linear operator, then the operator $A + B$ also generates a C_0-semigroup.

A C_0-**semigroup of contractions** is a C_0-semigroup $T(t)$ which satisfies $\| T(t) \| \leq 1$ (i.e., $M = 1$ and $w = 0$ in Theorem 5.2.4). Such semigroups are of particular interest in Hilbert spaces.

5.2.6. Definition. A linear operator $A: \mathrm{D}(A) \to H$, $\mathrm{D}(A) \subset H$, on a Hilbert space H is called **dissipative** if $\mathrm{Re}\langle Ax, x \rangle \le 0$ for all $x \in \mathrm{D}(A)$.

For C_0-semigroups of contractions we now have the following.

5.2.7. Theorem. If A is the infinitesimal generator of a C_0-semigroup of contractions on a Hilbert space H, then A is dissipative and the range of $(A - \lambda I)$ is all of H for any $\lambda > 0$. Conversely, if A is dissipative and if the range of $(A - \lambda I)$ is H for at least one constant $\lambda_0 > 0$, then A is closed and A is the infinitesimal generator of a C_0-semigroup of contractions.

The above theorem is particularly useful in connection with parabolic partial differential equations.

5.3   Nonlinear Semigroups

A (nonlinear) semigroup or dynamical system is a generalization of the notion of C_0-semigroup. In arriving at this generalization, the linear initial value problem (L) is replaced by the nonlinear initial value problem

$$\dot{x}(t) = A(x(t)), \qquad x(0) = x_0, \tag{N}$$

where $A: \mathrm{D}(A) \to X$ is a nonlinear mapping. If A is continuously differentiable (or at least locally Lipschitz continuous), then the theory of existence, uniqueness, and continuation of solutions of (N) is the same as in the finite-dimensional case (see Dieudonné [1, Chapter X, Section 4]). If A is only continuous, then (N) need not have a solution (see Dieudonné [1, p. 287, Problem 5]). Since we wish to have a theory which includes nonlinear partial differential equations we must allow A to be only defined on a dense set $\mathrm{D}(A)$ and to be discontinuous. For such functions A the accretive property replaces (and generalizes) the Lipschitz property.

5.3.1. Definition. Assume that C is a subset of a Banach space X. A function $T(t): C \to C$ is said to be a **nonlinear semigroup** on C if $T(t)x$ is continuous in (t, x) on $R^+ \times C$, $T(0)x = x$, and $T(t+s)x = T(t)T(s)x$ when t and $s \in R^+$ for each fixed x in C.

A function $T(t)$ is called a **quasicontractive semigroup** if it is a nonlinear semigroup on C and if there is a number $w \in R$ such that $\|T(t)x - T(t)y\| \le e^{wt}\|x - y\|$ for all $t \in R^+$ and for all $x, y \in C$. If $w \le 0$, then $T(t)$ is called a **contraction semigroup**. Note that $C = X$ is allowed as a special case.

The mapping A in (N) will sometimes be multivalued (i.e., a relation) and in general must be extended to be multivalued if it is to generate a quasicontractive semigroup. Thus we shall assume that $A(x)$ is a subset of X and identify A with its graph

$$\text{Gr}(A) = \{(x, y): x \in X \text{ and } y \in A(x)\} \subset X \times X.$$

In this case the **domain** of A, written $D(A)$, is the set of all $x \in X$ for which $A(x) \neq \varnothing$, the **range** of A is the set

$$\text{Ra}(A) = \bigcup \{A(x): x \in D(A)\},$$

and the **inverse** of A at any point y is defined as the set

$$A^{-1}(y) = \{x \in X : y \in A(x)\}.$$

Let λ be a real or complex scalar. Then λA is defined by

$$(\lambda A)(x) = \{\lambda y: y \in A(x)\}$$

and $A + B$ is defined by

$$A(x) + B(x) = \{y + z : y \in A(x), \ z \in B(x)\}.$$

5.3.2. Definition. A multivalued operator A is said to **generate a nonlinear semigroup** $T(t)$ on C if

$$T(t)x = \lim_{n \to \infty} (I - (t/n)A)^{-n}(x)$$

for all x in C.

The **infinitesimal generator** A_s of a semigroup $T(t)$ is still defined by

$$A_s(x) = \lim_{t \to 0^+} \frac{T(t)x - x}{t}, \qquad x \in D(A_s)$$

for all x such that the limit exists. The operator A and the infinitesimal generator A_s are generally different operators.

5.3.3. Definition. A multivalued function A on X is called **w-accretive** if

$$\|(x_1 - \lambda y_1) - (x_2 - \lambda y_2)\| \geq (1 - \lambda w)\|x_1 - x_2\| \qquad (5.3.4)$$

for all $\lambda \geq 0$ and for all $x_i \in D(A)$ and $y_i \in A(x_i)$, $i = 1, 2$.

If X is a Hilbert space, then inequality (5.3.4) reduces to

$$\langle (wx_1 - y_1) - (wx_2 - y_2), x_1 - x_2 \rangle \geq 0.$$

This property for the nonlinear case is analogous to $(A - wI)$ being dissipative in the linear symmetric case.

5.3.5. Theorem. Assume that A is w-accretive and that the range of $(I - \lambda A) \supset C = \overline{D(A)}$, the closure of $D(A)$, for each λ in the interval $(0, \lambda_0)$.

Then A generates a quasicontractive semigroup $T(t)$ on C with

$$\| T(t)x - T(t)y \| \le e^{wt} \| x - y \|$$

for all $t \in R^+$ and for all $x, y \in C$.

In general the trajectories $T(t)x$ determined by the semigroup in Theorem 5.3.5 are generalized solutions of (N) which need not be differentiable. Indeed an example is discussed in Crandall and Liggett [1, Section 4] where $w = 0$, $\overline{D(A)} = X$, A generates a quasicontraction $T(t)$ but the infinitesimal generator A_s has empty domain. This means that not even one trajectory $T(t)x$ is differentiable at even one time t. If the graph of A is closed, then A is always an extension of the infinitesimal generator A_s. So whenever $x(t) = T(t)x$ has a derivative, then $\dot{x}(t)$ must be in $A(x(t))$.

The situation is more reasonable in the setting of a Hilbert space H. If A is w-accretive and closed (i.e., its graph is a closed subset of $H \times H$), then for any $x \in D(A)$ the set $A(x)$ is closed and convex. Thus, there is an element $A^0(x) \in A(x)$ such that $A^0(x)$ is the element of $A(x)$ closest to the origin. Given a trajectory $x(t) = T(t)x$, the right derivative

$$D^+ x(t) = \lim_{h \to 0^+} \frac{x(t+h) - x(t)}{h}$$

must exist at all points $t \in R^+$ and be continuous except possibly at a countably infinite set of points. The derivative $\dot{x}(t)$ exists and is equal to $D^+ x(t)$ at all points where $D^+ x(t)$ is continuous. Furthermore,

$$D^+ x(t) = A^0(x(t))$$

for all $t \ge 0$. These results can be generalized to any space X which is uniformly convex. (Refer to Dunford and Schwartz [1, p. 74] for the definition of a uniformly convex space. In particular, any L_p space for $1 < p < \infty$ is uniformly convex.)

5.3.6. Definition. A trajectory $x(t) = T(t)x_0$ is called a **strong solution** of (N) if $x(t)$ is absolutely continuous on any bounded subset of R^+ (so that $\dot{x}(t)$ exists almost everywhere), if $x(t) \in D(A)$ and if $\dot{x}(t) \in A(x(t))$ almost everywhere on R^+.

We also have

5.3.7. Definition. Initial value problem (N) is called **well posed** on C if there is a semigroup $T(t)$ such that for any $x_0 \in D(A)$, $T(t)x_0$ is a strong solution of (N) and if $\overline{D(A)} = C$.

Thus, if X is a Hilbert space or a uniformly convex Banach space and if A is w-accretive and closed, then initial value problem (N) is well posed on $C = \overline{D(A)}$ and $\dot{x}(t) = A^0(x(t))$ almost everywhere on R^+.

5.4 Lyapunov Stability of Dynamical Systems

Throughout the present section $T(t)$ denotes either a linear or a nonlinear semigroup on a subset C of a Banach space X. We shall always assume that the origin 0 (the null vector of X) is in the interior of C and that $T(t)$ has the **trivial solution** $T(t) x = 0$ for all $t \in R^+$ when $x = 0$.

5.4.1. Definition. The trivial solution of $T(t)$ is said to be **stable** if for any $\mathscr{E} > 0$ there is a $\delta > 0$ such that $\|T(t)x\| < \mathscr{E}$ for all $t \in R^+$ whenever $\|x\| < \delta$ and $x \in C$.

In the above definition $\delta = \delta(\mathscr{E})$ depends on the choice of $\mathscr{E}$. In contrast to earlier stability definitions (see Chapter II) there is no mention here of an initial time t_0. This is because semigroups (as defined above) correspond to time-invariant (autonomous) processes. The initial time can always be taken to be $t_0 = 0$. Since stability and uniform stability are equivalent for time-invariant systems, we will not need to give a separate definition for uniform stability. Indeed this lack of need for uniformity will occur again when results concerning boundedness are considered.

5.4.2. Definition. The trivial solution of $T(t)$ is said to be **asymptotically stable** if it is stable and there is a $\delta_0 > 0$ such that $\|T(t)x\| \to 0$ as $t \to \infty$ whenever $\|x\| < \delta_0$.

5.4.3. Definition. The trivial solution of $T(t)$ is **uniformly asymptotically stable** if it is asymptotically stable and if $\|T(t)x\| \to 0$ as $t \to \infty$ uniformly in $\|x\| < \delta_0$.

5.4.4. Definition. The trivial solution of $T(t)$ is **exponentially stable** if it is stable and if in addition there are positive constants M and α such that $\|T(t)x\| \le Me^{-\alpha t}$ for all $t \ge 0$ and all x satisfying $\|x\| < \delta_0$.

5.4.5. Definition. The trivial solution of $T(t)$ is **unstable** if it is not stable. In this case there is an $\mathscr{E} > 0$, a sequence $\{x_n\}$ in C and a sequence $\{t_n\}$ such that $\|x_n\| \to 0$, $t_n \to \infty$, and $\|T(t_n)x_n\| \ge \mathscr{E}$ for all n.

When the trivial solution of $T(t)$ is asymptotically stable, then the **domain of attraction** of the trivial solution is the set $\{x \in C : T(t)x \to 0 \text{ as } t \to \infty\}$. The trivial solution of $T(t)$ is called **asymptotically stable in the large** if $C = X$ and if the domain of attraction is all of X. It is **uniformly asymptotically stable in the large** if it is uniformly asymptotically stable with domain of attraction X and if for any $N > 0$, $\|T(t)x\| \to 0$ as $t \to \infty$ uniformly on the set $\|x\| < N$. It is **exponentially stable in the large** if it is exponentially stable with domain of attraction X and if for any $N > 0$ there are constants α and M (depending only on N) such that $\|T(t)x\| \le Me^{-\alpha t}$ for all $t \ge 0$ and all x such that $\|x\| < N$.

5.4.6. Definition. The semigroup $T(t)$ is said to be **uniformly bounded** if given $\alpha > 0$ there is a constant M such that $\|T(t)x\| < M$ for all $t \in R^+$ whenever $\|x\| < \alpha$.

5.4.7. Definition. The semigroup $T(t)$ is **uniformly ultimately bounded**, with bound M, if given $\alpha > 0$ there exists $N = N(\alpha) > 0$ such that $\|x\| < \alpha$ implies that $\|T(t)x\| < M$ for all $t \geq N$.

In the subsequent results we will find the following convention convenient.

5.4.8. Definition. A function v is called a **Lyapunov function** if there is an open neighborhood $U \subset X$ of the origin such that $v: U \to R$ is continuous and such that $v(0) = 0$. For such a function

$$Dv(x) = \lim_{t \to 0^+} \sup \frac{v(T(t)x) - v(x)}{t}$$

is the upper right Dini derivative along trajectories.

We now summarize the basic theorems of Lyapunov's direct method for semigroups defined on Banach spaces. As in the preceding chapters, we phrase these results in terms of comparison functions of class K and class KR.

5.4.9. Theorem. If v is a Lyapunov function, if $\varphi_1 \in K$ satisfies $v(x) \geq \varphi_1(\|x\|)$ in a neighborhood U of the origin, and if $Dv(x) \leq 0$ for all $x \in U$, then the trivial solution of $T(t)$ is **stable**.

5.4.10. Theorem. If there exists a Lyapunov function v and three functions $\varphi_1, \varphi_2, \varphi_3 \in K$ such that for all x in a neighborhood U of the origin

$$\varphi_1(\|x\|) \leq v(x) \leq \varphi_2(\|x\|), \qquad Dv(x) \leq -\varphi_3(\|x\|)$$

then the trivial solution of $T(t)$ is **uniformly asymptotically stable**.

5.4.11. Theorem. If in Theorem 5.4.10 $U = C = X$ and if $\varphi_1 \in KR$, then the trivial solution of $T(t)$ is **uniformly asymptotically stable in the large**.

5.4.12. Theorem. If in Theorem 5.4.10 $\varphi_1, \varphi_2, \varphi_3 \in K$ are of the same order of magnitude, then the trivial solution of $T(t)$ is **exponentially stable**.

5.4.13. Theorem. If in Theorem 5.4.10 $U = C = X$, if $\varphi_1, \varphi_2, \varphi_3 \in KR$, and if $\varphi_1, \varphi_2, \varphi_3$ are of the same order of magnitude, then the trivial solution of $T(t)$ is **exponentially stable in the large**.

5.4.14. Theorem. Suppose v is a Lyapunov function and $\varphi \in K$ is a function satisfying $-Dv(x) \geq \varphi(\|x\|)$ on some neighborhood $U \subset C$ of the origin. If in every neighborhood $W \subset C$ of the origin there is at least one point $x_0 \in W$ for which $v(x_0) < 0$, then the trivial solution of $T(t)$ is **unstable**.

The next results yield conditions for uniform boundedness and uniform ultimate boundedness.

5.4.15. Theorem. Assume that $C = X$ and let $S = \{x \in X : \|x\| \geq R\}$ (where R may be large). If there exists a continuous function $v: S \to R$ and two functions $\varphi_1, \varphi_2 \in KR$ such that

$$\varphi_1(\|x\|) \leq v(x) \leq \varphi_2(\|x\|)$$

for all $x \in S$ and if $Dv(x) \leq 0$ for all $x \in S$, then the semigroup $T(t)$ is **uniformly bounded**.

5.4.16. Theorem. Assume that the hypotheses of Theorem 5.4.15 hold and that there exists a function $\varphi_3 \in K$ such that $Dv(x) \leq -\varphi_3(\|x\|)$ for all $x \in S$. Then $T(t)$ is **uniformly ultimately bounded**.

The computation of $Dv(x)$ can be a significant problem in some cases. However, if the semigroup $T(t)$ is a C_0-semigroup or a quasicontractive semigroup on a Hilbert space or on a uniformly convex Banach space, then the infinitesimal generator A_s of $T(t)$ exists on a set $D(A_s)$ which is dense in C. For such cases the computation of Dv can be simplified.

5.4.17. Definition. A pair (v, T) is called **admissible** if v is a Lyapunov function for $T(t)$, if the infinitesimal generator A_s of $T(t)$ is defined on a set $D_0 \subset D(A_s)$ dense in C, and if there is a function ∇v defined on $\{U \cap D_0\} \times X$ with values in R such that

(i) $v(y) - v(x) \leq \nabla v(x, y-x) + o(\|y-x\|)$ for all $x, y \in D_0 \cap U$, and
(ii) $\nabla v(x, h)$ is, for each fixed x, a bounded linear operator in $h \in X$.

We have in mind using the Frechet derivative of v at x for the operator $\nabla v(x, \cdot)$ (see Dieudonné [1, Chapter 8] for the definition and properties of such derivatives).

We now state and prove a result which we require in our subsequent development.

5.4.18. Theorem. If (v, T) is an admissible pair and if there exists a function $\varphi_3 \in K$ such that $\nabla v(x, A_s x) \leq -\varphi_3(\|x\|)$ for all $x \in D_0 \cap U$, then $Dv(x) \leq -\varphi_3(\|x\|)$ for all $x \in U$.

Proof. If $x \in D_0 \cap U$ then

$$Dv(x) = \lim_{t \to 0^+} \sup (1/t)\{v(T(t)x) - v(x)\}$$

$$\leq \lim_{t \to 0^+} \sup (1/t)\{\nabla v(x, T(t)x - x) + o(\|T(t)x - x\|)\}$$

$$= \lim_{t \to 0^+} \sup \nabla v(x, (T(t)x - x)/t) = \nabla v(x, A_s x) \leq -\varphi_3(\|x\|).$$

If $x \notin D_0$, choose a sequence $\{x_n\}$ in D_0 such that $x_n \to x$ as $n \to \infty$. Since each $x_n \in D_0$ we have by the result proved above,

$$v(T(t)x_n) - v(x_n) \le - \int_0^t \varphi_3(\|T(s)x_n\|)\, ds$$

for all $t \in R^+$. Since all of the functions involved are continuous, this inequality remains true in the limit, i.e.,

$$v(T(t)x) - v(x) \le - \int_0^t \varphi_3(\|T(s)x\|)\, ds.$$

Therefore,

$$\lim_{t \to 0^+} \sup (1/t)\{v(T(t)x) - v(x)\} \le \lim_{t \to 0^+} \sup (-1/t) \int_0^t \varphi_3(\|T(s)x\|)\, ds$$

$$= -\varphi_3(\|x\|).$$

Thus, $Dv(x) \le -\varphi_3(\|x\|)$ for all $x \in U$. ∎

For linear semigroups with generator A one can often establish stability by determining the spectrum of A. When $X = R^n$, A must be an $n \times n$ matrix whose spectrum is the set of eigenvalues $\{\lambda\}$ of A. In this case $\text{Re } \lambda < 0$ for all $\lambda \in \sigma(A)$ is certainly sufficient for stability. Slemrod [2] has studied the analogous problem for systems defined on infinite-dimensional spaces. He points out that there are C_0-semigroups of the type characterized by the following theorem.

5.4.19. Theorem. Given any two real numbers α and β with $\alpha < \beta$ there exists a C_0-semigroup $T(t)$ on a Hilbert space H such that $\text{Re } \lambda \le \alpha$ for all $\lambda \in \sigma(A)$ and in addition $\|T(t)\| = e^{\beta t}$ for all $t \ge 0$.

Slemrod [2] shows that a generalization of the stability result for the finite-dimensional case is possible for the following class of semigroups.

5.4.20. Definition. A C_0-semigroup $T(t)$ is called **differentiable** for $t > r$ if for each $x \in X$, $T(t)x$ is continuously differentiable on $r < t < \infty$.

For example, a system of functional differential equations with delay $[-r, 0]$ (as discussed in the next section) determines a semigroup which is differentiable for $t > r$. Also, systems of parabolic partial differential equations (as discussed in the next section) normally generate semigroups which are differentiable for $t > 0$. Following Slemrod [2] we have the following result.

5.4.21. Theorem. If $T(t)$ is a C_0-semigroup which is differentiable for $t > r$, if A is its generator and if $\text{Re } \lambda \le -\alpha_0$ for all $\lambda \in \sigma(A)$, then given any positive $\alpha < \alpha_0$ there is a constant $K(\alpha) > 0$ such that $\|T(t)\| \le K(\alpha)e^{-\alpha t}$ for all $t \in R^+$.

5.5 Examples of Semigroups

At this point it is appropriate to consider several examples of important semigroups which arise in applications and to provide related background material. Because of the great diversity of the material involved, our presentation is by necessity brief. However, we point to several references at the end of this chapter, where the proofs of the results cited in this section can be found.

A. Ordinary Differential Equations

A simple example of a semigroup is the solution of an autonomous (i.e., time independent) ordinary differential equation defined on R^n. Thus, if $g: R^n \to R^n$ is smooth enough so that the initial value problem

$$\dot{x} = g(x), \qquad x(0) = x_0$$

has a unique solution $\varphi(t, x_0)$ defined for all $t \in R^+$, then $T(t) x_0 = \varphi(t, x_0)$ is a semigroup on $X = R^n$. If g satisfies a Lipschitz condition

$$|g(x) - g(y)| < L|x - y|$$

for all $x, y \in R^n$, then g is w-accretive with $w = L$. In this case $T(t)$ is a quasi-contractive semigroup.

B. Functional Differential Equations

Differential equations with time delays and functional differential equations can also be used to generate semigroups. We have in mind equations such as

$$\dot{x}(t) = g(x(t), x(t-r))$$

or

$$\dot{x}(t) = g\left(x(t), \int_{-r}^{0} k(s, \dot{x}(t+s))\, ds\right)$$

for $t > 0$ with $x(t)$ equal to a given initial function ψ, i.e.,

$$x(t) = \psi(t) \qquad \text{on} \quad -r \le t \le 0.$$

Initial value problems of this type are special cases of a more general class of problems, discussed below.

Let C_r be the set of all continuous functions $\varphi: [-r, 0] \to R^n$ with norm defined by

$$\|\varphi\| = \max\{|\varphi(t)|: -r \le t \le 0\}.$$

Given a function $x(t)$ defined on $-r \le t < t_1$, let x_t be the function determined by $x_t(s) = x(t+s)$ for $-r \le s \le 0$. Then clearly, x_t is a mapping of the interval $[0, t_1)$ into C_r. A functional differential equation (with delay r) is an equation of the form

$$\dot{x}(t) = g(x_t), \qquad x_0 = \psi \tag{F}$$

where $g: C_r \to R^n$ and where ψ is the initial condition. We denote a solution of (F) with this initial condition by $\varphi(t, \psi)$. We assume here that for (F) solutions exist, are unique, and depend continuously on (t, ψ) (a sufficient condition for this is that $g(\varphi)$ be locally Lipschitz continuous), and that all solutions are defined for all $t \in R^+$. In this case $T(t) \psi = \varphi_t(\psi)$, or equivalently $(T(t) \psi)(s) = \varphi(t+s, \psi)$, defines a nonlinear semigroup on C_r.

If g is Lipschitz continuous, then $T(t)$ is a quasicontractive semigroup. For such a g define $A: D(A) \to C_r$ by

$$A\psi = \dot{\psi}, \qquad D(A) = \{\psi \in C_r : \dot{\psi} \in C_r \text{ and } \dot{\psi}(0) = g(\psi)\}.$$

Then the domain $D(A)$ is dense in C_r, A is the generator (and also the infinitesimal generator) of $T(t)$, and $T(t)$ is differentiable for $t > r$. (For proofs of the above assertions, see Webb [1].)

The most general **linear functional differential equation** has a right-hand side of the form

$$g(\varphi) = \int_{-r}^{0} dB(s)\,\varphi(s)$$

where $B(s) = [b_{ij}(s)]$ is an $n \times n$ matrix whose entries are of bounded variation on $[-r, 0]$. Such a functional is Lipschitz continuous on C_r with Lipschitz constant L less or equal to the variation of B. In this case the semigroup $T(t)$ is a C_0-semigroup and its stability can be ascertained by determining the spectrum of the generator. This spectrum consists of all solutions of the equation

$$\det\left(\int_{-r}^{0} e^{\lambda s} dB(s) - \lambda I\right) = 0. \tag{5.5.1}$$

If all solutions of Eq. (5.5.1) satisfy the relation $\operatorname{Re} \lambda \le -\gamma$ for some $\gamma > 0$, then the semigroup $T(t)$ is exponentially stable.

In the case of functional differential equations (F), the Lyapunov theorems can assume a special form. We give one example (see Yoshizawa [1, p. 191]).

5.5.2. Theorem. In Eq. (F) assume that $g(0) = 0$ and assume that $g(\psi)$ is locally Lipschitz continuous on C_r. Suppose there is a locally Lipschitz continuous function $v(\psi)$ and three comparison functions $\varphi_1, \varphi_2, \varphi_3 \in K$ such that

$$\varphi_1(|\psi(0)|) \le v(\psi) \le \varphi_2(\|\psi\|)$$

and

$$Dv_{(F)}(\psi) \le -\varphi_3(|\psi(0)|)$$

for all ψ in a neighborhood of the origin, where $Dv_{(F)}$ denotes the derivative of v with respect to (F). Then the trivial solution of (F) is **uniformly asymptotically stable**.

Modifications can be made in an obvious way to the above result to ensure uniform asymptotic stability in the large, exponential stability, uniform boundedness, uniform ultimate boundedness, and the like.

C. Volterra Integrodifferential Equations

Research on this topic is still in progress. Some work has been done on nonlinear equations but we shall restrict our attention to the linear case.

Volterra integral equations can be thought of as functional differential equations of the form (F) where the delay $[-r, 0]$ is replaced by the delay interval $(-\infty, 0]$. In this case the space X is a **fading memory space** as defined in Coleman and Mizel [1, 2]. If $h \geq 0$ is a given constant and $p(t)$ a positive, continuously differentiable function such that $\dot{p}(t) \geq 0$ on $-\infty < t < -h$, then X should consist of all functions $\varphi: (-\infty, 0] \to R^n$ such that φ is continuous on $-h \leq t \leq 0$ and such that

$$\|\varphi\| = \sup\{|\varphi(t)|: -h \leq t \leq 0\} + \int_{-\infty}^{-h} p(t)|\varphi(t)|\, dt$$

is finite. Another possible norm is

$$\|\varphi\| = \sup\{|\varphi(t)|: -h \leq t \leq 0\} + \left\{\int_{-\infty}^{-h} p(t)|\varphi(t)|^2\, dt\right\}^{1/2}.$$

If $h = 0$, this is equivalent to the inner product

$$\|\varphi\|^2 = \langle \varphi, \varphi \rangle = \langle \varphi(0), \varphi(0) \rangle + \int_{-\infty}^{0} p(t)\langle \varphi(t), \varphi(t) \rangle\, dt.$$

Most of the linear theory in the last section is also true for functional differential equations with infinite delay on a space of fading memory (see Hale [3, 4]).

We shall develop an alternate approach for Volterra equations of the form

$$\dot{x}(t) = Ax(t) + \int_{-\infty}^{t} K(s-t)x(s)\, ds$$

or equivalently

$$\dot{x}(t) = Ax_t(0) + \int_{-\infty}^{0} K(s)x_t(s)\, ds \tag{5.5.3}$$

for $t \geq 0$ with $x(t) = \psi(t)$ given on $-\infty < t < 0$ and $x(0) = z$. Given $p \in [1, \infty)$, the space X is in this case the set

$$X_p = \{(z, \psi): z \in R^n \text{ and } \psi \in L_p((-\infty, 0), R^n)\}$$

with the norm defined by

$$\|(z,\psi)\| = |z| + \left\{ \int_{-\infty}^{0} |\varphi(s)|^p \, dt \right\}^{1/p}.$$

The matrix A in Eq. (5.5.3) is a given constant $n \times n$ matrix and $K(t)$ is an $n \times n$ matrix whose entries $k_{ij} \in L_1(-\infty, 0)$.

If $x(t)$ is the solution of Eq. (5.5.3) for a given pair of initial conditions $(z, \psi) \in X_p$, we put

$$T(t)(z, \psi) = (x(t), x_t)$$

for $t \geq 0$, where as usual $x_t(s) = x(t+s)$ on $-\infty < s \leq 0$. Thus, $T(t)$ is a C_0-semigroup on X (see Barbu and Grossman [1] and Burns and Herdman [1]) with infinitesimal generator

$$\tilde{A}(z, \psi) = \left(Az + \int_{-\infty}^{0} K(s)\psi(s) \, ds, \, \dot{\psi} \right)$$

and domain

$$D(\tilde{A}) = \left\{ (z, \psi) : \dot{\psi} \in L_p^n \text{ and } \psi(t) = z + \int_0^t \dot{\psi}(s) \, ds \text{ for all } t \leq 0 \right\}.$$

If $\mathrm{Re}\,\lambda > 0$ then $\lambda \in \sigma(\tilde{A})$ if and only if

$$\det\left(A + \int_{-\infty}^{0} e^{\lambda s} K(s) \, ds - \lambda I \right) = 0. \tag{5.5.4}$$

If $\mathrm{Re}\,\lambda \leq 0$, then λ is always in $\sigma(\tilde{A})$.

Solutions of the Volterra equation

$$\dot{x}(t) = Ax(t) + \int_0^t B(t-u)x(u) \, du + \psi(t), \qquad x(0) = z \tag{5.5.5}$$

can also be used to construct C_0-semigroups (see Miller [3]). Let

$$X_u = \{(z, \psi) : z \in R^n, \, \psi \text{ bounded and uniformly continuous on } [0, \infty)\}$$

with norm

$$\|(z, \psi)\| = |z| + \sup\{|\psi(t)| : 0 \leq t < \infty\}.$$

If $x(t)$ is the solution of Eq. (5.5.5) and if

$$y^t(s) = \int_0^t B(t+s-u)x(u) \, du + \psi_t(s)$$

then

$$T(t)(z, \psi) = (x(t), y^t)$$

determines a C_0-semigroup on X_u with infinitesimal generator

$$\bar{A}(z, \psi) = (Az + \psi(0), B(\cdot)z + \psi(\cdot))$$

and

$$D(\bar{A}) = \{(z, \psi): \psi \text{ is the integral of } \dot{\psi} \text{ and } B(t)z + \dot{\psi}(t)$$

is bounded and uniformly continuous on $[0, \infty)\}$.

It can be shown (see Burns and Herdman [1]) that this semigroup is the adjoint of the one defined earlier on X_p for Eq. (5.5.3) if $p = 1$ and if $B(t) = K(-t)$ on $0 \le t < \infty$.

D. Partial Differential Equations

In our discussion of partial differential equations we require the following additional notation. A vector index or exponent is a vector $\alpha^T = (\alpha_1, \alpha_2, ..., \alpha_n)$ whose components are nonnegative integers, $|\alpha| = \sum_{j=1}^{n} \alpha_j$, and for any $x \in R^n$,

$$x^\alpha = (x_1, x_2, ..., x_n)^\alpha = x_1^{\alpha_1} x_2^{\alpha_2} ... x_n^{\alpha_n}.$$

Let $D_k = i(\partial/\partial x_k)$ for $k = 1, 2, ..., n$, where $i = (-1)^{1/2}$ and let $D = (D_1, D_2, ..., D_n)$ so that

$$D^\alpha = D_1^{\alpha_1} D_2^{\alpha_2} \cdots D_n^{\alpha_n}.$$

In the sequel G will be a fixed closed bounded subset of R^n with boundary ∂G. We will assume that ∂G is smooth, at least piecewise k-times continuously differentiable for some large enough k. This smoothness is easily seen to be true for the type of regions which normally occur in applications. Also, $H^l(B)$ will denote all functions $\varphi \in L_2(B)$ such that for $|\alpha| \le l$ the distribution derivatives $D^\alpha \varphi \in L_2(B)$ (see Zemanian [1]) with norm

$$\|\varphi\| = \left\{ \sum_{|\alpha| \le l} \int_B |D^\alpha \varphi(x)|^2 \, dx \right\}^{1/2}.$$

If $C_0^l(B)$ is the set of all l-times continuously differentiable functions φ on B such that $D^\alpha \varphi = 0$ on ∂B for all $|\alpha| \le l$ then $H_0^l(B)$ is the closure of $C_0^l(B)$ in $H^l(B)$. This construction builds "zero boundary conditions" into the space $H_0^l(B)$.

Given $m \times m$ constant square matrices A_α, let

$$A(D) = \sum_{|\alpha| \le r} A_\alpha D^\alpha$$

and consider the differential equation

$$\frac{\partial u}{\partial t} = A(D)u, \qquad u(0, x) = \psi(x) \tag{P}$$

for the unknown vector valued function $u(t, x)$. Here $t \geq 0$, $x \in R^n$, and $\psi \in L_2(R^n)$ is a given initial function. Proceeding intuitively for the moment, apply L_2-Fourier transforms to Eq. (P) to obtain

$$\frac{\partial \tilde{u}(t, \omega)}{\partial t} = A(\omega)\tilde{u}(t, \omega), \qquad \tilde{u}(0, \omega) = \tilde{\psi}(\omega)$$

where $A(\omega) = \Sigma_{|\alpha| \leq r} A_\alpha \omega^\alpha$ for all $\omega \in R^n$. In order to have a solution such that $u(t, x)$ and $\partial u(t, x)/\partial t$ are in L_2 over $x \in R^n$, it is necessary that $A(\omega)\tilde{u}(t, \omega)$ be in L_2 over $\omega \in R^n$. This places some restrictions on $A(\omega)$. For the proof of the next result, refer to Krein [1, p. 163].

5.5.6. Theorem. The mapping $T(t)\psi = u(t, \cdot)$ defined by the solutions $u(t, x)$ of Eq. (P) determines a C_0-semigroup on $X = L_2(R^n)$ if and only if there is a nonsingular matrix $S(\omega)$ and a constant $K > 0$ such that for all ω in R^n

(i) $|S(\omega)| \leq K$ and $|S(\omega)^{-1}| \leq K$,
(ii) $S(\omega)A(\omega)S(\omega)^{-1} = [C_{ij}(\omega)]$ is upper triangular,
(iii) $\operatorname{Re} C_{mm}(\omega) \leq \cdots \leq \operatorname{Re} C_{22}(\omega) \leq \operatorname{Re} C_{11}(\omega) \leq K$, and
(iv) $|C_{ik}(\omega)| \leq K(1 + |\operatorname{Re} C_{ii}(\omega)|)$ for $k = i+1, i+2, \ldots, m$.

Parabolic equations often satisfy these conditions while hyperbolic equations do not. For example, the equation

$$\frac{\partial u}{\partial t} = \frac{\partial^2 u}{\partial x^2} + \frac{\partial^2 u}{\partial y^2} + a\frac{\partial u}{\partial x} + b\frac{\partial u}{\partial y} + cu$$

yields $m = 1$, $r = n = 2$, $\omega = (\omega_1, \omega_2)$, and

$$A(\omega) = -\omega_1^2 - \omega_2^2 + ia\omega_1 + ib\omega_2 + c = C_{11}(\omega).$$

Clearly $\operatorname{Re} A(\omega) = -\omega_1^2 - \omega_2^2 + c \leq c$ on R^2. On the other hand the equation

$$\frac{\partial^2 u}{\partial t^2} = \frac{\partial^2 u}{\partial x^2}$$

or equivalently, the set of equations

$$\frac{\partial u_1}{\partial t} = u_2, \qquad \frac{\partial u_2}{\partial t} = \frac{\partial^2 u_1}{\partial x^2},$$

with $u_1 = u$ and $u_2 = \partial u/\partial t$, yields $m = 2$, $r = 2$, $n = 1$, and

$$A(\omega) = \begin{bmatrix} 0 & 1 \\ -\omega^2 & 0 \end{bmatrix}.$$

The eigenvalues of $A(\omega)$ are $C_{11}(\omega) = i\omega$ and $C_{22}(\omega) = -i\omega$. If $SAS^{-1} = C$, then

$$S^{-1} = \begin{bmatrix} x_1(\omega) & y_1(\omega) \\ x_2(\omega) & y_2(\omega) \end{bmatrix}, \qquad AS^{-1} = S^{-1}\begin{bmatrix} i\omega & C_{12}(\omega) \\ 0 & -i\omega \end{bmatrix}.$$

A straightforward computation yields

$$S(\omega)^{-1} = \begin{bmatrix} x_1(\omega) & y_1(\omega) \\ i\omega x_1(\omega) & C_{12}(\omega)x_1(\omega) - i\omega y_1(\omega) \end{bmatrix}$$

and

$$S(\omega) = \begin{bmatrix} C_{12}(\omega)x_1(\omega) - i\omega y_2(\omega) & -y_1(\omega) \\ -i\omega x_1(\omega) & x_1(\omega) \end{bmatrix}$$
$$\times \left[C_{12}(\omega)x_1^2(\omega) - 2i\omega x_1(\omega)y_1(\omega) \right]^{-1}.$$

If the hypotheses of Theorem 5.5.6 were true, then there would exist a constant K_1 such that $|C_{12}(\omega)| \leq K_1$ and all components of $S(\omega)$ and $S(\omega)^{-1}$ are bounded by K_1. In particular

$$|C_{12}(\omega)x_1(\omega) - i\omega y_1(\omega)| \leq K_1$$

and

$$|\omega| |C_{12}(\omega)x_1(\omega) - 2i\omega y_1(\omega)|^{-1} \leq K$$

can be combined to yield

$$|\omega|/K_1 \leq |C_{12}(\omega)x_1(\omega) - i\omega y_1(\omega)| + |i\omega y_1(\omega)| \leq K_1 + |i\omega y_1(\omega)|.$$

The first and last of these four inequalities imply that

$$(|\omega|/K_1) - 2K_1 \leq |C_{12}(\omega)x_1(\omega)|.$$

Since $|C_{12}(\omega)| \leq K_1$ and $|x_1(\omega)| \leq K_1$ for all $\omega \in R$, this is impossible. Thus no matrix S as asserted above exists.

Many hyperbolic problems can be treated by choosing X in a different fashion.

5.5.7. Theorem. In order that there exist integers $l_i \geq 0$ such that Eq. (P) is well posed on the product space $X = \bigtimes_{i=1}^{m} H^{l_i}(R^n)$ it is necessary and sufficient that there is a constant C such that $\operatorname{Re} \lambda(\omega) \leq C$ for all $\omega \in R^n$ and all eigenvalues $\lambda(\omega)$ of $A(\omega)$.

For example, if $X = H^1(R^n) \times L_2(R^n)$ for

$$\frac{\partial u_1}{\partial t} = u_2, \qquad \frac{\partial u_2}{\partial t} = \sum_{i=1}^{n} \frac{\partial^2 u_1}{\partial x_i^2}$$

then the above theorem applies.

Next, let us consider Eq. (P) on the bounded set G,

$$\frac{\partial u}{\partial t} = A(D)u, \qquad u(0, x) = \psi(x) \tag{P}$$

for $t > 0$, $x \in G$ and for $\psi \in L_2(G)$. Let $A_r(D) = \Sigma_{|\alpha|=r} A_\alpha D^\alpha$ be the part of $A(D)$ with the highest order derivatives. Then Eq. (P) will be called **strongly parabolic** if A is strongly elliptic. Here, A is **strongly elliptic** means that $r = 2s$ is an even integer, $s \geq 1$, and Re $\langle A_r(\omega) z, z \rangle < 0$ for all $\omega \in R^n$ and all $z \in R^m$. Let A_0 be the operator defined by $A_0 u = A(D) u$ on the domain

$$D(A_0) = \left\{ u \in L_2(G) : \text{the normal derivatives satisfy } u = \frac{\partial u}{\partial n} = \frac{\partial^2 u}{\partial n^2} \right.$$

$$= \cdots = \frac{\partial^{s-1} u}{\partial n^{s-1}} = 0 \text{ on } \partial G \left. \right\}.$$

Here $n(x)$ is the outer normal to a point $x \in \partial G$. Then A_0 can be extended to a closed linear mapping A on $L_2(G)$ with domain $H^r(G) \cap H_0^s(G)$ and $(A - \lambda I)$ will be dissipative on $L_2(G)$ when $\lambda \geq 0$ is chosen large enough. In general we have

5.5.8. Theorem. If $A(D)$ is strongly elliptic and ∂G is smooth, then Eq. (P) is well posed on $L_2(G)$.

Other boundary conditions can sometimes be used in Eq. (P). For example, if a_{ij}, b_{ij}, and c are real, if

$$A(D) u = \sum_{i,j=1}^{n} a_{ij} \frac{\partial^2 u}{\partial x_i \partial x_j} + \sum_{i=1}^{n} b_i \frac{\partial u}{\partial x_i} + cu$$

is strongly elliptic and if for some $\sigma > 0$,

$$\frac{\partial u}{\partial n} = \sigma u \quad \text{on} \quad \partial G$$

then Eq. (P) is well posed on $L_2(G)$.

Many nonlinear partial differential equations generate quasicontractive semigroups. We shall mention only one example here. Consider the partial differential equation

$$\frac{\partial u}{\partial t} = A(D) u - f(u), \qquad u(0, x) = \psi(x) \tag{5.5.9}$$

for $t > 0$ and $x \in G$ and with boundary conditions

$$u = \frac{\partial u}{\partial n} = \frac{\partial^2 u}{\partial n^2} = \cdots = \frac{\partial^{s-1} u}{\partial n^{s-1}} = 0$$

on ∂G. Here $A(D) = A_r(D)$ is an $r = 2s$th-order strongly elliptic linear operator and f is a nonlinear Lipschitz continuous function which maps $L_2(G)$ into itself.

We shall now consider the problem of constructing a Lyapunov function for Eq. (5.5.9) when $A(D) = \sum_{i=1}^{n} \partial^2/\partial x_i^2 = \Delta$ is the Laplacian and $zf(z) \geq \beta z^2$ for all $z \neq 0$. We have

$$\frac{\partial u}{\partial t} = \Delta u - f(u) \quad \text{in} \quad R^+ \times G, \qquad u(0, x) = \psi(x) \quad \text{in} \quad G,$$

and

$$u(t, x) = 0 \quad \text{on} \quad R^+ \times \partial G.$$

From the divergence theorem (also called Gauss's theorem, see e.g., Apostol [1, p. 339] or Buck [1, p. 350]) it follows that

$$\int_G (\nabla w \cdot \nabla w + w \Delta w)\, dx = \int_{\partial G} w \frac{\partial w}{\partial n}\, ds = 0$$

for $w \in H_0^1(G) \cap H^2(G)$ so that

$$\int_G w \Delta w\, dx = -\int_G |\nabla w|^2\, dx$$

where $dx = dx_1\, dx_2 \cdots dx_n$. For smooth u we define

$$v(u) = \int_G |u(x)|^2\, dx.$$

Using Theorem 5.4.18 we obtain

$$Dv(u) = \int_G u(\Delta u - f(u))\, dx \leq \int_G u \Delta u\, dx - \beta \int_G |u|^2\, dx$$

$$= -\int_G |\nabla u|^2\, dx - \beta \int_G |u|^2\, dx.$$

By Poincaré's inequality (see Bers, John and Schechter [1, p. 194, Lemma 5]) we have

$$\int_G |u(x)|^2\, dx \leq \gamma^2 \int_G |\nabla u(x)|^2\, dx,$$

where the constant γ can be chosen as $\delta/\sqrt{n}$ when n is the dimension of the X-space and G can be put into a cube of length δ. We therefore have,

$$Dv(u) \leq -(\beta + \gamma^{-2}) \int_G |u|^2\, dx = -(\beta + (1/\gamma^2)) v(u).$$

E. Stochastic Differential Equations

Stochastic differential equations (see Chapter IV) can also be used to generate semigroups. Here we consider random processes $\{x_t(\omega) \in R^n, t \in R^+\}$ which are defined on a probability space $(\Omega, \mathcal{A}, \mathcal{P})$. In this case Ω is the event space, $\mathscr{A}$ is a σ-algebra of events in Ω and $\mathscr{P}$ is a probability measure on $\mathscr{A}$. (Henceforth we simply write x_t in place of $\{x_t(\omega) \in R^n, t \in R^+\}$.)

The random behavior of x_t is characterized by the **distribution function**

$$P(t, B) = \mathscr{P}\{x_t \in B\}$$

and the **transition function**

$$P_a(t, B) = \mathscr{P}\{x_{t+\tau} \in B \mid x_\tau = a\},$$

where the latter expression represents a conditional probability. The evolution of the distribution function is completely determined by the transition function via the relation

$$P(t+\tau, B) = \int_{a \in R^n} P_a(t, B) \, P(\tau, da).$$

Now let us define the operators $T(t)$, $t \in R^+$, on the functionals of R^n, as the conditional expectation

$$u_t(a) = T(t) U(a) \triangleq E_a U(x_t) = \int_{b \in R^n} U(b) \, P_a(t, db).$$

If x_t is a homogeneous Markov process, then $T(t)$ will be a semigroup. If A is the infinitesimal generator of $T(t)$, then

$$\frac{\partial u}{\partial t} = Au, \qquad u_0(a) = U(a)$$

is called the **backward diffusion equation** of x_t.

In practice there are several specific types of Markov processes of interest. For example, $x_t(\omega)$ may be defined as the solution of the **Ito differential equation**

$$dx_t = f(x_t)\, dt + \sigma(x_t)\, d\xi_t$$

where $f: R^n \to R^n$ and $\{\xi_t \in R^m, t \in R^+\}$ is a normalized Gaussian random process with independent increments. If in particular ξ_t is also a normalized Wiener process, then the infinitesimal generator A can be computed as

$$AU(a) = f(a)^{\mathrm{T}} \nabla U(a) + \tfrac{1}{2} \operatorname{tr}\{\sigma(a)^{\mathrm{T}} \nabla_{aa} U(a) \sigma(A)\}$$

where the symbol ∇_{aa} is as defined in Chapter IV. This corresponds to the case when the disturbance is "white noise."

Other types of stochastic disturbances can also be considered, as well as certain other types of nonlinear stochastic differential equations.

For further details on stochastic processes and stochastic differential equations, refer to the books by Wong [1], Arnold [1], Doob [1], and Kushner [1, 3].

F. Hybrid Systems

It is possible to construct examples of semigroups which are generated by a mixture of different types of equations. Indeed systems which may be viewed as an interconnection of subsystems modeled by different types of equations (e.g., partial differential equations, ordinary differential equations, integrodifferential equations, functional differential equations, and the like) are quite common in practice. Such systems, consisting of distributed parameter and lumped parameter components are called **hybrid systems**. The class of sampled-data systems considered in Chapter III is an example of a hybrid system described on finite-dimensional spaces. For examples of hybrid dynamical systems described on infinite-dimensional spaces, see Wang [3, 4]. In Sections 6 and 7 of the present chapter we consider some theorems and examples pertaining to hybrid systems.

G. Other Types of Systems

As a final note, we point out that the theory of semigroups can be modified to include also **discrete-time systems** (such as those considered in Chapter III) and **time-varying systems** (i.e., nonautonomous systems) of various types. Since such modifications are even more mathematically complicated, while applications in the setting of infinite-dimensional spaces appear to be few, these modifications are not considered here.

5.6 Stability of Large Scale Systems Described on Banach Spaces

Now that all the required background material has been established, we are in a position to consider the qualitative analysis of large scale systems described on Banach spaces. We first formulate the class of composite systems which we will consider. Whereas the problem of well posedness presents no difficulties in the case of dynamical systems described on finite-dimensional spaces, great care must be taken in insuring existence and uniqueness of solutions of interconnected dynamical systems described on infinite-dimensional spaces. For this reason we present some well posedness results for the classes of problems considered herein. Finally, we state and prove several results for uniform asymptotic stability, exponential stability, instability, and uniform ultimate boundedness.

A. Interconnected Systems

We begin by considering l **isolated subsystems** $(\mathscr{S}_i)$ described by equations of the form

$$\dot{z}_i = f_i(z_i), \qquad (\mathscr{S}_i)$$

$i = 1, \ldots, l$, which are assumed to be **well posed** on Banach spaces Z_i with respective norms $\|\cdot\|_i$. By this we mean that there is a C_0-semigroup $T_i(t)$ or a nonlinear semigroup $T_i(t)$ defined on a set $C_i \subset Z_i$ where the origin is contained in the interior of C_i and $C_i = Z_i$ is allowed. The domain $D(f_i)$ must be dense in C_i and the function f_i must be the generator of the semigroup $T_i(t)$. We note that for a given integer i system $(\mathscr{S}_i)$ might represent a system of ordinary differential equations, a system of delay differential equations, a system of linear partial differential equations, or perhaps a nonlinear partial differential equation. In the case where $T_i(t)$ is a nonlinear semigroup recall that we must allow the possibility that f_i is multivalued, i.e., a relation. In this case system $(\mathscr{S}_i)$ should actually be replaced by

$$\dot{z}_i \in f_i(z_i).$$

Next, we interconnect these subsystems to form a system (Σ_i) described by equations of the form

$$\dot{z}_i = f_i(z_i) + g_i(x), \qquad (\Sigma_i)$$

$i = 1, \ldots, l$, where the operators g_i, representing the **interconnecting structure** of (Σ_i) are defined on $D(g_i) \subset X$ and have range in Z_i. The hypervector $x^T = (z_1, \ldots, z_l)$ is a point in the product space

$$X = \mathop{\text{\Large X}}_{i=1}^{l} Z_i$$

with the norm specified by

$$\|x\| = \sum_{i=1}^{l} \|z_i\|_i.$$

Letting $f(x)^T = [f_1(z_1), \ldots, f_l(z_l)]$, $g(x)^T = [g_1(x), \ldots, g_l(x)]$, we can express Eq. (Σ_i) equivalently as

$$\dot{x} = f(x) + g(x) \triangleq A(x). \qquad (\mathscr{S})$$

System $(\mathscr{S})$ is clearly a special case of system (N). As in the preceding chapters, we speak of **composite system** $(\mathscr{S})$ or **interconnected system** $(\mathscr{S})$ with decomposition (Σ_i). Note that $D(f+g) = D(f) \cap D(g) = D(f) \cap D(g_1) \cdots \cap D(g_l)$ is the domain for $(\mathscr{S})$. Henceforth, each subsystem $(\mathscr{S}_i)$ as well as composite system $(\mathscr{S})$ is assumed to have the trivial solution $z_i(t) \equiv 0$ and $x(t) \equiv 0$, respectively, as trajectories. In addition, the composite system ($\mathscr{S}$) is assumed to be well posed, that is, $f+g$ generates a semigroup $T(t)$. Moreover, $D_0 \triangleq D(f+g) \cap D(f_s) \cap D((f+g)_s)$ is dense in X.

B. Well Posedness

In most cases the well posedness of composite system ($\mathscr{S}$) must be verified, using the type of theory outlined in Sections 2, 3, and 5 of this chapter. However, in the next two theorems we present results for two special cases where the well posedness of the isolated subsystems and the form of the interconnecting structure guarantee the well posedness of system ($\mathscr{S}$).

5.6.1. Theorem. Let $A_i: D(A_i) \to Z_i$ be a linear operator with domain $D(A_i)$ dense in Z_i. If $f_i(z_i) \triangleq A_i z_i$ generates a C_0-semigroup on Z_i for each $i = 1, ..., l$ and if $D(g_i) = X$ and if $g_i: X \to Z_i$ is a bounded linear operator for each i, then composite system ($\mathscr{S}$) with decomposition (Σ_i) is well posed on X.

Proof. Since each subsystem ($\mathscr{S}_i$) determines a C_0-semigroup on Z_i, it follows from Theorem 5.2.4 that there are constants M_i and w such that

$$\| R(\lambda, A_i)^n \|_i \leq M_i (\lambda - w)^{-n}$$

for all $\lambda > w$ and for $i = 1, 2, ..., l$. If Ax is defined by

$$(Ax)^T = [(A_1 z_1), (A_2 z_2), ..., (A_l z_l)]$$

on $D(A) = \chi^l_{i=1} D(A_i)$, then A is closed and $D(A)$ is dense in X. Moreover, when $\lambda > w$, we have

$$\| R(\lambda, A)^n \| \leq \sum_{i=1}^{l} \| R(\lambda, A_i)^n \|_i \leq \left(\sum_{i=1}^{l} M_i \right) (\lambda - w)^{-n}.$$

From Theorem 5.2.4 it now follows that A generates a C_0-semigroup on X. If $(Bx)^T = [g_1(x), g_2(x), ..., g_l(x)]$, then B is a bounded linear operator on X. It follows from Theorem 5.2.5 that $A + B$ generates a C_0-semigroup on X. ∎

5.6.2. Theorem. If $f_i(z_i)$ is w_i-accretive on Z_i, if $\mathrm{Ra}(I - \lambda f_i) = Z_i = \overline{D(f_i)}$ for $0 < \lambda < \lambda_0$, if $D(g_i) = X$ with g_i Lipschitz continuous, and if $g_i(0) = 0$, $i = 1, ..., l$, then composite system ($\mathscr{S}$) with decomposition (Σ_i) is well posed and generates a quasicontraction on X.

Proof. For each i there is a w_i such that

$$\| (z_{1i} - \lambda y_{1i}) - (z_{2i} - \lambda y_{2i}) \|_i \geq (1 - \lambda w_i) \| z_{1i} - z_{2i} \|_i$$

for all (z_{1i}, y_{1i}) and (z_{2i}, y_{2i}) in the graph of f_i and for all $\lambda \geq 0$. Let f be the function with graph

$$f = \{((z_1, \ldots, z_l), (y_1, \ldots, y_l)) : (z_i, y_i) \in f_i \text{ for } i = 1, \ldots, l\}.$$

If $(x_1, y_1) \in f$ and $(x_2, y_2) \in f$, if $\lambda \geq 0$ and $w = \max_i w_i$, then

$$\|(x_1 - \lambda y_1) - (x_2 - \lambda y_2)\| = \sum_{i=1}^{l} \|(z_{1i} - \lambda y_{1i}) - (z_{2i} - \lambda y_{2i})\|_i$$

$$\geq \sum_{i=1}^{l} (1 - \lambda w_i) \|z_{1i} - z_{2i}\|_i$$

$$\geq (1 - \lambda w) \sum_{i=1}^{l} \|z_{1i} - z_{2i}\|_i$$

$$= (1 - \lambda w) \|x_1 - x_2\|.$$

Thus, f is w-accretive. Clearly the range $\mathrm{Ra}(I - \lambda f) = X$ for $0 \leq \lambda < \lambda_0$ since this is true by coordinates. By Theorem 5.3.5 it follows that f generates a quasicontractive semigroup on $C = X$.

Since $g(x)$ is Lipschitz continuous with some constant L, we have, given $(x_i, y_i + g(x_i)) \in f + g$,

$$\|[x_1 - \lambda(y_1 + g(x_1))] - [x_2 - \lambda(y_2 + g(x_2))]\|$$

$$\geq \|(x_1 - \lambda y_1) - (x_2 - \lambda y_2)\| - \lambda \|g(y_1) - g(y_2)\|$$

$$\geq (1 - \lambda w) \|x_1 - x_2\| - \lambda L \|x_1 - x_2\|$$

$$= [1 - \lambda(w + L)] \|x_1 - x_2\|.$$

Therefore, $f + g$ is $(w + L)$-accretive. To see that the range $\mathrm{Ra}(I - \lambda(f + g)) = X$, fix $z \in X$ and $\lambda \in (0, \lambda_0)$. The equation $[I - \lambda(f + g)] x = z$ for $x \in D(f + g)$ is equivalent to

$$(\lambda^{-1} I - f) x = z \lambda^{-1} + g(x) \qquad \text{or} \qquad x = R(\lambda^{-1}, f)(z \lambda^{-1} + g(x))$$

where $R(\mu, f) = (\mu I - f)^{-1}$ is the resolvent operator for f. Since $\dot{x} = f(x)$ is well posed and accretive, it follows from Eq. (5.3.4) that $|R(\mu, f)| \leq (\mu - w)^{-1}$ when $\mu > w = \max_i w_i$. This means that the right-hand side of the above equation is Lipschitz continuous, i.e.,

$$\|R(\lambda^{-1}, f)(z\lambda^{-1} + g(x)) - R(\lambda^{-1}, f)(z\lambda^{-1} + g(\bar{x}))\| \leq (\lambda^{-1} - w)^{-1} L \|x - \bar{x}\|.$$

By the contraction mapping theorem this means that the equation $x - \lambda(f(x) + g(x)) = z$ has a solution x for $\lambda < (L + w)^{-1} = \lambda_1$, that is, for $(\lambda^{-1} - w)^{-1} L < 1$. Since $z \in X$ is arbitrary, we have $\mathrm{Ra}[I - \lambda(f + g)] = X$ if $0 < \lambda < \lambda_1$. By Theorem 5.3.5, it now follows that $(f + g)$ generates a quasicontractive semigroup on X. ■

In addition to the well posedness assumption at the end of Section A, the domain hypothesis that the set $D_0 \triangleq D(f+g) \cap D(f_s) \cap D((f+g)_s)$ be dense can be verified in the settings of the last two theorems. The linear case is completely trivial since $A = A_s$, $A + B = (A+B)_s$ (see Theorems 5.2.3 and 5.2.4), while $D(A) = D(A+B)$ since B is bounded. In the nonlinear case it is necessary to make the additional assumption that all of the spaces Z_i are uniformly convex. If we use the equivalent norm

$$\||x\|| = \left\{ \sum_{i=1}^{l} (\|z_i\|_i)^2 \right\}^{1/2}$$

for all $x \in X$, then X will also be a uniformly convex Banach space. (The proof of Theorem 5.6.2 is easily modified to accommodate this change in norm.) By the last statement of Section 5.3, $D(\bar{f}) = D(f_s)$ where $\bar{f}$ is the closure of f. Since $D(g) = X$, it follows that $D_0 \triangleq D(f_s) \cap D((f+g)_s) \cap D(f+g) = D(f)$ is dense in X.

C. Uniform Asymptotic Stability

In characterizing the qualitative properties of subsystems $(\mathscr{S}_i)$, we use the following convention.

5.6.3. Definition. Isolated subsystem $(\mathscr{S}_i)$ with corresponding semigroup $T_i(t)$ is said to possess **Property A** if there exists a Lyapunov function v_i such that (v_i, T_i) is admissible in the sense of Definition 5.4.17, if there exist three functions $\varphi_{i1}, \varphi_{i2}, \varphi_{i3} \in K$, and if there exist real constants σ_i and $m_i > 0$ such that

  (i) $\varphi_{i1}(\|z_i\|_i) \le v_i(z_i) \le \varphi_{i2}(\|z_i\|_i)$ for all $z_i \in Z_i$ such that $\|z_i\|_i < m_i$, and
  (ii) $\nabla v_i(z_i, f_{is}(z_i)) \le \sigma_i \varphi_{i3}(\|z_i\|_i)$ for all $z_i \in D(f_{is})$ such that $\|z_i\|_i < m_i$.
Here f_{is} is the infinitesimal generator of $T_i(t)$.

5.6.4. Definition. Isolated subsystem $(\mathscr{S}_i)$ is said to possess **Property B** if it has Property A with $m_i = +\infty$ and if $\varphi_{i1}, \varphi_{i2} \in KR$.

If $\sigma_i < 0$ in Definition 5.6.3, then it follows from Theorem 5.4.10 that the trivial solution of $(\mathscr{S}_i)$ is uniformly asymptotically stable. If $\sigma_i < 0$ in Definition 5.6.4, then in accordance with Theorem 5.4.11, the trivial solution of $(\mathscr{S}_i)$ is uniformly asymptotically stable in the large. In the latter case we always assume that the semigroup generated by $(\mathscr{S}_i)$ is defined on the entire Banach space Z_i. If $\sigma_i > 0$ then $T_i(t)$ may possibly be unstable.

As in the preceding chapters, σ_i may be viewed as a measure of the degree of stability of subsystem $(\mathscr{S}_i)$. As such, σ_i will be useful in studying the qualitative effects of the isolated subsystems on the behavior of the entire interconnected system $(\mathscr{S})$.

We are now in a position to state and prove several stability results for composite system $(\mathscr{S})$.

5.6.5. Theorem. Assume that for composite system $(\mathscr{S})$ with decomposition (Σ_i) the following conditions hold.

(i)   Each isolated subsystem $(\mathscr{S}_i)$ possesses Property A;

(ii)   given the Lyapunov functions v_i for $(\mathscr{S}_i)$ and the corresponding comparison functions $\varphi_{i3} \in K$, $i = 1, \ldots, l$, there exist real constants b_{ij}, $i, j = 1, \ldots, l$, such that

$$\nabla v_i(z_i, g_i(x)) \leq \varphi_{i3}(\|z_i\|_i)^{1/2} \sum_{j=1}^{l} b_{ij} \varphi_{j3}(\|z_j\|_j)^{1/2}$$

for all $x^{\mathrm{T}} = (z_1, \ldots, z_l) \in D(f+g)$ with $\|z_i\|_i < m_i$; and

(iii)   there exist positive constants α_i, $i = 1, \ldots, l$, such that the test matrix $R = [r_{ij}]$ defined by

$$r_{ij} = \begin{cases} \alpha_i(\sigma_i + b_{ii}), & \text{if } i = j \\ \tfrac{1}{2}(\alpha_i b_{ij} + \alpha_j b_{ji}), & \text{if } i \neq j \end{cases}$$

is negative definite.

Then the trivial solution of composite system $(\mathscr{S})$ is **uniformly asymptotically stable**.

Proof. Let $Q = \{x^{\mathrm{T}} = (z_1, \ldots, z_l) : \|z_i\|_i < m_i \text{ for all } i\}$. On Q we define

$$v(x) = \sum_{i=1}^{l} \alpha_i v_i(z_i),$$

where the constants $\alpha_i > 0$ are as given in hypothesis (iii). Clearly, $v(x)$ is continuous and $v(0) = 0$, since each $v_i(z_i)$ satisfies these conditions. Since each subsystem $(\mathscr{S}_i)$ possesses Property A, it follows that

$$\sum_{i=1}^{l} \alpha_i \varphi_{i1}(\|z_i\|_i) \leq v(x) \leq \sum_{i=1}^{l} \alpha_i \varphi_{i2}(\|z_i\|_i)$$

for all $x \in Q$. The two summations above are both positive definite, decrescent functions, i.e., there are functions $\varphi_1, \varphi_2 \in K$ such that

$$\varphi_1(\|x\|) \leq \sum_{i=1}^{l} \alpha_i \varphi_{i1}(\|z_i\|_i) \quad \text{and} \quad \varphi_2(\|x\|) \geq \sum_{i=1}^{l} \alpha_i \varphi_{i2}(\|z_i\|_i).$$

Thus,

$$\varphi_1(\|x\|) \leq v(x) \leq \varphi_2(\|x\|) \tag{5.6.6}$$

for all $x \in Q$.

To compute the derivative of v along solutions of $(\mathscr{S})$, note that for $x \in D_0$

$$v(x+h) - v(x) = \sum_{i=1}^{l} \alpha_i \{v_i(z_i+h_i) - v_i(z_i)\}$$

$$\leq \sum_{i=1}^{l} \alpha_i \{\nabla v_i(z_i, h_i) + o(\|h_i\|_i)\}$$

$$= \sum_{i=1}^{l} \alpha_i \nabla v_i(z_i, h_i) + o(\|h\|),$$

so that

$$\nabla v(x, h) = \sum_{i=1}^{l} \alpha_i \nabla v_i(z_i, h_i).$$

Since each $\nabla v_i(z_i, h_i)$ is continuous and linear in h_i it follows that $\nabla v(x, h)$ is continuous and linear in h for each fixed value x. By this linearity and in view of hypothesis (ii), we have

$$\nabla v(x, f(x)+g(x)) = \sum_{i=1}^{l} \alpha_i \nabla v_i(z_i, f(x)+g(x))$$

$$= \sum_{i=1}^{l} \alpha_i \nabla v_i(z_i, f(x)) + \sum_{i=1}^{l} \alpha_i \nabla v_i(z_i, g(x))$$

$$\leq \sum_{i=1}^{l} \alpha_i \left[\sigma_i \varphi_{i3}(\|z_i\|_i) + \varphi_{i3}(\|z_i\|_i)^{1/2} \sum_{j=1}^{l} b_{ij} \varphi_{j3}(\|z_j\|_j)^{1/2} \right]$$

$$= u^{\mathsf{T}} R u$$

where R is the test matrix given in hypothesis (iii) and

$$u^{\mathsf{T}} = [\varphi_{13}(\|z_1\|_1)^{1/2}, \varphi_{23}(\|z_2\|_2)^{1/2}, \ldots, \varphi_{l3}(\|z_l\|_l)^{1/2}].$$

Since R is a negative definite symmetric matrix, its largest eigenvalue $\lambda_M(R)$ is negative and we have

$$\nabla v(x, f(x)+g(x)) \leq u^{\mathsf{T}} R u \leq \lambda_M(R) |u|^2.$$

Also, since

$$|u|^2 = \sum_{i=1}^{l} \varphi_{i3}(\|z_i\|_i) \geq \varphi_3(\|x\|)$$

for some function $\varphi_3 \in K$, it follows that

$$\nabla v(x, f(x)+g(x)) \leq \lambda_M(R) \varphi_3(\|x\|) \tag{5.6.7}$$

for all $x \in Q \cap D_0$. Applying Theorem 5.4.18 to inequality (5.6.7) we obtain

$$Dv_{(\mathscr{S})}(x) \leq \lambda_M(R) \varphi_3(\|x\|). \tag{5.6.8}$$

In view of inequalities (5.6.6) and (5.6.8), the hypotheses of Theorem 5.4.10 are satisfied and the trivial solution of composite system $(\mathscr{S})$ is uniformly asymptotically stable. ∎

For global asymptotic stability we have the following result.

5.6.9. Theorem. Assume that for composite system $(\mathscr{S})$ with decomposition (Σ_i) the following conditions hold.

(i)　Each isolated subsystem $(\mathscr{S}_i)$ possesses Property B;

(ii)　given the Lyapunov functions v_i for $(\mathscr{S}_i)$ and the corresponding comparison functions $\varphi_{i3} \in K$, $i = 1, \ldots, l$, there exist real constants b_{ij}, $i, j = 1, \ldots, l$, such that

$$\nabla v_i(z_i, g_i(x)) \le \varphi_{i3}(\|z_i\|_i)^{1/2} \sum_{j=1}^{l} b_{ij}\, \varphi_{j3}(\|z_j\|_j)^{1/2}$$

hold for all $x \in D(f+g)$ where $x^{\mathrm{T}} = (z_1, \ldots, z_l)$; and

(iii)　there exist positive constants α_i for $i = 1, \ldots, l$ such that the test matrix $R = [r_{ij}]$ defined by

$$r_{ij} = \begin{cases} \alpha_i(\sigma_i + b_{ii}), & \text{if } i = j \\ \tfrac{1}{2}(\alpha_i b_{ij} + \alpha_j b_{ji}), & \text{if } i \ne j \end{cases}$$

is negative definite.

Then the trivial solution of composite system $(\mathscr{S})$ with decomposition (Σ_i) is **uniformly asymptotically stable in the large**.

Proof. As in the proof of Theorem 5.6.5, we choose $v(x) = \sum_{i=1}^{l} \alpha_i v_i(z_i)$. It is an easy matter to show that

$$\varphi_1(\|x\|) \le \sum_{i=1}^{l} \alpha_i \varphi_{i1}(\|z_i\|_i) \quad \text{and} \quad \varphi_2(\|x\|) \ge \sum_{i=1}^{l} \alpha_i \varphi_{i2}(\|z_i\|_i)$$

can be chosen so that $\varphi_1, \varphi_2 \in KR$, and that the inequalities

$$\varphi_1(\|x\|) \le v(x) \le \varphi_2(\|x\|)$$

are satisfied for all $x \in X = \underset{i=1}{\overset{l}{\bigtimes}} Z_i$. In addition, the inequality

$$Dv_{(\mathscr{S})}(x) \le \lambda_M(R)\, \varphi_3(\|x\|)$$

with

$$\varphi_3(\|x\|) \le \sum_{i=1}^{l} \varphi_{i3}(\|z_i\|_i)$$

is satisfied for all $x \in X$. It now follows from Theorem 5.4.11 that the trivial solution of composite system $(\mathscr{S})$ is uniformly asymptotically stable in the large. ∎

D. Exponential Stability

For exponential stability of interconnected system $(\mathscr{S})$ we have the following theorem.

5.6.10. Theorem. Assume that for composite system $(\mathscr{S})$ with decomposition (Σ_i) the following conditions hold.

(i)   Each isolated subsystem $(\mathscr{S}_i)$ possesses Property A;

(ii)   all functions in the set $\{\varphi_{ij}: i = 1, 2, ..., l, \text{ and } j = 1, 2, 3\}$ are of the same order of magnitude; and

(iii)   there exist real constants b_{ij}, $i, j = 1, ..., l$, such that the inequalities

$$Vv_i(z_i, g_i(x)) \leq \varphi_{i3}(\|z_i\|_i)^{1/2} \sum_{j=1}^{l} b_{ij}\,\varphi_{j3}(\|z_j\|_j)^{1/2}$$

hold for all $x \in D(f+g)$ with $\|z_i\|_i < m_i$; and

(iv)   there exist positive constants α_i, $i = 1, ..., l$, such that the test matrix $R = [r_{ij}]$ defined by

$$r_{ij} = \begin{cases} \alpha_i(\sigma_i + b_{ii}), & \text{if } i = j \\ \tfrac{1}{2}(\alpha_i b_{ij} + \alpha_j b_{ji}), & \text{if } i \neq j \end{cases}$$

is negative definite.

Then the trivial solution of composite system $(\mathscr{S})$ is **exponentially stable.**

Proof. As in the proof of Theorem 5.6.5, we choose

$$v(x) = \sum_{i=1}^{l} \alpha_i v_i(z_i),$$

so that

$$\sum_{i=1}^{l} \alpha_i \varphi_{i1}(\|z_i\|_i) \leq v(x) \leq \sum_{i=1}^{l} \alpha_i \varphi_{i2}(\|z_i\|_i)$$

and

$$Dv_{(\mathscr{S})}(x) \leq \lambda_M(R) \sum_{i=1}^{l} \varphi_{i3}(\|z_i\|_i)$$

where $\lambda_M(R) < 0$ denotes again the largest eigenvalue of the test matrix R.

To complete the proof we must show that there are functions $\varphi_1, \varphi_2, \varphi_3 \in K$ which are of the same order of magnitude such that

$$\varphi_1(\|x\|) \leq v(x) \leq \varphi_2(\|x\|)$$

for all $x \in Q = \{x^T = (z_1, ..., z_l) : \|z_i\|_i < m_i \text{ for all } i\}$, and

$$Dv_{(\mathscr{S})}(x) \leq \lambda_M(R)\varphi_3(\|x\|)$$

for all $x \in Q$.

By hypothesis, there is a function φ, e.g., $\varphi = \varphi_{11}$, and positive constants k_{ij} such that

$$\varphi_{i1}(r) \geq k_{i1}\,\varphi(r), \qquad \varphi_{i2}(r) \leq k_{i2}\,\varphi(r), \qquad \varphi_{i3}(r) \geq k_{i3}\,\varphi(r)$$

for all r in a region $0 \leq r < r_0$ and for $i = 1, 2, ..., l$. We define

$$\|x\|_\infty = \max_i \|z_i\|_i$$

$$\varphi_1(r) = \min_i (\alpha_i k_{i1})\,\varphi(r)$$

$$\varphi_2(r) = \left(\sum_{i=1}^{l} \alpha_i k_{i2} \right) \varphi(r)$$

and

$$\varphi_3(r) = \min_i (k_{i3})\,\varphi(r).$$

Clearly, each $\varphi_i \in K$. Moreover, since φ is strictly increasing, we have

$$\sum_{i=1}^{l} \alpha_i \varphi_{i1}(\|z_i\|_i) \geq \sum_{i=1}^{l} \alpha_i k_{i1}\,\varphi(\|z_i\|_i)$$

$$\geq \min_i (\alpha_i k_{i1}) \sum_{i=1}^{l} \varphi(\|z_i\|_i)$$

$$\geq \min_i (\alpha_i k_{i1}) \max_i \varphi(\|z_i\|_i)$$

$$= \min_i (\alpha_i k_{i1})\,\varphi\!\left(\max_i \|z_i\|_i \right)$$

$$= \min_i (\alpha_i k_{i1})\,\varphi(\|x\|_\infty) = \varphi_1(\|x\|_\infty).$$

Also,

$$\sum_{i=1}^{l} \alpha_i \varphi_{i2}(\|z_i\|_i) \leq \sum_{i=1}^{l} \alpha_i k_{i2}\,\varphi(\|z_i\|_i)$$

$$\leq \left(\sum_{i=1}^{l} \alpha_i k_{i2} \right) \max_i \varphi(\|z_i\|_i)$$

$$= \left(\sum_{i=1}^{l} \alpha_i k_{i2} \right) \varphi\!\left(\max_i \|z_i\|_i \right) = \varphi_2(\|x\|_\infty)$$

and

$$\sum_{i=1}^{l} \varphi_{i3}(\|z_i\|_i) \geq \sum_{i=1}^{l} k_{i3}\,\varphi(\|z_i\|_i) \geq \left(\min_i k_{i3} \right) \sum_{i=1}^{l} \varphi(\|z_i\|_i)$$

$$\geq \left(\min_i k_{i3} \right) \max_i \varphi(\|z_i\|_i)$$

$$= \left(\min_i k_{i3} \right) \varphi\!\left(\max_i \|z_i\|_i \right) = \varphi_3(\|x\|_\infty).$$

By Theorem 5.4.12 it follows that the trivial solution of composite system ($\mathscr{S}$) is exponentially stable in the norm $\|\cdot\|_\infty$. Since

$$\|x\| = \sum_{i=1}^{l} \|z_i\|_i \leq l \|x\|_\infty$$

we conclude that the trivial solution of composite system ($\mathscr{S}$) is exponentially stable in the norm $\|x\|$. ∎

For exponential stability in the large we have the following result.

5.6.11. Theorem. Assume that for composite system ($\mathscr{S}$) with decomposition (Σ_i) the following conditions hold.

(i)   Each isolated subsystem ($\mathscr{S}_i$) possesses Property B;

(ii)   all functions in the set $\{\varphi_{ij}: i = 1, 2, ..., l \text{ and } j = 1, 2, 3\}$ belong to class KR and are of the same order of magnitude; and

(iii)   there exist real constants b_{ij}, $i, j = 1, ..., l$, such that the inequalities

$$\nabla v_i(z_i, g_i(x)) \leq \varphi_{i3}(\|z_i\|_i)^{1/2} \sum_{j=1}^{l} b_{ij}\, \varphi_{j3}(\|z_j\|_j)^{1/2}$$

hold for all $x \in D(f+g)$; and

(iv)   there exist positive constants α_i, $i = 1, ..., l$, such that the test matrix $R = [r_{ij}]$ defined by

$$r_{ij} = \begin{cases} \alpha_i(\sigma_i + b_{ii}), & \text{if } i = j \\ \tfrac{1}{2}(\alpha_i b_{ij} + \alpha_j b_{ji}), & \text{if } i \neq j \end{cases}$$

is negative definite.

Then the trivial solution of composite system ($\mathscr{S}$) is **exponentially stable in the large**.

Proof. The proof is similar to the proof of Theorem 5.6.10. ∎

Consistent with the method of analysis advanced in the preceding chapters, the above results express stability properties of composite system ($\mathscr{S}$) in terms of the qualitative properties of the lower order subsystems ($\mathscr{S}_i$) and in terms of the characteristics of the interconnecting structure. As before, the confining relationships between these properties are determined by the test matrix R. Concerning test matrix R, several observations are once more in order.

Since a necessary condition for the negative definiteness are the relations $\sigma_i + b_{ii} < 0$, $i = 1, ..., l$, each subsystem must either be stable or else the interconnecting structure must provide local stabilizing feedback for each unstable subsystem.

The nature of the bounds on the interconnecting structure expresses the strength (and the direction) of coupling relative to the degree of stability of each subsystem. The negative definiteness condition on the test matrix R has the effect of limiting the degree to which the interconnecting structure is allowed to affect the qualitative behavior of system $(\mathscr{S})$ by constraining the size and/or sign of the coupling terms.

As in the finite-dimensional case, the above observations suggest a systematic procedure for the stabilization of unstable composite systems by utilizing local stabilizing feedback for the subsystems, and also, by using local feedback associated with various interconnecting terms which has the effect of decreasing the strength of coupling.

E. Application of M-Matrices

In all of the results above it is necessary to find weighting factors $\alpha_i > 0$ such that matrix R is negative definite. Although the choice of such constants is not unique, it is not always evident that these constants exist. In the next results we show that in the special case when $b_{ij} \geq 0$ for all $i \neq j$, the necessity of choosing a weighting vector $\alpha^T = (\alpha_1, ..., \alpha_l)$ can be eliminated. It is emphasized that these results are at best only equivalent to the ones presented thus far in this section. Indeed, since no sign restrictions on the b_{ij}, $i \neq j$, exist in the above theorems, these stability results remain important due to their greater generality.

5.6.12. Theorem. Consider the $l \times l$ matrix $S = [s_{ij}]$ defined by

$$s_{ij} = \begin{cases} -(\sigma_i + b_{ii}), & \text{if } i = j \\ -b_{ij}, & \text{if } i \neq j \end{cases} \tag{5.6.13}$$

where σ_i, $i = 1, ..., l$, and b_{ij}, $i, j = 1, ..., l$, are as defined in Theorems 5.6.5, 5.6.9, 5.6.10, and 5.6.11.

(i) Assume that hypotheses (i) and (ii) of Theorem 5.6.5 are satisfied with $b_{ij} \geq 0$ for all $i \neq j$. If all successive principal minors of test matrix S are positive, then the trivial solution of composite system $(\mathscr{S})$ is **uniformly asymptotically stable**.

(ii) Assume that hypotheses (i) and (ii) of Theorem 5.6.9 are satisfied with $b_{ij} \geq 0$ for all $i \neq j$. If all successive principal minors of test matrix S are positive, then the trivial solution of composite system $(\mathscr{S})$ is **uniformly asymptotically stable in the large**.

(iii) Assume that hypotheses (i), (ii), and (iii) of Theorem 5.6.10 are satisfied with $b_{ij} \geq 0$ for all $i \neq j$. If all successive principal minors of test matrix S are positive, then the trivial solution of composite system $(\mathscr{S})$ is **exponentially stable**.

(iv) Assume that hypotheses (i), (ii), and (iii) of Theorem 5.6.11 are satisfied with $b_{ij} \geq 0$ for all $i \neq j$. If all successive principal minors of test matrix S are positive, then the trivial solution of composite system $(\mathscr{S})$ is **exponentially stable in the large**.

Proof. Since the proofs of the four parts follow along similar lines, we prove only part (i).

As in the proof of Theorem 5.6.5, we choose

$$v(x) = \sum_{i=1}^{l} \alpha_i v_i(z_i)$$

where the constants $\alpha_i > 0$, $i = 1, ..., l$, will further be specified later. As before, there are functions $\varphi_1, \varphi_2 \in K$ such that

$$\varphi_1(\|x\|) \leq v(x) \leq \varphi_2(\|x\|) \tag{5.6.14}$$

for all $x \in Q$.

Also if $x \in Q$, then

$$Dv_{(\mathscr{S})}(x) \leq \sum_{i=1}^{l} \alpha_i \left\{ \sigma_i \varphi_{i3}(\|z_i\|_i) + \varphi_{i3}(\|z_i\|_i)^{1/2} \sum_{j=1}^{l} b_{ij} \varphi_{j3}(\|z_j\|_j)^{1/2} \right\}$$

$$= -\tfrac{1}{2} u^T (AS + S^T A) u,$$

where $S = [s_{ij}]$ is the matrix defined in (5.6.13),

$$u^T = (\varphi_{13}(\|z_1\|_1)^{1/2}, \varphi_{23}(\|z_2\|_2)^{1/2}, ..., \varphi_{l3}(\|z_l\|_l)^{1/2}),$$

and A is the diagonal matrix specified by $A = \text{diag}[\alpha_1, ..., \alpha_l]$. Now by Corollary 2.5.6, the condition on the successive principal minors of matrix S is equivalent to the existence of a diagonal matrix A with positive elements such that the matrix $(AS + S^T A)$ is positive definite. Thus, if λ_M is the largest eigenvalue of the matrix $-(AS + S^T A)/2$, then $\lambda_M < 0$ and

$$Dv_{(\mathscr{S})}(x) \leq \lambda_M \sum_{i=1}^{l} \varphi_{i3}(\|z_i\|_i) \leq \lambda_M \varphi_3(\|x\|) \tag{5.6.15}$$

for all $x \in Q$, where φ_3 is some function of class K. It now follows from inequalities (5.6.14) and (5.6.15) and from Theorem 5.4.10 that the trivial solution of composite system $(\mathscr{S})$ is uniformly asymptotically stable. ∎

It is emphasized that the condition on the test matrix S in the above results can be replaced by any of the other equivalent M-matrix conditions discussed in Chapter II which insure that $-S$ is a stable matrix. In particular, $S - \mu I$ will also be an M-matrix if and only if $\mu < \min \text{Re} \, \lambda(S)$. This means that any modification of the subsystems $(\mathscr{S}_i)$, or their local feedback, which increases each diagonal term $\sigma_i + b_{ii}$ by less than μ will leave composite system $(\mathscr{S})$ asymptotically stable (or exponentially stable). In this case μ can again be interpreted as a margin of stability for system $(\mathscr{S})$.

F. Instability

In discussing instability results for composite system $(\mathscr{S})$, we find it convenient to employ the following convention.

5.6.16. Definition. Isolated subsystem $(\mathscr{S}_i)$ is said to possess **Property C** if there exists a Lyapunov function $v_i(z_i)$ such that (v_i, T_i) is admissible in the sense of Definition 5.4.17, two functions $\varphi_{i2}, \varphi_{i3} \in K$ and real constants σ_i and $m_i > 0$ such that

$$\varphi_{i2}(\|z_i\|_i) \leq -v_i(z_i)$$

$$\nabla v_i(z_i, f_{is}(z_i)) \leq \sigma_i \varphi_{i3}(\|z_i\|_i)$$

for all $z_i \in D(f_{is})$, where $D(f_{is})$ denotes the domain of the infinitesimal generator of $T_i(t)$, with $\|z_i\|_i < m_i$.

5.6.17. Definition. Isolated subsystem $(\mathscr{S}_i)$ is said to possess **Property C′** if there exists a Lyapunov function $v_i(z_i)$ with (v_i, T_i) admissible in the sense of Definition 5.4.17, a function $\varphi_{i3} \in K$, and real constants σ_i and $m_i > 0$ such that

$$\nabla v_i(z_i, f_{is}(z_i)) \leq \sigma_i \varphi_{i3}(\|z_i\|_i)$$

for all $\|z_i\|_i < m_i$, $z_i \in D(f_{is})$.

If in Definition 5.6.16 $\sigma_i < 0$, then the trivial solution of subsystem $(\mathscr{S}_i)$ is unstable (in fact, completely unstable).

We now have the following instability result.

5.6.18. Theorem. Let N be a nonempty subset of $L = \{1, 2, ..., l\}$. Assume that for composite system $(\mathscr{S})$ with decomposition (Σ_i) the following conditions hold.

(i) If $i \in N$, then $(\mathscr{S}_i)$ possesses Property C and if $i \notin N$, $i \in L$, then $(\mathscr{S}_i)$ possesses Property C';

(ii) there exist real constants b_{ij} such that the inequalities

$$\nabla v_i(z_i, g_i(x)) \leq \varphi_{i3}(\|z_i\|_i)^{1/2} \sum_{j=1}^{l} b_{ij} \varphi_{j3}(\|z_j\|_j)^{1/2}$$

hold for all $x^T = (z_1, ..., z_l) \in D(f+g)$ with $\|z_i\|_i < m_i$ for all $i \in L$;

(iii) there exist positive constants α_i for all $i \in L$ such that the test matrix $R = [r_{ij}]$ defined by

$$r_{ij} = \begin{cases} \alpha_i(\sigma_i + b_{ii}), & \text{if } i = j \\ \tfrac{1}{2}(\alpha_i b_{ij} + \alpha_j b_{ji}), & \text{if } i \neq j \end{cases}$$

is negative definite.

Then the trivial solution of composite system $(\mathscr{S})$ is **unstable**.

Proof. Given the constants $\alpha_i > 0$, $i \in L$, choose

$$v(x) = \sum_{i=1}^{l} \alpha_i v_i(z_i).$$

As in the proof of Theorem 5.6.5, we can show that v is a Lyapunov function such that

$$Dv_{(\mathscr{S})}(x) \leq \lambda_M(R) \, \varphi_3(\|x\|) \tag{5.6.19}$$

for all $x \in Q$, where $Q = \{x^{\mathrm{T}} = (z_1, ..., z_l) : \|z_i\|_i < m_i \text{ for all } i \in L\}$, where $\lambda_M(R) < 0$ is the largest eigenvalue of matrix $R = [r_{ij}]$ and where φ_3 is a function of class K such that

$$\varphi_3(\|x\|) \leq \sum_{i=1}^{l} \varphi_{i3}(\|z_i\|_i).$$

Now let $X_u = \{x \in Q : x^{\mathrm{T}} = (z_1, ..., z_l) \text{ and } z_i = 0 \text{ whenever } i \notin N\}$. Since $v(0) = 0$, then for $x \in X_u$, $x \neq 0$, we have

$$v(x) = \sum_{i \in N} \alpha_i v_i(z_i) \leq -\sum_{i \in N} \alpha_i \varphi_{i2}(\|z_i\|_i) < 0, \tag{5.6.20}$$

i.e., in every neighborhood of the origin there is at least one point x_0 for which $v(x_0) < 0$.

It now follows from inequalities (5.6.19), (5.6.20), and Theorem 5.4.14 that the trivial solution of composite system $(\mathscr{S})$ is unstable. ∎

The proof of Theorem 5.6.18 can be modified along the lines outlined in the proof of Theorem 5.6.12 to show that the following result is true.

5.6.21. Theorem. Assume that for composite system $(\mathscr{S})$ with decomposition (Σ_i) the following conditions hold.

(i) Hypotheses (i) and (ii) of Theorem 5.6.18 are satisfied with $b_{ij} \geq 0$ for all $i \neq j$;

(ii) all successive principal minors of the test matrix $S = [s_{ij}]$ are positive, where

$$s_{ij} = \begin{cases} -(\sigma_i + b_{ii}), & \text{if } i = j \\ -b_{ij}, & \text{if } i \neq j. \end{cases}$$

Then the trivial solution of composite system $(\mathscr{S})$ is **unstable**.

G. Uniform Ultimate Boundedness

Next, we consider the boundedness of solutions of system $(\mathscr{S})$.

5.6.22. Definition. Isolated subsystem $(\mathscr{S}_i)$ is said to possess **Property D** if there exists a Lyapunov function v_i defined for all $z_i \in Z_i$ with (v_i, T_i) admissible in the sense of Definition 5.4.17, three functions $\varphi_{i1}, \varphi_{i2}, \varphi_{i3} \in KR$, a constant $\sigma_i \in R$, and positive constants M_i and R_i, such that the inequalities

$$\varphi_{i1}(\|z_i\|_i) \leq v_i(z_i) \leq \varphi_{i2}(\|z_i\|_i)$$

and

$$\nabla v_i(z_i, f_{is}(z_i)) \leq \sigma_i \varphi_{i3}(\|z_i\|_i)$$

hold when $z_i \in D(f_{is})$ and $\|z_i\|_i > R_i$, and if $|v_i(z_i)| \leq M_i$ and $|\nabla v_i(z_i, f_{is}(z_i))| \leq M_i$ when $\|z_i\|_i \leq R_i$ and $z_i \in D(f_{is})$. (As before, f_{is} is the infinitesimal generator of $T_i(t)$.)

If in Definition 5.6.22 $\sigma_i < 0$, then isolated subsystem $(\mathcal{S}_i)$ is uniformly ultimately bounded.

5.6.23. Theorem. Assume that for composite system $(\mathcal{S})$ with decomposition (Σ_i) the following conditions hold.
  (i)   Each isolated subsystem $(\mathcal{S}_i)$ possesses Property D;
  (ii)   there exist real constants b_{ij}, $i, j = 1, \ldots, l$, such that the inequalities

$$\nabla v_i(z_i, g_i(x)) \leq \varphi_{i3}(\|z_i\|_i)^{1/2} \sum_{j=1}^{l} b_{ij} \varphi_{j3}(\|z_j\|_j)^{1/2}$$

hold for all $x^T = (z_1, \ldots, z_l) \in D(f+g)$; and
  (iii)   there exist positive constants α_i, $i = 1, \ldots, l$, such that the test matrix $R = [r_{ij}]$ defined by

$$r_{ij} = \begin{cases} \alpha_i(\sigma_i + b_{ii}), & \text{if} \quad i = j \\ \tfrac{1}{2}(\alpha_i b_{ij} + \alpha_j b_{ji}), & \text{if} \quad i \neq j \end{cases}$$

is negative definite.
  Then composite system $(\mathcal{S})$ is **uniformly ultimately bounded**.

Proof. Given positive constants α_i, $i = 1, \ldots, l$, we choose

$$v(x) = \sum_{i=1}^{l} \alpha_i v_i(z_i).$$

Let $\overline{B_i(R_i)} = \{z_i \in Z_i : \|z_i\|_i \leq R_i\}$. Following the procedure in the proof of Theorem 5.6.5, it is clear that

$$\sum_{i=1}^{l} \alpha_i \varphi_{i1}(\|z_i\|_i) \leq v(x) \leq \sum_{i=1}^{l} \alpha_i \varphi_{i2}(\|z_i\|_i)$$

and

$$Dv_{(\mathcal{S})}(x) \leq \lambda_M(R) \sum_{i=1}^{l} \varphi_{i3}(\|z_i\|_i)$$

for all $x \in X - \bigtimes_{i=1}^{l} \overline{B_i(R_i)}$.

To complete the proof we must consider the situation where some but not all z_i satisfy the condition $\|z_i\|_i \le R_i$. First we suppose that $\|z_i\|_i > R_i$ for $i = 1, 2, \dots, r$ and $|z_i| \le R_i$ for $i = r+1, \dots, l$ and $x^T = (z_1, \dots, z_l) \in D(f+g)$. For such values of x we have

$$\sum_{i=1}^{r} \alpha_i \varphi_{i1}(\|z_i\|_i) + \sum_{i=r+1}^{l} \alpha_i v_i(z_i) \le v(x) \le \sum_{i=1}^{r} \alpha_i \varphi_{i2}(\|z_i\|_i) + \sum_{i=r+1}^{l} \alpha_i v_i(z_i)$$

or

$$\sum_{i=1}^{r} \alpha_i \varphi_{i1}(\|z_i\|_i) - \sum_{i=r+1}^{l} \alpha_i M_i \le v(x) \le \sum_{i=1}^{r} \alpha_i \varphi_{i2}(\|z_i\|_i) + \sum_{i=r+1}^{l} \alpha_i M_i.$$

The functions on the left and right of the above inequality are dominated by two functions $\varphi_1{}^*, \varphi_2{}^* \in KR$ such that

$$\varphi_1{}^*\left(\sum_{i=1}^{r} \|z_i\|_i \right) \le \sum_{i=1}^{r} \alpha_i \varphi_{i1}(\|z_i\|_i) - \sum_{i=r+1}^{l} \alpha_i M_i$$

and

$$\varphi_2{}^*\left(\sum_{i=1}^{r} \|z_i\|_i \right) \ge \sum_{i=1}^{r} \alpha_i \varphi_{i2}(\|z_i\|_i) + \sum_{i=r+1}^{l} \alpha_i M_i.$$

Similarly, for such values of x we also have

$$\nabla v(x, f(x)+g(x)) = \sum_{i=1}^{l} \alpha_i \{ \nabla v_i(z_i, f_{is}(z_i)) + \nabla v_i(z_i, g_i(x)) \}$$

$$\le \sum_{i=1}^{r} \Bigg(\alpha_i \sigma_i \varphi_{i3}(\|z_i\|_i)$$

$$+ \alpha_i \varphi_{i3}(\|z_i\|_i)^{1/2} \sum_{j=1}^{r} b_{ij} \varphi_{j3}(\|z_j\|_j)^{1/2} \Bigg)$$

$$+ \sum_{i=1}^{r} \alpha_i \varphi_{i3}(\|z_i\|_i)^{1/2} \sum_{j=r+1}^{l} b_{ij} \varphi_{j3}(\|z_j\|_j)^{1/2}$$

$$+ \sum_{i=r+1}^{l} \alpha_i \nabla v_i(z_i, f_{is}(z_i))$$

$$+ \sum_{i=r+1}^{l} \alpha_i \varphi_{i3}(\|z_i\|_i)^{1/2} \sum_{j=1}^{r} b_{ij} \varphi_{j3}(\|z_j\|_j)^{1/2}$$

$$+ \sum_{i=r+1}^{l} \alpha_i \varphi_{i3}(\|z_i\|_i)^{1/2} \sum_{j=r+1}^{l} b_{ij} \varphi_{j3}(\|z_j\|_j)^{1/2}.$$

Now if $u = (\varphi_{13}(\|z_1\|_1)^{1/2}, \varphi_{23}(\|z_2\|_2)^{1/2}, \dots, \varphi_{r3}(\|z_r\|_r)^{1/2})$ and if R^* denotes the $r \times r$ matrix $R^* = [r_{ij}]^r_{i,j=1}$, then we have

$$\nabla v(x, f(x)+g(x)) \leq u^{\mathrm{T}} R^* u + \sum_{i=1}^{r} \alpha_i \varphi_{i3}(\|z_i\|_i)^{1/2} \left[\sum_{j=r+1}^{l} b_{ij} \varphi_{j3}(R_j)^{1/2} \right]$$

$$+ \sum_{i=r+1}^{l} \alpha_i M_i + \sum_{i=r+1}^{l} \alpha_i \varphi_{i3}(R_i)^{1/2} \sum_{j=1}^{r} b_{ij} \varphi_{j3}(\|z_j\|_j)^{1/2}$$

$$+ \sum_{i=r+1}^{l} \alpha_i \varphi_{i3}(R_i)^{1/2} \sum_{j=r+1}^{l} b_{ij} \varphi_{j3}(R_j)^{1/2}.$$

We thus have an inequality of the form

$$\nabla v(x, f(x)+g(x)) \leq u^{\mathrm{T}} R^* u + u^{\mathrm{T}} L_0 + L_1$$

for some constant $L_1 > 0$ and some vector $L_0 \in R^r$. Since the test matrix R is negative definite, the submatrix R^* is also negative definite. Thus, there is a constant $\lambda_M(R^*) < 0$, a real constant r^*, and a function $\varphi_3^* \in KR$ such that

$$\nabla v(x, f(x)+g(x)) \leq \lambda_M(R^*)|u|^2 + u^{\mathrm{T}} L_0 + L_1$$

$$\leq \tfrac{1}{2}\lambda_M(R^*)|u|^2$$

$$= \tfrac{1}{2}\lambda_M(R^*) \sum_{i=1}^{r} \varphi_{i3}(\|z_i\|_i) \leq \tfrac{1}{2}\lambda_M(R^*) \varphi_3^* \left(\sum_{i=1}^{r} \|z_i\|_i \right)$$

for all $x^{\mathrm{T}} = (z_1, ..., z_l) \in D(f+g)$ such that $\|z_i\|_i > r^*$ for $i = 1, ..., r$ and $\|z_i\|_i \leq R_i$ for $i = r+1, ..., l$.

Above we have assumed that $\|z_i\|_i > R_i$ for $i = 1, 2, ..., r$ and $\|z_i\|_i \leq R_i$ for $i = r+1, ..., l$. The identical argument works for any other combination of indices.

In view of Theorem 5.4.18 and the above, we can now apply Theorem 5.4.16, which completes the proof. ∎

Using the modifications outlined in the proof of Theorem 5.6.12, we can also prove the following result.

5.6.24. Theorem. Assume that for composite system $(\mathscr{S})$ with decomposition (Σ_i) the following conditions hold.

(i)   Hypotheses (i) and (ii) of Theorem 5.6.23 are satisfied with $b_{ij} \geq 0$ for all $i \neq j$; and

(ii) all successive principal minors of the test matrix $S = [s_{ij}]$ defined by

$$s_{ij} = \begin{cases} -(\sigma_i + b_{ii}), & \text{if } i = j \\ -b_{ij}, & \text{if } i \neq j \end{cases}$$

are positive.

Then composite system $(\mathscr{S})$ is **uniformly ultimately bounded**.

5.7 Some Examples and Applications

In the present section we consider two specific dynamical systems to demonstrate the applicability and usefulness of the results of Section 5.6. The first of these is a simple hybrid system, while the second involves the point kinetics model of a coupled core nuclear reactor.

5.7.1. Example. (*Hybrid System.*) Consider a hybrid system described by the set of equations

$$\dot{z}_1(t) = Az_1(t) + b \int_G H_1(y, z_2(t, y)) \, dy$$

$$\frac{\partial z_2(t, y)}{\partial t} = \alpha \Delta z_2(t, y) - H_2(z_2(t, y)) + h_2(y) c^T z_1(t) \tag{5.7.2}$$

with boundary conditions

$$z_2(t, y) = 0 \qquad \text{for all} \quad (t, y) \in R^+ \times \partial G \tag{5.7.3}$$

and initial conditions $z_1(0) = z_{10}$ given and $z_2(0, y) \triangleq \psi(y)$ given for $y \in G$. Here A is an $n \times n$ matrix, b and c are given n-vectors, α and L are given positive numbers, Δ is the Laplacian in m-space R^m, G is an open subset of R^m with smooth boundary ∂G, H_1 and H_2 are given functions which satisfy the conditions

$$H_1(y, 0) = 0 \qquad \text{for all} \quad y \in G, \qquad H_2(0) = 0$$

$$|H_1(y, z) - H_1(y, z^*)| \le |h_1(y)| \, |z - z^*|$$

for all $y \in G$, and $z, z^* \in R$, and

$$|H_2(u) - H_2(w)| \le L|u - w|$$

for all $u, w \in R$, and $h_i \in L_2(G)$, $i = 1, 2$, are given functions.

Hybrid system (5.7.2) may be viewed as an interconnection of two isolated subsystems $(\mathscr{S}_1)$, $(\mathscr{S}_2)$ described by the equations

$$\dot{z}_1(t) = f_1(z_1) = Az_1 \tag{$\mathscr{S}_1$}$$

$$\frac{\partial z_2}{\partial t} = f_2(z_2) = \alpha \Delta z_2 - H_2(z_2). \tag{$\mathscr{S}_2$}$$

The interconnecting structure is specified in this case by

$$g_1(z_1, z_2) = b \int_G H_1(y, z_2(y)) \, dy$$

and

$$g_2(z_1, z_2) = h_2(y) c^T z_1.$$

In the present case $Z_1 = R^n$, $Z_2 = L_2(G)$, and $X = R^n \times L_2(G)$. As usual, the norms for R^n and $L_2(G)$ are denoted by $|\cdot|$ and $\|\cdot\|_2$, respectively.

Since $(\mathscr{S}_1)$ is an ordinary differential equation, it is certainly well posed on Z_1. Moreover, if A is a stable matrix, there exists a positive definite symmetric matrix P such that

$$A^{\mathrm{T}}P + PA = -Q$$

is negative definite. Choosing

$$v_1(z_1) = z_1{}^{\mathrm{T}} P z_1$$

and using the notation of Section 5.6, we obtain

$$\varphi_{11}(|z_1|) = \lambda_m(P)|z_1|^2 \le v_1(z_1) \le \lambda_M(P)|z_1|^2 = \varphi_{12}(|z_1|)$$

and

$$\nabla v_1(z_1, f_1(z_1)) = -z_1{}^{\mathrm{T}} Q z_1 \le -\lambda_m(Q)|z_1|^2$$

so that $\sigma_1 = -\lambda_m(Q)$ and $\varphi_{13}(r) = r^2$.

Subsystem $(\mathscr{S}_2)$ is the type of system discussed at the end of Section 5.5, Part D (see Eq. (5.5.9)). The Laplacian Δ is a second order, strongly elliptic operator while H_2 is Lipschitz continuous. Thus, subsystem $(\mathscr{S}_2)$ determines an w-accretive operator on $L_2(G)$ and by Theorem 5.6.2 is well posed. If H_2 satisfies the condition

$$uH_2(u) \ge \beta u^2 \qquad \text{for all} \quad u \in R \tag{5.7.4}$$

for some fixed constant $\beta \ge 0$, then we can choose

$$v_2(z_2) = \int_G |z_2(y)|^2 \, dy = (\|z_2\|_2)^2.$$

In the notation of Section 5.6 we have in this case

$$\varphi_{21}(\|z_2\|_2) = v_2(z_2) = \varphi_{22}(\|z_2\|_2).$$

Also, using the notation of Section 5.5, Part D,

$$\nabla v_2(z_2, f_2(z_2)) = -\alpha \int_G |\nabla z_2|^2 \, dy - \int_G z_2 \, H_2(z_2) \, dy$$

$$\le -\alpha \int_G |\nabla z_2|^2 \, dy - \beta \int_G |z_2|^2 \, dy$$

$$\le -\alpha\gamma^{-2} \int_G |z_2|^2 \, dy - \beta \int_G |z_2|^2 \, dy$$

$$= -(\alpha\Gamma + \beta)(\|z_2\|_2)^2.$$

Using the notation of Section 5.6 we have $\sigma_2 = -(\alpha\Gamma + \beta)$ and $\varphi_{32}(r) = r^2$. The constant $\Gamma = \gamma^{-2}$ can be estimated by $\gamma \leq \delta/n^{1/2}$, where

$$\delta = \sup\{|u - v|: u, v \in G \text{ and for some } j (1 \leq j \leq n),$$

$$u_i = v_i \text{ for } i = 1, ..., n, \; i \neq j\}.$$

Thus it is clear that isolated subsystems $(\mathscr{S}_1)$ and $(\mathscr{S}_2)$ are well posed and that they satisfy Property A as well as Property B if matrix A is stable and if condition (5.7.4) is true.

The interconnecting terms are Lipschitz continuous. Indeed,

$$|g_1(z_1, z_2) - g_1(u_1, u_2)| \leq |b| \int_G |h_1(y)| \, |z_2(y) - u_2(y)| \, dy$$

$$\leq |b| \, \|h_1\|_2 \, \|z_2 - u_2\|_2$$

by the above assumptions and the Schwarz inequality. Also,

$$\int_G |g_2(z_1, z_2) - g_2(u_1, u_2)|^2 \, dy = \int_G |h_2(y) \, c^T (z_1 - u_1)|^2 \, dy$$

$$\leq (\|h_2\|_2 |c| \, |z_1 - u_1|)^2.$$

It now follows from Theorem 5.6.2 that composite system (5.7.2) is well posed on $X = R^n \times L_2(G)$.

Next, we check hypothesis (ii) of Theorem 5.6.5. We have

$$\nabla v_1(z_1, g_1(x)) = 2z_1^T P g_1(x)$$

$$= 2z_1^T P \left\{ b \int_G H_1(y, z_2(y)) \, dy \right\}$$

$$\leq 2|z_1| \lambda_M(P) |b| \int_G |h_1(y)| \, |z_2(y)| \, dy$$

$$\leq 2|z_1| \lambda_M(P) |b| \, \|h_1\|_2 \, \|z_2\|_2$$

with the last inequality following from the Schwarz inequality. Using the notation of Theorem 5.6.5 we have $b_{11} = 0$ and $b_{12} = 2\lambda_M(P) |b| \, \|h_1\|_2$. We also have

$$\nabla v_2(z_2, g_2(x)) = 2 \int_G z_2(y) h_2(y) c^T z_1 \, dy \leq 2 \|z_2\|_2 \|h_2\|_2 |c| \, |z_1|.$$

In the notation of Theorem 5.6.5 we have $b_{22} = 0$ and $b_{21} = 2 \|h_2\|_2 |c|$. The test matrix R of this theorem assumes the form

$$R = \begin{bmatrix} -\alpha_1 \lambda_m(Q) & [\alpha_1 \lambda_M(P) |b| \, \|h_1\|_2 + \alpha_2 \|h_2\|_2 |c|] \\ [\alpha_1 \lambda_M(P) |b| \, \|h_1\|_2 + \alpha_2 \|h_2\|_2 |c|] & -\alpha_2(\beta + \alpha\Gamma) \end{bmatrix}.$$

If we choose $\alpha_1 = b_{12}^{-1}$ and $\alpha_2 = b_{21}^{-1}$, matrix R is negative definite if and only if

$$(\beta + \alpha\Gamma)\left(\frac{\lambda_m(Q)}{\lambda_M(P)}\right) > 4\,|b|\,|c|\,\|h_1\|_2\,\|h_2\|_2. \qquad (5.7.5)$$

Therefore, if matrix A is stable and if inequalities (5.7.4) and (5.7.5) are satisfied, the trivial solution of composite system (5.7.2) is **uniformly asymptotically stable**, by Theorem 5.6.5, and **exponentially stable**, by Theorem 5.6.10. As a matter of fact, since subsystems $(\mathscr{S}_1)$ and $(\mathscr{S}_2)$ satisfy Property B under the above conditions, it follows that the trivial solution of composite system (5.7.2) is **uniformly asymptotically stable in the large** by Theorem 5.6.9 and **exponentially stable in the large** by Theorem 5.6.11.

5.7.6. Remark. Since $b_{12} \geq 0$ and $b_{21} \geq 0$ we can also apply Theorem 5.6.12. In this case the test matrix S is specified by

$$S = \begin{bmatrix} \lambda_m(Q) & -2\lambda_M(P)\,|b|\,\|h_1\|_2 \\ -2\,\|h_2\|_2\,|c| & (\beta + \alpha\Gamma) \end{bmatrix}.$$

Matrix S has positive successive principal minors if and only if inequality (5.7.5) is satisfied. We have thus arrived at the same stability conditions as before.

5.7.7. Remark. Next, let us replace condition (5.7.4) by the condition

$$uH_2(u) \geq \beta u^2 \qquad \text{if} \quad |u| \geq R_1. \qquad (5.7.8)$$

Then Theorem 5.6.23 can be applied. Indeed, if matrix A is stable and if conditions (5.7.5) and (5.7.8) are true, then composite system (5.7.2) is **uniformly ultimately bounded**, by Theorem 5.6.23.

5.7.9. Remark. Next, let us assume that $H_1(y, z_2)$ and $H_2(z_2)$ are linear, having the special form

$$H_1(y, z_2) = f_1(y)z_2, \qquad H_2(z_2) = \beta z_2 \qquad (5.7.10)$$

where $\beta > 0$ is a constant and f_1 is a real function such that $\|f_1\|_2 < \infty$. In this case, composite system (5.7.2) determines a C_0-semigroup on $X = R^n \times L_2(G)$. If in particular G is the one-dimensional interval $G = [a, b]$, then Γ in (5.7.5) can be replaced by $(b-a)^{-2}$.

5.7.11. Remark. For composite system (5.7.2) boundary conditions different from (5.7.3) can also be used. For example, let us consider for Eq. (5.7.2) the boundary conditions

$$\frac{\partial z_2}{\partial n}(t, y) = 0 \qquad \text{for all} \quad (t, y) \in R^+ \times \partial G$$

where n denotes the outward unit normal on ∂G. For these boundary conditions we can compute

$$\nabla v_2(z_2, f_2(z_2)) \leq -\alpha \int_G |\nabla z_2|^2 \, dy - \beta \int_G |z_2|^2 \, dy \leq -\beta \int_G |z_2|^2 \, dy$$

so that $\sigma_2 = -\beta$, $\Gamma = 0$, and $\varphi_{23}(r) = r^2$. The rest of the analysis is the same as before. Thus, if for composite system (5.7.2) the boundary condition (5.7.3) is replaced by the last boundary condition above, then the stability condition (5.7.5) is replaced by the condition

$$\beta \left(\frac{\lambda_m(Q)}{\lambda_M(P)} \right) > 4 |b| \, |c| \, \|h_1\|_2 \, \|h_2\|_2. \tag{5.7.12}$$

5.7.13. Remark. Similarly, if we replace G by R^m, $Z_2 = L_2(R^m)$, and if we assume (5.7.4), then condition (5.7.12) will still be the stability condition for (5.7.2).

5.7.14. Example. Let us reconsider hybrid system (5.7.2) with boundary conditions (5.7.3). However, in the present case we assume that A is an unstable matrix. Specifically, we assume there is a nonsingular real matrix B such that

$$BAB^{-1} = \begin{bmatrix} A_1 & 0 \\ 0 & A_2 \end{bmatrix} \tag{5.7.15}$$

where $-A_1$ is a stable $(k \times k)$ matrix, A_2 is a stable $(j \times j)$ matrix, $k + j = n$, and $1 \leq k \leq n$. This transformation is always possible if A has at least one eigenvalue with positive real part and has no eigenvalue with zero real part. We allow the possibility that $k = n, j = 0$, and $B = I$, i.e., that A is completely unstable, and will discuss it separately later. Let $w = Bz_1$ so that

$$\dot{w}(t) = \begin{bmatrix} A_1 & 0 \\ 0 & A_2 \end{bmatrix} w(t) + Bb \int_G H_1(y, z_2(t, y)) \, dy.$$

If we put

$$w = \begin{bmatrix} w_1 \\ w_2 \end{bmatrix}, \qquad Bb = \begin{bmatrix} b_1 \\ b_2 \end{bmatrix}, \qquad (c^T B^{-1})^T = \begin{bmatrix} c_1 \\ c_2 \end{bmatrix}, \qquad w_3 = z_2$$

then Eq. (5.7.2) can be replaced by

$$\dot{w}_1 = A_1 w_1 + b_1 \int_G H_1(y, w_3(t, y)) \, dy$$

$$\dot{w}_2 = A_2 w_2 + b_2 \int_G H_1(y, w_3(t, y)) \, dy \tag{5.7.16}$$

$$\frac{\partial w_3}{\partial t} = \alpha \Delta w_3 - H_2(w_3) + h_2(y)[c_1^T w_1 + c_2^T w_2]$$

with $w_3(t, y) = 0$ for all $(t, y) \in R^+ \times \partial G$. System (5.7.16) may now be viewed as an interconnection of three isolated subsystems of the form

$$\dot{w}_1 = f_1(w_1) = A_1 w_1 \tag{$\mathscr{S}_1$}$$

$$\dot{w}_2 = f_2(w_2) = A_2 w_2 \tag{$\mathscr{S}_2$}$$

$$\frac{\partial w_3}{\partial t} = f_3(w_3) = \alpha \Delta w_3 - H_2(w_3). \tag{$\mathscr{S}_3$}$$

Since the trivial solution of $(\mathscr{S}_1)$ is completely unstable, there exists a symmetric positive definite matrix P_1 such that the matrix

$$A_1^T P_1 + P_1 A_1 = Q_1$$

is positive definite. Let

$$v_1(w_1) = -w_1^T P_1 w_1.$$

Then

$$-v_1(w_1) \geq \lambda_m(P_1)|w_1|^2 \triangleq \varphi_{12}(|w_1|)$$

and

$$\nabla v_1(w_1, f_1(w_1)) = -w_1^T Q w_1 \leq -\lambda_m(Q_1)|w_1|^2 \triangleq \sigma_1 \varphi_{13}(|w_1|).$$

Also, since the trivial solution of $(\mathscr{S}_2)$ is asymptotically stable, there exists a positive definite symmetric matrix P_2 such that the matrix

$$A_2^T P_2 + P_2 A_2 = -Q_2$$

is negative definite. Let

$$v_2(w_2) = w_2^T P_2 w_2.$$

Then

$$\nabla v_2(w_2, f_2(w_2)) \leq -\lambda_m(Q_2)|w_1|^2 \triangleq \sigma_2 \varphi_{23}(|w_2|).$$

Next, choose

$$v_3(w_3) = (\|w_3\|_2)^2.$$

Then the same calculation as in Example 5.7.1 can be used to see that

$$\nabla v_3(w_3, f_3(w_3)) \leq -\alpha \int_G |\nabla w_3|^2 \, dy - \beta \int_G |w_3|^2 \, dy$$

$$\leq -(\alpha \Gamma + \beta)(\|w_3\|_2)^2 \triangleq \sigma_3 \varphi_{33}(\|w_3\|_2).$$

Using the notation of Section 5.6 we have the estimate

$$\nabla v_1(w_1, g_1(x)) = 2w_1^T P_1 b_1 \int_G H_1(y, w_3(y)) \, dy$$

$$\leq 2|w_1| \lambda_M(P_1)|b_1| \int_G |h_1(y)| \, |w_3(y)| \, dy$$

$$\leq 2|w_1| \lambda_M(P_1)|b_1| \, \|h_1\|_2 \|w_3\|_2$$

so that $b_{11} = b_{12} = 0$ and $b_{13} = 2\lambda_M(P_1)|b_1| \, \|h_1\|_2$.

Similarly, we have the estimate

$$\nabla v_2(w_2, g_2(x)) = 2w_2^T P_2 b_2 \int_G H_1(y, w_3(y))\, dy$$

$$\leq 2|w_2|\lambda_M(P_2)|b_2| \int_G |h_1(y)|\,|w_3(y)|\, dy$$

$$\leq 2|w_2|\lambda_M(P_2)|b_2|\,\|h_1\|_2\,\|w_3\|_2$$

so that $b_{21} = b_{22} = 0$ and $b_{23} = 2\lambda_M(P_2)|b_2|\,\|h_1\|_2$.

Also, we have the estimate

$$\nabla v_3(w_3, g_3(x)) = 2 \int_G w_3(y) h_2(y)[c_1^T w_1 + c_2^T w_2]\, dy$$

$$\leq 2\|h_2\|_2(|c_1|\,|w_1| + |c_2|\,|w_2|)\|w_3\|_2,$$

so that $b_{31} = 2\|h_2\|_2|c_1|$, $b_{32} = 2\|h_2\|_2|c_2|$, and $b_{33} = 0$.

We are now in a position to apply either Theorem 5.6.18 or Theorem 5.6.21. The test matrix of the latter assumes the form

$$S = \begin{bmatrix} \lambda_m(Q_1) & 0 & -2\lambda_M(P_1)|b_1|\,\|h_1\|_2 \\ 0 & \lambda_m(Q_2) & -2\lambda_M(P_2)|b_2|\,\|h_1\|_2 \\ -2\|h_2\|_2|c_1| & -2\|h_2\|_2|c_2| & \beta + \alpha\Gamma \end{bmatrix}.$$

The successive principal minors of matrix S are positive if and only if the inequality

$$\lambda_m(Q_1)\lambda_m(Q_2)(\beta + \alpha\Gamma) > 4\|h_1\|_2\|h_2\|_2[\lambda_M(P_1)\lambda_m(Q_2)|b_1|\,|c_1|$$

$$+ \lambda_M(P_2)\lambda_m(Q_1)|b_2|\,|c_2|] \qquad (5.7.17)$$

is satisfied. Thus, it follows from Theorem 5.6.21 that the trivial solution of composite system (5.7.2) is **unstable** if inequality (5.7.4), transformation (5.7.15), and inequality (5.7.17) hold.

Finally note that in case $A = A_1$, so that $B = I$ in (5.7.15), then there is no w_2-component and inequality (5.7.17) reduces to inequality (5.7.5).

5.7.18. Example. (*Point Kinetics Model of a Coupled Core Nuclear Reactor.*) In the present case we consider a somewhat more complicated example, motivated by a physical problem. Among other items, this problem illustrates the fact that the choice of subsystems and interconnecting structure is often obvious in applications.

We consider the point kinetics model of a coupled core nuclear reactor with l cores (see Akcasu, Lillouche, and Shotkin [1], Plaza and Kohler [1]) described by the set of equations of the form

$$\Lambda_i \dot{p}_i(t) = [\rho_i(t) - \mathscr{E}_i - \beta_i] \, p_i(t) + \rho_i(t) + \sum_{k=1}^{6} \beta_{ki} \, c_{ki}(t)$$

$$+ \sum_{j=1}^{l} \mathscr{E}_{ji}(P_{j0}/P_{i0}) \int_{-\infty}^{t} h_{ji}(t-s) \, p_j(s) \, ds \qquad (5.7.19)$$

$$\dot{c}_{ki}(t) = \lambda_{ki}[p_i(t) - c_{ki}(t)], \qquad i = 1, 2, ..., l,$$

$$k = 1, 2, ..., 6,$$

where $p_i: R \to R$ and $c_{ki}: R \to R$ represent the power in the ith core and the concentration of the kth precursor in the ith core, respectively. The constants Λ_i, $\mathscr{E}_i$, β_{ki}, $\mathscr{E}_{ji}$, P_{i0}, and λ_{ki} are all positive and

$$\beta_i = \sum_{k=1}^{6} \beta_{ki}.$$

The functions $h_{ji} \in L_1(R^+, R)$. They determine the coupling between cores due to neutron migration from the jth to the ith core. The function ρ_i represents the reactivity of the ith core which we assume to have the form

$$\rho_i(t) = \int_{-\infty}^{t} w_i(t-s) \, p_i(s) \, ds \qquad (5.7.20)$$

where $w_i \in L_1(R^+, R)$. The functions $p_i(t)$ and $c_{ki}(t)$ are assumed to be known, bounded continuous functions on $-\infty \le t \le 0$. The problem is to determine these functions and their stability for $t > 0$.

If we make the physically realistic assumption that $c_{ki}(t) \, e^{\lambda_{ki} t} \to 0$ as $t \to -\infty$, then we can solve for c_{ki} in terms of p_i, obtaining

$$c_{ki}(t) = \int_{-\infty}^{t} \lambda_{ki} \, e^{-\lambda_{ki}(t-s)} p_i(s) \, ds. \qquad (5.7.21)$$

Using Eqs. (5.7.20) and (5.7.21) to eliminate ρ_i and c_{ki} from Eq. (5.7.19), we are left with l Volterra integrodifferential equations for $p_i(t)$, $i = 1, ..., l$. In order to write these equations in a more convenient and compact form, we use the notation

$$F_i(t) = \Lambda_i^{-1} \left[w_i(t) + \sum_{k=1}^{6} \beta_{ki} \lambda_{ki} \, e^{-\lambda_{ki} t} + \mathscr{E}_{ii} h_{ii} \right],$$

$$K_i = \Lambda_i^{-1}[\mathscr{E}_i + \beta_i], \qquad n_i(t) = \Lambda_i^{-1} w_i(t),$$

$$G_{ij} = \frac{\mathscr{E}_{ji} P_{j0} h_{ji}(t)}{\Lambda_i P_{i0}},$$

and $z^i(t) = p_i(t)$ on $-\infty < t < \infty$. We have

$$\dot{p}_i(t) = -K_i p_i(t) + \int_{-\infty}^t F_i(t-s) p_i(s) \, ds + p_i(t) \int_{-\infty}^t n_i(t-s) p_i(s) \, ds$$

$$+ \sum_{j=1, i \neq j}^l \int_{-\infty}^t G_{ij}(t-s) p_j(s) \, ds, \qquad i = 1, \ldots, l,$$

for $t \geq 0$ and $p_i(t) = \varphi_i(t)$ given on $-\infty < t \leq 0$.

Now let Z_i be the **fading memory space** of all measurable functions φ_i such that

$$\|\varphi_i\|^2 = |\varphi_i(0)|^2 + \int_{-\infty}^0 |\varphi_i(s)|^2 e^{L_i s} \, ds$$

where $L_i > 0$ is some constant which will be specified later. Let $X = \mathcal{X}_{i=1}^l Z_i$. Then Eq. (5.7.19) can be expressed as

$$\dot{z}^i(t) = -K_i z_t^i(0) + \int_{-\infty}^0 F_i(-s) z_t^i(s) \, ds + z_t^i(0) \int_{-\infty}^0 n_i(-s) z_t^i(s) \, ds$$

$$+ \sum_{j=1, i \neq j}^l \int_{-\infty}^0 G_{ij}(-s) z_t^j(s) \, ds, \qquad i = 1, \ldots, l, \qquad (5.7.22)$$

with $z_0^i = \varphi_i \in Z_i$ given. (Refer to Section 5.5, Part C, for a brief discussion of Volterra integrodifferential equations.)

Composite system (5.7.22) may be viewed as an interconnection of l isolated subsystems $(\mathcal{S}_i)$ described by equations of the form

$$\dot{z}_t^i(u) = f_i(z_t^i) \triangleq \begin{cases} -K_i z_t^i(0) + \int_{-\infty}^0 F_i(-s) z_t^i(s) \, ds \\ \quad + z_t^i(0) \int_{-\infty}^0 n_i(-s) z_t^i(s) \, ds, \qquad u = 0 \\ \dot{z}_t^i(u), \qquad \qquad \text{on} \quad -\infty < u < 0 \end{cases} \qquad (\mathcal{S}_i)$$

with interconnecting structure characterized by

$$g_i(x) = \sum_{j=1, i \neq j}^l \int_{-\infty}^0 G_{ij}(-s) z_t^j(s) \, ds.$$

For each subsystem $(\mathcal{S}_i)$ we choose

$$v_i(z^i) = z^i(0)^2 + K_i \int_{-\infty}^0 z^i(u)^2 e^{L_i u} \, du.$$

Then

$$\min(1, K_i) \|z^i\|^2 \leq v_i(z^i) \leq \max(1, K_i) \|z^i\|^2$$

and

$$\nabla v_i(z^i, f_i(z^i)) = 2z^i(0) f_i(z^i) + 2K_i \int_{-\infty}^0 z^i(s) f_i(z^i) e^{L_i s} \, ds.$$

If $\varphi_i \in D(f_i)$, then $\dot\varphi_i \in Z_i$, and

$$\lim_{t \to -\infty} \varphi_i(t)^2 e^{L_i t} = 0.$$

Integrating by parts, we obtain

$$\int_{-\infty}^0 z^i(u) f_i(z^i)(u) e^{L_i u} \, du = \int_{-\infty}^0 z^i(s) \dot z^i(s) e^{L_i s} \, ds$$

$$= \tfrac{1}{2} z^i(0)^2 - (L_i/2) \int_{-\infty}^0 [z^i(s)]^2 e^{L_i s} \, ds$$

$$= \tfrac{1}{2} [z^i(0)]^2 - (L_i/2) b_i^2$$

where b_i is defined in an obvious way. Now assume that $L_i > 0$ can be chosen so that

$$c_i \triangleq \left(\int_0^\infty [F_i(u)]^2 e^{L_i u} \, du \right)^{1/2} < \infty$$

and

$$d_i \triangleq \left(\int_0^\infty [n_i(u)]^2 e^{L_i u} \, du \right)^{1/2} < \infty.$$

The definitions for c_i and d_i, integration by parts, and the Schwarz inequality can now be used to show that

$$\nabla v_i(z^i, f_i(z^i)) = 2z^i(0) \left[-K_i z^i(0) + \int_{-\infty}^0 F_i(-s) z^i(s) \, ds \right.$$

$$\left. + z^i(0) \int_{-\infty}^0 n_i(-s) z^i(s) \, ds \right]$$

$$+ 2K_i [\tfrac{1}{2} (z^i(0))^2 - \tfrac{1}{2} L_i b_i^2]$$

$$= -2K_i [z^i(0)]^2 + 2z^i(0) \int_{-\infty}^0 [F_i(-s) e^{-L_i(s/2)}][z^i(s) e^{L_i(s/2)}] \, ds$$

$$+ [2z^i(0)]^2 \int_{-\infty}^0 [n_i(-s) e^{-L_i(s/2)}][z^i(s) e^{L_i(s/2)}] \, ds$$

$$+ 2K_i [\tfrac{1}{2} (z^i(0))^2 - \tfrac{1}{2} L_i b_i^2]$$

$$\leq -K_i [z^i(0)]^2 + 2c_i z^i(0) b_i - K_i L_i b_i^2 + 2d_i [z^i(0)]^2 b_i.$$

$$(5.7.23)$$

The first three terms in (5.7.23) form a quadratic form which is negative definite if

$$K_i\sqrt{L_i} > c_i. \tag{5.7.24}$$

The fourth term in (5.7.23) is cubic so that in a neighborhood of the origin,

$$\nabla v_i(z^i, f_i(z^i)) < \sigma_i[(z^i(0))^2 + b_i^2]$$

for

$$\sigma_i = -\frac{K_i(L_i+1)}{2} + \left[\frac{K_i(L_i-1)^2}{4} + c_i^2\right]^{1/2}. \tag{5.7.25}$$

Next, we consider the interconnecting structure. We assume that

$$c_{ij} = \left(\int_0^\infty [G_{ij}(s)]^2 e^{L_j s}\, ds\right)^{1/2} < \infty.$$

Then

$$\nabla v_i(z^i, g_i(x)) = 2z^i(0) \sum_{j=1,\, i\neq j}^{l} \int_{-\infty}^{0} (G_{ij}(-s)e^{-L_j(s/2)})(z^j(s)e^{L_j(s/2)})\, ds$$

$$\leq z^i(0) \sum_{j=1,\, i\neq j}^{l} 2c_{ij}b_j \leq \|z^i\| \sum_{j=1,\, i\neq j}^{l} 2c_{ij}\|z^j\| \triangleq \|z^i\| \sum_{j=1,\, i\neq j}^{l} b_{ij}\|z^j\|.$$

Thus, the hypotheses of Theorem 5.6.12, part (iii), are satisfied, provided that the test matrix $S = [s_{ij}]$ has positive successive principal minors, where

$$s_{ij} = \begin{cases} -\sigma_i, & i = j \\ -2c_{ij}, & i \neq j \end{cases}$$

and where σ_i is specified by Eq. (5.7.25).

For example, if $l = 2$, then the trivial solution of composite system (5.7.19) is **exponentially stable** if

$$\sigma_1\sigma_2 > 4c_{12}c_{21}.$$

Thus, if in a two core coupled nuclear reactor each core is exponentially stable and if the coupling between cores via neutron migration is sufficiently weak, then the reactor is exponentially stable.

5.8 Functional Differential Equations — Special Results

Functional differential equations with finite delays determine semigroups in the space C_r, as outlined in Section 5.5, Part B. As such, the general stability results for composite systems, presented in Section 5.6, can be applied to systems described by functional differential equations. However, equations of this type have a special form and as such, they have special Lyapunov theorems and comparison theorems which are not applicable to general dynamical systems. For further details and examples, refer to Yoshizawa [1, Chapter 8], Lakshmikantham and Leela [2, Chapters 6–8], and Driver [1]. In the process of characterizing the qualitative properties of isolated subsystems, we give two examples of Lyapunov theorems for functional differential equations, which are useful in studying composite systems. Then we present two sample stability theorems for interconnected systems.

In the present section the notation of Section 5.5, Part B, is employed. In particular we let $C_r^{n_i}$ denote the set of all continuous functions $\psi^i: [-r, 0] \to R^{n_i}$ with norm defined by

$$\|\psi^i\| = \max\{|\psi^i(t)|: -r \le t \le 0\}.$$

We consider systems which may be described by equations of the form

$$\dot{z}^i(t) = f_i(z_t^i) + \sum_{j=1, i \ne j}^{l} g_{ij}(z_t^j), \qquad i = 1, \dots, l \qquad (5.8.1)$$

where

$$f_i: C_r^{n_i} \to C_r^{n_i} \quad \text{and} \quad g_{ij}: C_r^{n_j} \to C_r^{n_i}.$$

Letting $\sum_{i=1}^{l} n_i = n$,

$$x^T = [(z^1)^T, \dots, (z^l)^T] \in R^n,$$

$$[f(x)]^T = [(f_1(z_1))^T, \dots, (f_l(z_l))^T],$$

$$g_i(x) = \sum_{j=1, i \ne j}^{l} g_{ij}(z^j),$$

and

$$[g(x)]^T = [(g_1(x))^T, \dots, (g_l(x))^T],$$

system (5.8.1) may be rewritten as

$$\dot{x}(t) = f(x_t) + g(x_t) \triangleq h(x_t). \qquad (5.8.2)$$

Composite system (5.8.2) with decomposition (5.8.1) may be viewed as an interconnection of l isolated subsystems $(\mathscr{S}_i)$ described by equations of the form

$$\dot{z}^i(t) = f_i(z_t^i) \qquad (\mathscr{S}_i)$$

with interconnecting strucure characterized by

$$g_i(x) = \sum_{j=1, i \ne j}^{l} g_{ij}(z^j).$$

Henceforth we assume that the right-hand sides of Eqs. (5.8.1), (5.8.2), and $(\mathscr{S}_i)$ are sufficiently smooth so that their initial value problems have unique solutions. We also assume that these equations possess the trivial solution.

5.8.3. Definition. Isolated subsystem $(\mathscr{S}_i)$ is said to have **Property F1** if there is a continuous functional $v_i(\psi^i)$ defined on $C_r^{n_i}$, three functions $\varphi_{i1}, \varphi_{i2}, \varphi_{i3} \in K$, and constants $c_i > 0$ and $L_i > 0$, such that the inequalities

(i) $\varphi_{i1}(\|\psi^i\|) \le v_i(\psi^i) \le \varphi_{i2}(\|\psi^i\|)$,
(ii) $Dv_{i(\mathscr{S}_i)}(\psi^i) \le -c_i\varphi_{i3}(\|\psi^i\|)$, and
(iii) $|v_i(\psi_1^i) - v_i(\psi_2^i)| \le L_i\|\psi_1^i - \psi_2^i\|$

hold for all $\psi^i, \psi_1^i, \psi_2^i$ in a neighborhood $\|\psi^i\| < H_i$ of $C_r^{n_i}$.

If isolated subsystem $(\mathscr{S}_i)$ possesses Property F1, then, as shown in Yoshizawa [1, pp. 189–192], the trivial solution of subsystem $(\mathscr{S}_i)$ is uniformly asymptotically stable. (Note that if composite system (5.8.2) possesses Property F1, where this property is rephrased for the appropriate space C_r^n, then the trivial solution of system (5.8.2) is uniformly asymptotically stable.)

5.8.4. Theorem. Assume that for composite system (5.8.2) with decomposition (5.8.1), the following conditions hold.

(i) Each isolated subsystem $(\mathscr{S}_i)$ possesses Property F1;
(ii) for each $i, j = 1, \ldots, l$, $i \ne j$, there are constants $k_{ij} \ge 0$ such that $|g_{ij}(\psi^j)| \le k_{ij}\varphi_{j3}(\|\psi^j\|)$ for all $\|\psi^j\| < H_j$; and
(iii) the successive principal minors of the test matrix $S = [s_{ij}]$ defined by

$$s_{ij} = \begin{cases} c_i, & \text{if } i = j \\ -L_i k_{ij}, & \text{if } i \ne j \end{cases}$$

are all positive.

Then the trivial solution of composite system (5.8.2) is **uniformly asymptotically stable**.

Proof. Given v_i of hypothesis (i), let $\alpha_i > 0$, $i = 1, \ldots, l$, be arbitrary constants and choose as a Lyapunov function for system (5.8.2),

$$v(\psi) = \sum_{i=1}^{l} \alpha_i v_i(\psi^i)$$

for $\psi^T = [(\psi^1)^T, \ldots, (\psi^l)^T]$ and $\|\psi^i\| < H_i$. As before, there exist functions $\varphi_1, \varphi_2 \in K$ such that

$$\varphi_1(\|\psi\|) \le \sum_{i=1}^{l} \alpha_i \varphi_{i1}(\|\psi^i\|) \le v(x) \le \sum_{i=1}^{l} \alpha_i \varphi_{i2}(\|\psi^i\|) \le \varphi_2(\|x\|).$$

$$(5.8.5)$$

Moreover, the derivative $Dv(\psi)$ along solutions of Eq. (5.8.2) can be estimated as (see Yoshizawa [1, pp. 186–189])

$$
\begin{aligned}
Dv(\psi) &= \sum_{i=1}^{l} \alpha_i \, Dv(\psi^i) \\
&\leq \sum_{i=1}^{l} \alpha_i \left\{ -c_i \varphi_{i3}(\|\psi^i\|) + L_i \left| \sum_{j=1,\, i \neq j}^{l} g_{ij}(\psi^j) \right| \right\} \\
&\leq \sum_{i=1}^{l} \alpha_i \left\{ -c_i \varphi_{i3}(\|\psi^i\|) + \sum_{j=1,\, i \neq j}^{l} L_i k_{ij}\, \varphi_{j3}(\|\psi^j\|) \right\} \\
&= -\alpha^{\mathrm{T}} S u,
\end{aligned}
$$

where $\alpha^{\mathrm{T}} = (\alpha_1, \ldots, \alpha_l)$, $u^{\mathrm{T}} = (\varphi_{13}(\|\psi^1\|), \ldots, \varphi_{l3}(\|\psi^l\|))$, and S is the test matrix given in hypothesis (iii). The results in Chapter 2 on M-matrices imply that α can be picked in such a way that $\alpha^{\mathrm{T}} S > 0$ (see in particular the proof of Theorem 2.5.15). Thus, there is a function $\varphi_3 \in K$ such that

$$
Dv(\psi) \leq -\alpha^{\mathrm{T}} S u \leq -\varphi_3(\|\psi\|). \tag{5.8.6}
$$

It now follows from inequalities (5.8.5) and (5.8.6) (and from the remark following Definition 5.8.3) that the trivial solution of composite system (5.8.2) is uniformly asymptotically stable. ■

5.8.7. Definition. Isolated subsystem $(\mathcal{S}_i)$ is said to possess **Property F2** if there is a continuous function $v_i(\psi_i)$, three functions $\varphi_{i1}, \varphi_{i2}, \varphi_{i3} \in K$, and positive constants c_i and L_i, such that the inequalities

(i) $\varphi_{i1}(|\psi^i(0)|) \leq v_i(\psi^i) \leq \varphi_{i2}(\|\psi^i\|)$,
(ii) $Dv_{i(\mathcal{S}_i)}(\psi^i) \leq -c_i \varphi_{i3}(|\psi^i(0)|)$, and
(iii) $|v_i(\psi_1^i) - v_i(\psi_2^i)| \leq L_i \|\psi_1^i - \psi_2^i\|$

hold for all $\psi_1^i, \psi_2^i, \psi^i$ in a neighborhood $\|\psi^i\| < H_i$ of $C_r^{n_i}$.

If isolated subsystem $(\mathcal{S}_i)$ possesses Property F2, then the trivial solution of $(\mathcal{S}_i)$ is uniformly asymptotically stable (see Yoshizawa [1, pp. 189–192]). (When rephrased for the space C_r^n, Property F2 implies the uniform asymptotic stability of composite system (5.8.2).)

5.8.8. Definition. The interconnections g_{ij} in Eq. (5.8.1) are said to be **delay free** if there are functions G_{ij} such that

$$
g_{ij}(\psi^j) = G_{ij}(\psi^j(0)),
$$

$i, j = 1, \ldots, l$, $i \neq j$, for all $\|\psi^j\| < H_j$.

5.8.9. Theorem. The trivial solution of composite system (5.8.2) with decomposition (5.8.1) is **uniformly asymptotically stable** if the following conditions hold.

(i)   Each isolated subsystem ($\mathscr{S}_i$) possesses Property F2;
(ii)   all interconnections g_{ij} are delay free and satisfy the inequalities

$$|g_{ij}(\psi^j)| = |G_{ij}(\psi^j(0))| \le k_{ij}\,\varphi_{j3}(|\psi^j(0)|)$$

for $\|\psi^j\| < H_j$; and
(iii)   all successive principal minors of the test matrix $S = [s_{ij}]$ defined by

$$s_{ij} = \begin{cases} c_i, & \text{if } i = j \\ -L_i k_{ij}, & \text{if } i \ne j \end{cases}$$

are positive.

Proof. The proof is essentially the same as the proof of Theorem 5.8.4, once Theorem 5.5.2 is noted. ∎

5.9   Application of Comparison Theorems to Vector Lyapunov Functions

Comparison theorems can be applied to vector Lyapunov functions in the qualitative analysis of dyanmical systems defined on infinite-dimensional spaces. Since these comparison theorems can be formulated to make use of the **maximum principle for parabolic partial differential equations,** their use has the potential to yield significant improvement over the scalar and vector Lyapunov theory outlined thus far. However, at this time it is not yet clear whether the full power of the theory involving this maximum principle can be used to advantage in the qualitative theory of interconnected systems. Indeed, a good deal of research will be required in this area. For this reason we treat this subject very briefly. For further information on the maximum principle for parabolic partial differential equations, refer to Matrosov [1, 2], Walter [1, especially p. 269], and Lakshmikantham and Leela [1, 2].

In the present section the notation established in Section 5.5, Part D, is used. Let G be a bounded open subset of R^n with smooth boundary ∂G. Given any twice continuously differentiable function $u: R^+ \times G \to R^l$, with components u^k, we define u_x^k and u_{xx}^k as

$$u_x^k = (u_{x_1}^k, u_{x_2}^k, \ldots, u_{x_n}^k), \qquad u_x = (u_x^1, \ldots, u_x^l)$$

and

$$u_{xx}^k = [u_{x_i x_j}^k], \quad i, j = 1, \ldots, n, \qquad u_{xx} = [u_{xx}^k], \quad k = 1, \ldots, l,$$

where $u_{x_i}^k$ denotes the first partial derivative of u^k with respect to x_i while $u_{x_i x_i}^k$ is the second partial derivative of u^k with respect to x_i. For a vector

Lyapunov function $v(x, u(x))$ the v_x^k and v_{xx}^k are similarly defined via the chain rule. We now consider a partial differential operator

$$T(w) = w_t - F(x, w, w_x, w_{xx}) \qquad (5.9.1)$$

where $x \in G$ and $w \in R^l$ while F has components $F_i(x, w, w_x^i, w_{xx}^i)$ mapping $G \times R^l \times R^n \times R^{n^2}$ into R.

5.9.2. Definition. The function $F(x, w, w_x, w_{xx})$ is called **elliptic** in G if for any $x \in G$, any $w \in R^l$, any $P \in R^n$, and any symmetric matrices R and Q, where $R - Q$ is positive semidefinite, each component F_i of F satisfies the inequality

$$F_i(x, w, P, Q) \le F_i(x, w, P, R).$$

The operator T in Eq. (5.9.1) is called **parabolic** if F is elliptic.

Notice that these definitions differ from the earlier definitions of strongly elliptic and parabolic, given in Section 5.5, Part D.

We now consider the class of problems characterized by the set of equations

$$
\begin{aligned}
u_t &= F(x, u, u_x, u_{xx}) &&\text{for } (t, x) \in R^+ \times G \\
u(0, x) &= \psi_1(x) &&\text{for } x \in G, \quad \psi_1 \in C(\bar{G}) \qquad (5.9.3) \\
u(t, x) &= 0 &&\text{for } (t, x) \in R^+ \times \partial G.
\end{aligned}
$$

We also consider the ordinary differential equation

$$\dot{y} = g(y), \qquad y(0) = y_0 \qquad (5.9.4)$$

where $g: R^l \to R^l$ and g is sufficiently smooth for existence and uniqueness of solutions of (5.9.4). For the proof of the following comparison theorems and further details, see Lakshmikantham and Leela [2, Chapter 10].

5.9.5. Theorem. Assume that $\varphi_1, \varphi_2 \in K$, that $u(t, x)$ solves Eq. (5.9.3), that $y(t)$ solves Eq. (5.9.4), and that the following hypotheses hold.

  (i) F and H are defined and continuous on $G \times R^l \times R^{l \cdot n} \times R^{l \cdot n^2}$ into R^l;
  (ii) H is elliptic on G, $v: G \times R^l \to R^l$, and

$$Dv = \frac{\partial v}{\partial t} + \left(\frac{\partial v}{\partial u}\right)^{\mathsf{T}} F(x, u, u_x, u_{xx}) \le H(x, v, v_x, v_{xx})$$

where $\partial v / \partial u = [\partial v_i / \partial u_j]$, $i, j = 1, \ldots, l$; and
  (iii) g is quasimonotone, $g(0) = 0$, $H(x, v, 0, 0) \le g(v)$ for all $(x, v) \in G \times R^l$, $v \ge 0$, and $v(x, \psi_1(x)) \le y_0$ for all $x \in G$.

Then $v(x, u(t, x)) \le y(t)$ for all $t \ge 0$. Moreover, if in addition to hypotheses (i)–(iii) we assume that

(iv) $\varphi_1(|v|) \le v(x, v) \le \varphi_2(|v|)$ for all $x \in G$, $v \in R^l$, $v \ge 0$; and
(v)   the trivial solution of Eq. (5.9.4) is uniformly asymptotically stable,

then the trivial solution of Eq. (5.9.3) is **uniformly asymptotically stable** on the space $C(\bar{G})$.

Similar results, as the above, can be proved for the case when the trivial solution of Eq. (5.9.4) is uniformly asymptotically stable in the large or exponentially stable. Moreover, the boundary condition in Eq. (5.9.3) may be replaced by the conditions

$$u(t, x) = 0 \qquad \text{for} \quad x \in \partial G - \partial G_\alpha$$

and

$$\alpha(x) \frac{\partial u}{\partial n}(t, x) = Q(x, u) \qquad \text{on } \partial G_\alpha$$

where $\alpha \in C(\partial G)$, $\partial G_\alpha = \{x \in \partial G : \alpha > 0\}$, n denotes the outer normal to a point $x \in \partial G_\alpha$, and Q is quasimonotone in u. For these boundary conditions, we need in addition to (i), (ii), and (iii) of Theorem 5.9.5, the condition

$$\alpha(x) \frac{\partial u(t, x)}{\partial n} \le Q(x, u(t, x)) - Q(x, y(t)) \qquad \text{for} \quad (t, x) \in R^+ \times \partial G_\alpha.$$

As a simple example, let G_k be a bounded domain in R^n with smooth boundary ∂G_k and define two operators

$$L_k(u) \triangleq \sum_{i, j = 1}^{n} a_{ij}^k \frac{\partial^2 u}{\partial x_i \, \partial x_j} + \sum_{j=1}^{n} b_j^k \frac{\partial u}{\partial x_j}$$

for $k = 1, 2$, where $a_{ij}^k \in R$, $b_j^k \in R$, and L_k is strongly elliptic. Next, consider isolated subsystems described by equations of the form

$$\begin{aligned}
\partial u_k / \partial t &= L_k(u_k) - c_k u_k & &\text{on } R^+ \times G_k \\
u_k(0, x) &= \psi_k(x) & &\text{on } G_k & & (\mathscr{S}_k) \\
u_k(t, x) &= 0 & &\text{on } R^+ \times \partial G_k
\end{aligned}$$

with $\psi_k \in C(\bar{G}_k)$.

We now form an interconnected system by considering interconnecting terms of the form

$$g_k(u_1, u_2) = B_k u_j, \qquad j \ne k \quad \text{and} \quad k, j = 1, 2,$$

where $B_i \in R$. Choosing

$$v_k(u_k) = \tfrac{1}{2}(u_k)^2$$

the derivative of v_k for the resulting composite system can be obtained by the following series of computations.

$$Dv_k = u_k L_k(u_k) - c_k u_k{}^2 + B_k u_k u_j$$

$$= L_k(u_k{}^2/2) - c_k u_k{}^2 + B_k u_k u_j - \sum a_{ij}^k \frac{\partial u_k}{\partial x_i} \frac{\partial u_k}{\partial x_j}$$

$$\leq L_k(v_k) - 2c_k v_k + 2|B_k| \sqrt{v_k} \sqrt{v_j}.$$

Letting $x = (u_1, u_2)^{\mathrm{T}}$ and $v = (v_1, v_2)^{\mathrm{T}}$, we have

$$Dv \leq \left[\begin{array}{c} L_1(v_1) - 2c_1 v_1 + 2|B_1| \sqrt{v_1} \sqrt{v_2} \\ L_2(v_2) - 2c_2 v_2 + 2|B_2| \sqrt{v_1} \sqrt{v_2} \end{array} \right].$$

In accordance with Theorem 5.9.5, we can take

$$g(v) = -2 \left[\begin{array}{c} c_1 v_1 - |B_1| \sqrt{v_1} \sqrt{v_2} \\ -|B_2| \sqrt{v_1} \sqrt{v_2} + c_2 v_2 \end{array} \right].$$

For this g the transformation $w_i = \sqrt{v_i}$, $i = 1, 2$, linearizes $\dot{v} = g(v)$. The trivial solution of Eq. (5.9.4) is asymptotically stable if $c_1 > 0$, $c_2 > 0$, and $c_1 c_2 > |B_1| |B_2|$.

The same analysis works if one or both of the boundary conditions $u_k(t, x) = 0$ on ∂G_k is replaced by

$$\frac{\partial u_k(t, x)}{\partial n} + \sigma_k u_k = 0 \qquad \text{on } \partial G_k,$$

where $\sigma_k > 0$ is a constant and n is the outward normal on ∂G_k.

5.10   Notes and References

For the proofs of the results given in Section 5.2 and for additional results on linear semigroups refer to the book by Hille and Phillips [1] which is encyclopedic. Refer also to the book by Krein [1, Chapter 1] which has an especially good exposition from the point of view of abstract differential equations. Consult also Pazy's notes [1] which have many useful results including material on differentiable and on compact semigroups not found elsewhere in book form.

For a more complete discussion and for proofs of the results given in Section 5.3, see Crandall [1], Crandall and Liggett [1], Brezis [1, 2], and Kurtz [1]. For further results on existence and nonexistence of solutions of initial value problems (N), when A is continuous, see Godunov [1] where it is shown that for any given infinite-dimensional space X there is a continuous $A(x)$ and an initial condition x_0 such that (N) has no solution. Also, see Lasota and Yorke [1] where it is shown that for "most $A(x)$ and most x_0" there must be a solution.

General references on stability results presented in Section 5.4 include Hahn [2], Yoshizawa [1], and Krasovskii [1]. Some interesting material on stability theory in a Banach space setting can also be found in Lakshmikantham [1], Massera [1], and Massera and Schaeffer [1].

Webb [1] studied functional differential equations as quasicontinuous semigroups. Proofs and additional results on the Lyapunov theory for functional differential equations (Section 5.5, Part B) can be found in Driver [1] and in the books by Yoshizawa [1], Hale [2], and Krasovskii [1]. Hale [3, 4] studied functional differential equations with infinite delay when X is a fading memory space.

The results on Volterra integrodifferential equations in Section 5.5,. Part C, are based on the references by Barbu and Grossman [1], Burns and Herdman [1], and Miller [1].

For more information and examples of well posed linear partial differential equations (see Section 5.5, Part D), refer to Hille and Phillips [1], Krein [1, Chapter 1, Section 8], and Pazy [1, Chapter 5]. For nonlinear problems described by partial differential equations, see Crandall [1], Crandall and Liggett [1], Brezis [1, 2], and Kurtz [1]. For examples and hints on how to construct Lyapunov functions for partial differential equations, see Infante and Walker [1], Walker [1], Chaffee and Infante [1], Matrosov [2], Sirazetdinov [1], Zubov [1, Chapter 5], Wang [1, 2], Pao [1], and Plant and Infante [1]. Some results on combining Lyapunov functions and the idea of invariant sets can be found in Walker and Infante [1] and Defermos and Slemrod [1].

For general references on stochastic processes see Doob [1] and Wong [1]. For stability results of systems described by stochastic differential equations and stochastic difference equations (Section 5.5, Part E) see Kushner [1–3], Arnold [1], and Kozin [1].

The stability results for interconnected systems in Section 5.6 and the examples in Section 5.7 are expansions and modifications of work reported by Rasmussen and Michel [2, 4] and Rasmussen [1]. Matrosov [2] has surveyed some related literature. Theorem 5.6.2 is good enough for many applications, however, it would be interesting to obtain a theorem of this type which allows a mixed group of quasicontractive subsystems and linear C_0 subsystems. Note that Theorems 5.6.1 and 5.6.2 are stability theorems for composite system $(\mathcal{S})$ in the sense that well posedness is a weak type of stability.

The results in Section 5.8 are modifications of work by Michel [5, 7]. Many other modifications are possible for composite systems described by functional differential equations, along the lines of the results of Chapter II. Specifically, global asymptotic stability, exponential stability, instability, boundedness, as well as results for time varying systems can readily be established. Note also that Theorem 5.8.4 can be proved for a general dynamical system in a Banach space.

CHAPTER VI

Input–Output Stability
of Large Scale Systems

In the present chapter we consider the input–output stability and instability of time-varying nonlinear interconnected systems. As in the preceding chapters the objective is still to analyze large scale systems in terms of their lower order subsystems and in terms of their interconnecting structure.

By the term input–output stability we mean boundedness and/or continuity of the relations (describing a dynamical system) which connect system inputs to system outputs. Here we note that the motivation for input–output stability has its roots in systems engineering, where frequently systems are viewed as "black boxes" (i.e., relations) which connect input signals to output signals. In this context, stable systems are those for which bounded input signals result in bounded output signals, and/or small changes in input signals result in small changes in output signals.

Of particular interest in the present chapter are stability results in the setting of L_2-space, l_2-space, L_∞-space, and l_∞-space. Wherever possible, we emphasize results which allow analysis based on graphical methods.

The first section contains required background material. Specifically, we establish in this section the necessary notation and recall several useful general results from systems theory. In the second and third sections we present the main stability and instability results, respectively, in a rather general form. In the remainder of this chapter, as well as in the next chapter, we establish several special stability and instability results and we consider specific examples to illustrate applications and usefulness of the method of analysis advanced. Specifically, in the fourth section we combine a special transformation with some theorems from Section 6.2 to obtain several specific useful results involving sector and conicity conditions in a Hilbert space setting. The fifth section contains Popov-type results for composite dynamical systems while in the sixth section stability results for systems defined on L_∞- and l_∞-spaces are presented. Finally, in the seventh section a design and compensation procedure, using the viewpoint of treating dynamical systems in terms of lower order subsystems and interconnecting structure, is proposed. The chapter is concluded in the eighth section with a discussion of references.

6.1 Preliminaries

The present section consists of five parts. First we present several classes of useful function spaces. Next, we discuss properties of relations on extended spaces. Then we consider an important class of nonlinear operators on extended spaces. This is followed by a discussion of an important class of linear operators on extended spaces. The section is concluded with the statement of two types of input–output stability theorems. The first type involves gains of relations (small gain theorem) while the second involves sector conditions.

A. Extended Function Spaces

In this chapter we again let $L_p = L_p(R^+, R)$ denote the **space of Lebesgue measurable functions** $f: R^+ \to R$ with norm

$$\|f\|_{L_p} = \left\{ \int_0^\infty |f(t)|^p \, dt \right\}^{1/p} < \infty, \qquad 1 \le p < \infty,$$

while $L_p{}^m = L_p(R^+, R^m)$ is the corresponding space of vector valued functions $f: R^+ \to R^m$ with norm $\|f\|_{L_p}$. The spaces L_∞ and $L_\infty{}^m$ are defined in a similar manner (refer to Section 5.1).

The **truncation** of a function $f \in L_p$ or a function $f \in L_p{}^m$ at time T is denoted by f_T and is defined by

$$f_T(t) = \begin{cases} f(t), & \text{if } t \le T \\ 0, & \text{if } t > T. \end{cases}$$

The **extended space** L_{pe}^m is defined as

$$L_{pe}^m = \{f: R^+ \to R^m : f_T \in L_p^m \text{ for all } T > 0\}.$$

Given $f \in L_{pe}^m$, we let

$$\|f\|_T = \|f_T\|_{L_p}.$$

Note that $\|f\|_T = 0$ does not mean that $f(t) \equiv 0$ on R^+ since in this case $|f(t)|$ can be nonzero for $t > T$. Therefore, the function $\|\cdot\|_T$ is a **pseudonorm** or **seminorm** (see Hille and Phillips [1, p. 137]) rather than a norm.

Let C denote the **space of all continuous functions** $f: R^+ \to R^m$ and let $BC = C \cap L_\infty^m$ denote the **subspace of bounded and continuous functions**. If we put

$$\|f\|_T = \sup\{|f(t)|: 0 \le t \le T\}$$

for any $f \in C$ and

$$\|g\|_\infty = \sup\{|g(t)|: 0 \le t < \infty\}$$

for $g \in BC$, then it is clear that C can be thought of as an extension of BC. Other subspaces of C are also of interest. For example, the **space of continuous functions which have a finite limit at infinity**, C_l, is defined as

$$C_l = \{f \in C : f(t) \to f(\infty) \text{ as } t \to \infty, f(\infty) \in R^m\};$$

the **space of continuous functions which tend to zero**, C_0, is given by

$$C_0 = \{f \in C_l : f(\infty) = 0\};$$

the **space of asymptotically** (or **ultimately**) **periodic functions**, A_ω, is specified by

$$A_\omega = \{f = p + e: e \in C_0, p \in C, \text{ and } p(t+\omega) = p(t) \text{ for all } t \in R\};$$

or, the **space of asymptotically** (or **ultimately**) **almost periodic functions**, AP, is defined as

$$AP = \{f = p + e: e \in C_0, p \in C, \text{ and } p \text{ is almost periodic}\}$$

(see Besicovitch [1] or Fink [1] for a discussion of almost periodic functions). All of these spaces have the same norm as BC, namely $\|\cdot\|_\infty$.

Also of interest will be **weighted spaces**. Let X be any of the spaces L_p^m, BC, C_l, C_0, or A_ω and let $g \in C$ be a given positive scalar function. Then the weighted space X_g consists of all $f: R^+ \to R^m$ such that $f/g \in X$. In particular, let f/g be defined by $(f/g)(t) = f(t)/g(t)$ and let

$$(L_p^m)_g = \{f \in L_{pe}^m : f/g \in L_p^m\}$$

with norm

$$\|f\|_g = \left\{\int_0^\infty |f(t)/g(t)|^p \, dt\right\}^{1/p}. \tag{6.1.1}$$

Also,

$$(BC)_g = \{f \in C : f/g \in BC\}$$

with norm

$$\|f\|_g = \sup\{|f(t)|/g(t): 0 \le t < \infty\}.$$

Weighted spaces for the cases when $X = C_l$, C_0, or A_ω are defined similarly.

The natural extended space for the spaces $(BC)_g$, $(C_l)_g$, $(C_0)_g$, or $(A_\omega)_g$ is the space $(C)_g$, where $(C)_g$ is the space C with the new seminorms defined by

$$\|f\|_T = \sup\{|f(t)|/g(t): 0 \le t \le T\}.$$

The natural extended space for $(L_p{}^m)_g$ is the space L_{pe}^m with the new seminorms defined by

$$\|f\|_T = \left\{ \int_0^T [|f(t)|/g(t)]^p \, dt \right\}^{1/p}. \tag{6.1.2}$$

Note that in the above, $g(t) \equiv 1$ is allowed and that for this special case, $X_g = X$. Also note that for $p = 2$, $(L_p{}^m)_g$ has a **Hilbert space structure**, i.e., both the seminorms (6.1.2) and the norm (6.1.1) are induced by

$$\langle f_1, f_2 \rangle_T \triangleq \int_0^T f_1(t)^{\mathrm{T}} f_2(t) g(t)^{-2} \, dt$$

$$\langle h_1, h_2 \rangle_g \triangleq \int_0^\infty h_1(t)^{\mathrm{T}} h_2(t) g(t)^{-2} \, dt$$

while $(\|f\|_T)^2 = \langle f,f \rangle_T$ and $(\|f\|_g)^2 = \langle f,f \rangle_g$. In the special case where $g = e^a$ is defined by

$$g(t) = \exp(at)$$

for some real number a, $(L_2{}^m)_g$ is the well-known **shifted L_2-space**.

Next, let S_m denote the space of all R^m-sequences $\{a_n : n = 0, 1, 2, \ldots\} = \{a_n\} \subset R^m$. Then $l_p{}^m$ is the subset of S_m for which the norm, defined by

$$\|\{a_n\}\|_{l_p} = \left\{ \sum_{n=0}^\infty |a_n|^p \right\}^{1/p}$$

is finite. Let $g = \{g_n\}$ be a given real valued sequence with $g_n > 0$ for all n. We now define $(l_p{}^m)_g$ as the set of elements of S_m for which

$$\|\{a_n\}\|_g = \left\{ \sum_{n=0}^\infty [|a_n|/g_n]^p \right\}^{1/p} < \infty.$$

The extended space of $(l_p{}^m)_g$ is S_m with seminorms given by

$$\|\{a_n\}\|_T = \left\{ \sum_{n=0}^T [|a_n|/g_n]^p \right\}^{1/p}$$

for $T = 0, 1, 2, \ldots$. Similarly, we define c_l as

$$c_l = \left\{ \{a_n\} \in S_m : \lim_{n \to \infty} a_n = a_\infty \text{ exists as } n \to \infty, \ a_\infty \in R^m \right\}$$

and c_0 as

$$c_0 = \{\{a_n\} \in c_l : a_\infty = 0\}$$

with norm given by

$$\|\{a_n\}\|_\infty = \sup\{|a_n| : n = 0, 1, 2, \ldots\}.$$

The extended space for c_l and for c_0 is S_m with seminorms

$$\|\{a_n\}\|_T = \max\{|a_n| : n = 0, 1, 2, \ldots, T\}.$$

The spaces $(c_0)_g$ and $(c_l)_g$ can now be defined in an obvious way.

From the above discussion it is clear that the number of possible spaces which one may want to consider is rather large. This enables us to construct particular spaces to meet specific needs in various applications. For example, the space of measurable, asymptotically periodic functions

$$AM_\omega = \{f = p + e : e \in L_p^m, \ p \in L_p[0, \omega], \text{ and } p(t + \omega) = p(t)\}$$

is sometimes useful in applications, particularly when $p = 1, 2$. We also remark that the idea of weighted spaces can be generalized by replacing the scalar function $g(t)$ by a matrix weight function $G(t)$ (see Gollwitzer [1] and Dotseth [1]).

Using the spaces discussed thus far as motivation, we characterize an extended space as follows.

6.1.3. Definition. An **extended space** X_e is a linear space of functions $f : R^+ \to R^m$ (or sequences $f : \{0, 1, 2, \ldots\} \to R^m$) and a family of seminorms $\{\|\cdot\|_T\}$ for any $T \in R^+$ (or any $T \in \{0, 1, 2, \ldots\}$) such that
  (i) the set $(X_e)_T \triangleq \{f_T : f \in X_e\}$ is a Banach space under $\|\cdot\|_T$, and
  (ii) for each $f \in X_e$, $\|f\|_T$ is a nondecreasing function of T.

Note that for all spaces discussed above we have the following property.

6.1.4. Definition. A subspace X of an extended space X_e is called a **completely compatible subspace** of X_e if the norm $\|\cdot\|$ of X is determined from the seminorms of X_e by the relation

$$\|f\| = \lim_{T \to \infty} \|f\|_T.$$

These definitions allow for the possibility of mixing the basic extended spaces L_{pe}^m and C. For example, X_e could represent the four-dimensional vector valued functions whose coordinates $(f_1^T, f_2^T, f_3^T, f_4^T) \in X_e =$

$(L_{pe}^m)_g \times (L_{qe}^m)_h \times C \times C$. If $X_1 = (L_p^m)_g$, $X_2 = (L_q^m)_h$, $X_3 = BC$, and $X_4 = C_0$, then $X = X_1 \times X_2 \times X_3 \times X_4$ is a completely compatible subspace of X_e.

The spaces BC, C_l, C_0, A_ω, and AP could all be thought of as subspaces of L_{pe}^m for some $p \in [1, \infty]$, while the space $X = L_p^m \cap BC$ with norm

$$\|f\| = \sup\{|f(t)| : t \in R^+\} + \left\{\int_0^\infty |f(t)|^p\, dt\right\}^{1/p}$$

is a subspace of L_{pe}^m or of C. These subspaces are not completely compatible but they do possess the following property.

6.1.5. Definition. A subspace X of an extended space X_e is said to be **compatible** if the condition

$$\lim_{n \to \infty} \|f_n - f\|_X = 0$$

for f_n and f in X, implies that

$$\lim_{n \to \infty} \|f_n - f\|_T = 0$$

for all $T > 0$.

B. Relations on Extended Function Spaces

The ideas of a **multivalued function** H and of a **relation** H were discussed in Section 5.5 and we shall use the notation and definitions introduced there. In particular, if H is a relation on an extended space X_e (i.e., H is a subset of the product space $X_e \times X_e$) and if $(x, y) \in H$, we write $x \in D(H)$ and $y \in Hx \subset Ra(H)$. We often will need to consider the inverse of H, denoted by H^{-1} (H^{-1} always exists) and we often find it convenient to identify H with its graph.

A relation H defined on X_e is called **causal** if

$$(Hx)_T = (Hx_T)_T$$

for all $T > 0$ and all $x \in X_e$.

A relation H is **time invariant** if it commutes with all time delays and **memoryless** if the value of Hx at time t depends only on the value of x at time t.

Given a relation H on X_e and given a subspace X, we let $S(H)$ denote the **stable manifold** of H, defined as

$$S(H) = \{x \in X : Hx \subset X\}$$

and we say that H is **bounded with respect to** X (or for short, **bounded**) if $S(H) = X$.

The **conditional gain** of H (if it exists) is defined by

$$g_c(H) = \inf\{M : \|Hx\|_T \le M \|x\|_T \text{ for all } x \in S(H) \text{ and all } T > 0\}.$$

If H is bounded, then the **gain** of H (if it exists) is defined by

$$g(H) = \inf\{M: \|Hx\|_T \le M\|x\|_T \text{ for all } x \in X_e \text{ and all } T > 0\}.$$

In this case $S(H) = X$ and $g_c(H) = g(H)$.

We call a relation H on X_e **interior conic** (c, r) if there are real numbers $r \ge 0$ and c such that $\|y - cx\|_T \le r\|x\|_T$ for all $T > 0$, all $x \in X_e$, and all $(x, y) \in H$. In this case H is bounded and has gain $g(H) \le \max\{|c+r|, |c-r|\}$. The numbers c and r are called the **center** and **radius** of H, respectively. Similarly, a relation H is called **exterior conic** (c, r) if the inequality above is replaced by $\|y - cx\|_T \ge r\|x\|_T$ for all $T > 0$, all $x \in X_e$ and all $(x, y) \in H$.

Next, let us assume that the extended space X_e has Hilbert space structure. We say that a relation H is **inside the sector** $\{a, b\}$ if for all $(x, y) \in H$ and all $T > 0$, the inequality

$$\langle y - ax, y - bx \rangle_T \le 0$$

holds, where a and b are real constants. We say that H is **outside the sector** $\{a, b\}$ if for all $(x, y) \in H$ and $T > 0$, the inequality

$$\langle y - ax, y - bx \rangle_T \ge 0$$

is satisfied. The relation H is said to be **positive** if $\langle x, y \rangle_T \ge 0$ for all $(x, y) \in H$ and all $T > 0$. In these definitions no particular ordering of a and b is implied. Thus, the statement "H is inside the sector $\{a, b\}$" is equivalent to the statement "H is inside the sector $\{b, a\}$." If in particular $c = (a+b)/2$ and $r = |b-a|/2$, it is easy to see that H is inside (outside) the sector $\{a, b\}$ if and only if it is interior (exterior) conic (c, r).

It will sometimes be necessary to consider relations H with domain in one extended space X_{1e} and range in a second extended space X_{2e} (i.e., H is a subset of $X_{1e} \times X_{2e}$). In this case we will assume X_1 and X_2 are compatible or completely compatible subspaces. The **gain** of such a relation H is defined by

$$g(H) = \inf\{M: \|Hx\|_T \le M\|x\|_T \text{ for all } T > 0 \text{ and all } x \in X_{1e}\}.$$

Moreover, H is called **bounded with respect to** (X_1, X_2) if $Hx \in X_2$ whenever $x \in X_1$.

In order to discuss the continuity of dynamical systems, we introduce the corresponding incremental definitions, but only for relations which are single valued (i.e., H must be an operator). Thus, the **conditional incremental gain**, $\hat{g}_c(H)$, is the smallest $M \ge 0$ such that $\|Hx - Hy\|_T \le M\|x - y\|_T$ for all $T > 0$ and all x, y in the stable manifold $S(H)$. If in particular $S(H) = X$, then $\hat{g}_c(H) = \hat{g}(H)$ is the **incremental gain**. Also, H is said to be **incrementally interior conic** (c, r) if $r \ge 0$, $c \in R$, and $\|(Hx - Hy) - c(x - y)\|_T \le r\|x - y\|_T$ for all $T > 0$ and all $x, y \in X_e$. Similarly, H is said to be **incrementally exterior conic** (c, r) if $r \ge 0$, $c \in R$, and $\|(Hx - Hy) - c(x - y)\|_T \ge r\|x - y\|_T$ for all

$T > 0$ and all $x, y \in X_e$. If X_e has Hilbert space structure, we define relation H to be **incrementally inside** (or **outside**) **of the sector** $\{a, b\}$ if

$$\langle (Hx - Hy) - a(x - y), (Hx - Hy) - b(x - y) \rangle_T \leq 0$$

(or ≥ 0) for all $x, y \in X_e$ and all $T > 0$. In this case we also define H to be **incrementally positive** if $\langle x - y, Hx - Hy \rangle_T \geq 0$ for all $x, y \in X_e$ and all $T > 0$. Note that if $H0 = 0$ and H satisfies any of the incremental definitions, then it will also satisfy the corresponding nonincremental definitions.

Now let $H = [H_{ij}]$ be a matrix such that H_{ij} is a relation on X_{je} to X_{ie} (i.e., H_{ij} is a subset of $X_{je} \times X_{ie}$) with gain $g(H_{ij}) < \infty$. We define $G(H) = [g(H_{ij})]$ as the matrix of these gains. Similarly, we define $\hat{G}(H) = [g(H_{ij})]$ as the matrix of incremental gains.

C. A Class of Nonlinear Operators

Next, we characterize an important class of operators.

6.1.6. Definition. Let $\mathcal{N}$ denote the class of operators from X_e into X_e such that $\bar{N} \in \mathcal{N}$ means $(\bar{N}x)(t) \triangleq N(x(t), t)$ for all $x \in X_e$ and all $t \geq 0$, $N(0, t) = 0$ for all $t \geq 0$, and there is a constant $F \geq 0$ such that

$$|N(y, t)| \leq F|y| \tag{6.1.7}$$

for all $y \in R^m$ and all $t \geq 0$. (Subsequently we write N in place of $\bar{N}$.)

Any operator $N \in \mathcal{N}$ is causal and memoryless. It is time invariant if and only if the function $N(x, t)$ is time invariant and linear if and only if $N(x, t)$ is linear. If in particular $X = L_p{}^m$, BC, or C_0, then the gain with respect to X satisfies $g(N) \leq F$. If $X = C_l$ and N is time invariant, then N is bounded on X and $g(N) \leq F$.

We call $N \in \mathcal{N}$ **interior conic** (c, r) if $|N(x, t) - cx| \leq r|x|$ for all $x \in R^m$ and all $t \geq 0$, **exterior conic** (c, r) if $|N(x, t) - cx| \geq r|x|$ for all $x \in R^m$ and all $t \geq 0$, and **positive** if $x^T N(x, t) \geq 0$ for all $x \in R^m$ and all $t \geq 0$.

Replacing inequality (6.1.7) by the inequality

$$|N(x, t) - N(y, t)| \leq F|x - y|$$

for all $x, y \in R^m$ and all $t \geq 0$, we define and characterize the incremental counterpart of class $\mathcal{N}$, denoted by $\hat{\mathcal{N}}$, in an obvious way.

For the discrete time case we define the class of operators η by replacing $t \geq 0$ by $j = 0, 1, 2, \dots, N$ by n, and inequality (6.1.7) by

$$|n(x, j)| \leq F|x|$$

for all $x \in R^m$ and all $j = 0, 1, 2, \dots$. The incremental counterpart of class η, denoted by $\hat{\eta}$, is characterized by making obvious modifications.

D. A Class of Linear Operators

We now consider a large class of linear, causal, time invariant operators defined on various extended spaces X_e. We present sufficient conditions for boundedness of these operators and establish estimates of the gain of these operators defined on various spaces X_e.

6.1.8. Definition. Let $\mathcal{L}_e$ denote the set of all operators H of the following form. There is a sequence $\{h_j\}$ of $m \times m$ real matrices, a sequence of real numbers $\{t_j\}$, $0 = t_0 < t_1 < t_2 < \cdots < t_n \to \infty$ as $n \to \infty$, and an $m \times m$ matrix valued function $h(t) = [h_{ij}(t)]$ such that all $h_{ij} \in L_{1e}$ and such that for all $x \in X_e$,

$$Hx(t) = \sum_{j=0}^{\infty} h_j x(t - t_j) + \int_{-\infty}^{t} h(t - \tau) x(\tau)\, d\tau. \tag{6.1.9}$$

6.1.10. Definition. The set $\mathcal{L}$ is that subset of $\mathcal{L}_e$ such that

$$\sum_{j=0}^{\infty} |h_j| < \infty \quad \text{and} \quad \int_0^{\infty} |h(t)|\, dt < \infty. \tag{6.1.11}$$

The set $\mathcal{L}_{1e}$ is that subset of $\mathcal{L}_e$ such that $h_i = 0$ for $i > 0$ and $\mathcal{L}_{0e}$ is the subset of $\mathcal{L}_{1e}$ such that $h_0 = 0$. Also, $\mathcal{L}_1$ is defined as $\mathcal{L}_{1e} \cap \mathcal{L}$ and $\mathcal{L}_0$ is defined as $\mathcal{L}_0 = \mathcal{L} \cap \mathcal{L}_{0e}$.

The subscripts in $\mathcal{L}_1$ and $\mathcal{L}_0$ refer to the number of elements h_j which are allowed to be nonzero.

It will be convenient to assume that the functions x in any extended space X_e are defined as $x(t) = 0$ for all $t < 0$. The function $h(t)$ in Eq. (6.1.9) will also be taken as zero for all $t < 0$. With this proviso, the infinite sum in Eq. (6.1.9) reduces to a finite sum at any particular time $t \geq 0$ and is thus always defined. Furthermore, in this case the integral in Eq. (6.1.9) can be written as

$$\int_0^t h(t - \tau) x(\tau)\, d\tau = \int_{-\infty}^{\infty} h(t - \tau) x(\tau)\, d\tau$$

since $x(\tau) = 0$ if $\tau < 0$ and $h(t - \tau) = 0$ if $\tau > t$.

6.1.12. Theorem. If $x \in C$ and $H \in \mathcal{L}_{0e}$ then $Hx \in C$. If $x \in L_{pe}^m$ and $H \in \mathcal{L}_e$, then $Hx \in L_{pe}^m$.

Proof. Both results are known when $H \in \mathcal{L}_{0e}$ (see, e.g., Miller [1, pp. 165–167, 273]). To finish the proof assume that $x \in L_{pe}^m$ and $H \in \mathcal{L}_e$ with

$$Hx(t) = \sum_{j=0}^{\infty} h_j x(t - t_j) + \int_0^t h(t - \tau) x(\tau)\, d\tau.$$

In view of the first sentence of the proof we can now take $h(t) \equiv 0$ so that only the sum $\sum_{j=0}^{\infty} h_j x(t-t_j)$ is left to consider. Since $t_j \to \infty$ as $j \to \infty$ then on any finite interval $0 \le t \le T$ only a finite number of terms of the infinite sum are nonzero. Since each term is in $L_p([0,T], R^m)$, so is the finite sum and $\|Hx\|_T \le (\sum_{j=0}^{\infty} |h_j|) \|x\|_T$. ■

The above argument can also be used to prove the next result.

6.1.13. Theorem. If $H \in \mathscr{L}$ and $1 \le p \le \infty$, then H is bounded on L_p^m and

$$g(H) \le \sum_{j=0}^{\infty} |h_j| + \int_0^\infty |h(t)| \, dt.$$

The proof of the next theorem can be found in Corduneanu [2, Section 2.7] or in Miller [1, Chapter V].

6.1.14. Theorem. If $H \in \mathscr{L}_1$ and $X = BC$, C_l, C_0, A_ω, or AP, then H is bounded on X and satisfies the estimate

$$g(H) \le |h_0| + \int_0^\infty |h(t)| \, dt.$$

The next result is a special case of a theorem due to Corduneanu (see Corduneanu [2, Section 2.7] or Miller [1, pp. 261–262]).

6.1.15. Theorem. If $H \in \mathscr{L}_{0e}$ and X_1 is $(BC)_g$, and X_2 is the corresponding space $(BC)_G$, then H is bounded with respect to (X_1, X_2) if and only if there is a constant $M \ge 0$ such that

$$\int_0^t |h(t-\tau)| g(\tau) \, d\tau \le MG(t)$$

for all $t \ge 0$. In this case the gain of H satisfies the estimate $g(H) \le M$.

As a consequence of the above theorem, we have the following result.

6.1.16. Corollary. If $H \in \mathscr{L}_{1e}$ and if X is $(BC)_g$, then H is bounded on X if and only if there is a constant $M \ge 0$ such that

$$\int_0^t |h(t-\tau)| g(\tau) \, d\tau \le Mg(t)$$

for all $t \ge 0$. In this case $g(H) \le |h_0| + M$.

Proof. Since $h_0 x(t) \in X$ whenever $x \in X$ and since $|h_0 x(t)| \le |h_0| |x(t)|$, it is only necessary to apply Theorem 6.1.15 to the integral part of operator H. ■

The next result follows from the above corollary.

6.1.17. Corollary. If $h = H \in \mathscr{L}_{0e}$ and if $X = BC$, C_0, or C_l, then H is bounded on X if and only if $|h| \in L_1$. In this case $g(H) \le \|h\|_{L_1}$.

Proof. The proof follows from Corollary 6.1.16 with $g(t) \equiv 1$.   ∎

The arguments used in Corduneanu [2, Section 2.2] can be used to prove the next result.

6.1.18. Theorem. If $X = A_\omega$ for some $\omega > 0$ or $X = AP$ and if $H \in \mathscr{L}_1$, then H is bounded on X and $g(H) \leq |h_0| + \int_0^\infty |h(t)| \, dt$.

Little is known about when an operator $H \in \mathscr{L}_e$ is bounded on $(L_p{}^m)_g$, except for the shifted case, i.e., except when $g(t) = \exp(at)$ for some $a \in R$. Fortunately, the shifted L_2-case is of most interest in applications. Let $H^*(s)$ denote the **Laplace transform** of H, i.e.,

$$H^*(s) = \sum_{i=0}^\infty h_i \exp(-st_i) + \int_0^\infty h(t) \exp(-st) \, dt,$$

where $s = \sigma + j\omega$, $j = \sqrt{-1}$, and $\sigma, \omega \in R$. This representation is guaranteed to converge to the continuous function $H^*(s)$, when $\operatorname{Re} s = \sigma \geq a$ and $H_1 = e^{-a}H \in \mathscr{L}$, and may possibly be extendable to other complex values of s by analytic continuation. For $s = j\omega$ and $-\infty < \omega < \infty$, the function $H_1{}^*(j\omega) = H^*(a + j\omega)$ is essentially the **Fourier transform** or **frequency response** of the operator H_1. The expression

$$\bar{H}(j\omega) = \operatorname{Re} H^*(j\omega) + j\omega [\operatorname{Im} H^*(j\omega)]$$

is called the **modified frequency response** of H. The graph of $\{H^*(j\omega): -\infty < \omega < \infty\}$ in the complex plane is called the **Nyquist plot** of H. For the next three well-known results refer to Willems [1] and Desoer and Vidyasagar [1].

6.1.19. Theorem. If $X = (L_2)_g$, where $g = e^a$, and if $e^{-a}H \in \mathscr{L}$, then H is bounded on X and $g(H) = \operatorname{ess\,sup}_{\omega \in R^+} |H^*(a + j\omega)|$.

6.1.20. Theorem. If $X = (L_2{}^m)_g$, where $g = e^a$, and if $e^{-a}H \in \mathscr{L}$, then H is bounded on X and $g(H) = \operatorname{ess\,sup}_{\omega \in R^+} \lambda_M [H^*(a + j\omega)^\mathsf{T} H^*(a + j\omega)]^{1/2}$.

6.1.21. Theorem. If $e^{-a}H \in \mathscr{L}$, if $X = (L_2)_g$, if $X_e = (L_{2e})_g$ for $g = e^a$, and if $H_1 = e^{-a}H$, then

   (i) H is interior conic (c, r) if $|H_1{}^*(j\omega) - c| \leq r$ for all $\omega \in R^+$,
   (ii) H is exterior conic (c, r) if $|H_1{}^*(j\omega) - c| \geq r$ for all $\omega \in R^+$, and
   (iii) H is positive if $\operatorname{Re} H_1{}^*(j\omega) \geq 0$ for all $\omega \in R^+$.

Since H is a linear operator, it is not necessary to state separate incremental forms of any of the preceding definitions or theorems. For $m > 1$ a version of Theorem 6.1.21 can also be stated though it is not very useful, since in this case simple graphical interpretation is no longer possible. For example, if

$m > 1$, H is interior conic (c, r) on $(L_2{}^m)_g$ with $g = e^a$ if

$$\overline{[H_1{}^*(j\omega) - cI]^{\mathrm{T}} [H_1{}^*(j\omega) - cI]} - r^2 I$$

is a negative semidefinite $m \times m$ matrix for all $\omega \in R^+$.

We also wish to point out that if $H_1 = e^{-a}H$ with $a \neq 0$, then H_1 is called the **shifted operator**. Likewise, $H_1{}^*(j\omega) = H^*(a + j\omega)$ is called the **shifted Fourier transform**, the graph $H_1{}^*(j\omega)$, $-\infty < \omega < \infty$, is termed the **shifted Nyquist plot**, and so on.

6.1.22. Definition. An operator $H \in \mathcal{L}_e$ is said to have **decaying L_1-memory** if there is a nonnegative, nonincreasing function $m_1 \in L_1$ such that

$$|Hx(t)|^2 \leq \int_0^t |x(\tau)|^2 m_1(t - \tau)\, d\tau$$

for all $t \geq 0$ and all $x \in L_{2e}^m$.

For operators with this property we have the following result.

6.1.23. Theorem. If $a > 0$ and if $H \in \mathcal{L}_{0e}$ with $e^a h \in L_2{}^m$, then H has decaying L_1-memory.

Proof. By the Schwarz inequality we have

$$|Hx(t)|^2 = \left| \int_0^t h(t - \tau) x(\tau)\, d\tau \right|^2$$

$$\leq \int_0^t |x(\tau)|^2 \exp(-a(t - \tau))\, d\tau \int_0^t |h(t - \tau)|^2 \exp(a(t - \tau))\, d\tau$$

$$= \int_0^t |x(\tau)|^2 \exp(-a(t - \tau))\, d\tau \int_0^t |h(\tau)|^2 \exp(a\tau)\, d\tau$$

$$\leq \int_0^t |x(\tau)|^2 m_1(t - \tau)\, d\tau$$

where $m_1(t) = m_1(0) \exp(-at)$ and

$$m_1(0) = \int_0^\infty |h(\tau)|^2 \exp(a\tau)\, d\tau < \infty. \quad \blacksquare$$

We also have the following theorem.

6.1.24. Theorem. If H has decaying L_1-memory with memory function m_1 and if $x(t)$ satisfies

$$\exp(-ct) \int_0^t |x(\tau)|^2 \exp(c\tau)\, d\tau \leq a$$

for some constants $c > 0$, and $a > 0$, then for all $t \geq 0$

$$|Hx(t)|^2 \leq a\left[c\int_0^\infty m_1(\tau)\,d\tau + m_1(0)\right].$$

Proof. Let $J(t) = \int_0^t |x(\tau)|^2 \exp(-c(t-\tau))\,d\tau$, that is $(dJ(t)/dt) + cJ(t) = |x(t)|^2$ and $J(0) = 0$. Then $J(t) \leq a$ and

$$|Hx(t)|^2 \leq \int_0^t |x(\tau)|^2 m_1(t-\tau)\,d\tau = \int_0^t \left[\frac{dJ(\tau)}{d\tau} + cJ(\tau)\right]m_1(t-\tau)\,d\tau.$$

Using integration by parts we see that

$$|Hx(t)|^2 \leq c\int_0^t J(\tau)\,m_1(t-\tau)\,d\tau + J(\tau)\,m_1(t-\tau)\big|_0^t + \int_0^t J(\tau)\,dm_1(t-\tau)$$

$$\leq c\int_0^t J(\tau)\,m_1(t-\tau)\,d\tau + J(t)\,m_1(0)$$

$$\leq a\left[c\int_0^\infty m_1(\tau)\,d\tau + m_1(0)\right]. \quad\blacksquare$$

For shifted L_p spaces we have the following theorem.

6.1.25. Theorem. If $X = (L_p^m)_g$ and $g(t) = \exp(at)$ for some p, $1 \leq p \leq \infty$, and some real number a, if $H \in \mathcal{L}_{0e}$ and if $e^{-a}H \in \mathcal{L}_0$, then H is bounded on X with gain

$$g(H) \leq \int_0^\infty |h(t)|\,e^{-at}\,dt.$$

Proof. If $x \in X$ then $e^{-a}x \in L_p$. Let F be defined as

$$F(t) = \exp(-at)\int_0^t h(t-\tau)\,x(\tau)\,d\tau$$

$$= \int_0^t \{h(t-\tau)\exp(-a(t-\tau))\}\,x(\tau)\exp(-a\tau)\,dt.$$

By Theorem 6.1.13 it follows that $F \in L_p$ and

$$\|F\|_{L_p} \leq \|e^{-a}h\|_{L_1}\|e^{-a}x\|_{L_p}. \quad\blacksquare$$

Analogous to the above theorems we can also develop results for bounded operators for discrete time systems. In this case $\mathcal{L}_e$ is replaced by $l_e = S_{m\times m}$, the space of $m \times m$ matrix valued sequences, and l is the space $l_p^{m^2}$ for $p = 1$.

6.1.26. Definition. Let l_e be the space of all $m \times m$ matrix valued sequences $H = \{h_j\} = \{h_0, h_1, h_2, \ldots\}$. Given any sequence $\{x_j\} \subset S_m$, we define

$$(Hx)_j = \sum_{k=0}^j h_{j-k}\,x_k.$$

The following results are straightforward modifications of the results stated above for the continuous time case.

6.1.27. Theorem. If $H \in l$ and if $X = l_p^m$ for $1 \le p \le \infty$, or $X = c_l$, or $X = c_0$, then H is bounded on X and has gain

$$g(H) \le \sum_{i=0}^{\infty} |h_i|.$$

Next, for $H \in l_e$ let $H_1 = e^{-a}H = \{h_j e^{-aj}\}$. If $H_1 \in l$, we define the z-**transform** of H_1 as

$$H_1(z) = \sum_{j=0}^{\infty} (h_j e^{-aj}) z^{-j}.$$

This representation converges to a continuous function of z when $|z| \ge 1$. As is well known, when $X = (l_2^m)_g$, $m = 1$, $g = e^a$, and $H_1 = e^{-a}H \in l$, then H is bounded on X with gain

$$g(H) \le \max_{|z|=1} |H_1(z)|.$$

Note that $a = 0$ or $a < 0$ is allowed.

6.1.28. Theorem. If $H \in l$ and $X_e = l_{2e}^m$, then

    (i) H is interior conic (c, r) if $|H(z) - c| \le r$ for $|z| = 1$,
    (ii) H is exterior conic (c, r) if $|H(z) - c| \ge r$ for $|z| = 1$, and
    (iii) H is positive if $\operatorname{Re} H(z) \ge 0$ for $|z| = 1$.

As before, in the case of shifted operators, H is replaced by $H_1 = e^{-a}H$. Also, since the operators $H \in l$ are linear, it is not necessary to state separate incremental forms of the preceding definitions and theorems.

6.1.29. Definition. A relation $H \in l_e$ is said to have **decaying l_1-memory** if there is a nonnegative, nonincreasing sequence $M = \{m_j\} \in l_1$ such that

$$|(Hx)_j|^2 \le \sum_{k=0}^{j} m_{j-k} |x_j|^2, \qquad j = 0, 1, 2, \dots,$$

for all $\{x_j\} \in S_m$.

As in the continuous time case, if $e^a H \in l_2$ and if $a > 0$, then H has decaying l_1-memory. If H has decaying l_1-memory, if $c > 0$, and if there exists $a > 0$ such that

$$e^{-cj} \sum_{k=0}^{j} |x_j|^2 e^{cj} \le a$$

then

$$|(Hx)_j|^2 \le a \left[c \sum_{j=0}^{\infty} m_j + m_0 \right].$$

E. The Small Gain and the Sector Theorems

Let us now consider physical systems which may be represented by the familiar block diagram of Fig. 6.1 and which may be represented by equations of the form

$$e_1 = (a_1 x + w_1) + y_2, \qquad y_1 = H_1 e_1$$
$$e_2 = (a_2 x + w_2) + y_1, \qquad y_2 = H_2 e_2. \tag{6.1.30}$$

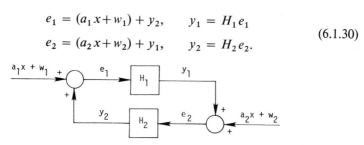

Figure 6.1   Block diagram of a single loop feedback system.

We call x an **input**, w_1 and w_2 **reference signals** (in the case of linear systems w_1 and w_2 are called **biases**), we call e_1 and e_2 **errors** and y_1 and y_2 **outputs**. For system (6.1.30), X_e is some fixed extended space and X is a completely compatible subspace. Also, H_1 and H_2 are two given relations (not necessarily operators) on X_e, a_1 and a_2 are real constants, $x \in X_e$, $w_1, w_2 \in X$, $e_1, e_2 \in X_e$, and $y_1, y_2 \in X_e$. Furthermore, for system (6.1.30), there are relations E_1, E_2, Y_1, and Y_2 on X_e which are defined by the errors and outputs corresponding to a given input, so that $E_i x = e_i$ and $Y_i x = y_i$, $i = 1, 2$.

Subsequently, we shall avoid all questions of existence by assuming that solutions for system (6.1.30) belonging to X_e exist. Also, we shall often avoid the question of uniqueness of solutions by the use of operators instead of relations.

The next result is called the **small gain theorem**.

6.1.31. Theorem. If $g(H_1) g(H_2) < 1$, then the relations E_1, E_2, Y_1, and Y_2 associated with system (6.1.30) are bounded.

The next theorem is called the **incremental form** of the small gain theorem.

6.1.32. Theorem. If $\hat{g}(H_1) \hat{g}(H_2) < 1$, then the relations E_1, E_2, Y_1, and Y_2 associated with system (6.1.30) are all continuous with respect to X.

When X_e has Hilbert space structure, the following result, called the **sector theorem** can be proved.

6.1.33. Theorem. Let γ and $\mathscr{E}$ be real constants, one positive and the other nonnegative, such that
(i) $-H_2$ is inside of the sector $\{a+\gamma, b-\gamma\}$, where $b > 0$, and
(ii) H_1 satisfies one of the following conditions:

(a) if $a > 0$, H_1 is outside the sector $\{-a^{-1} - \mathscr{E}, -b^{-1} + \mathscr{E}\}$,

(b) if $a < 0$, H_1 is inside the sector $\{-b^{-1} + \mathscr{E}, -a^{-1} - \mathscr{E}\}$,

(c) if $a = 0$, $H_1 + (b^{-1} - \mathscr{E})I$ is positive and if $\gamma = 0$, then also $g(H_1) < \infty$.

Then the relations E_1, E_2, Y_1, and Y_2 associated with system (6.1.30) are all bounded.

In closing we remark that the incremental form of Theorem 6.1.33 is also true.

6.2 Stability of Large Scale Systems: Results Involving Gains

In this section we establish sufficient conditions for input–output stability of a class of large-scale systems. All results are phrased in terms of gains of relations, and as in previous chapters, all results make use of test matrices. We consider two sets of results. The first of these is concerned with general multi input–multi output systems. In the second set of results we view certain classes of multi input–multi output systems as interconnected systems.

First we need to establish some additional notation and prove a preliminary result.

Let X_{ie} be a given extended space with completely compatible subspace X_i for $i = 1, ..., l$. Let $X_e = X_{1e} \times \cdots \times X_{le}$, let $X = X_1 \times \cdots \times X_l$, and let $M = [M_{ij}]$ denote an $l \times l$ matrix relation where M_{ij} is a relation on $X_{je} \times X_{ie}$ (i.e., M_{ij} is a subset of $X_{je} \times X_{ie}$). We use $\|x\| = (\sum_{i=1}^{l} \|x_i\|_i^2)^{1/2}$ as a norm on X, or any other convenient standard norm. As before, if a matrix A has only real eigenvalues, then $\lambda_M(A)$ denotes the largest eigenvalue of A.

6.2.1. Theorem. If $M = [M_{ij}]$ is a relation on X_e, if $g(M_{ij}) < \infty$ for all $i, j = 1, 2, ..., l$, and if $G(M) \triangleq [g(M_{ij})]$, then

$$g(M) \leq (\lambda_M[G(M)^\mathsf{T} G(M)])^{1/2}.$$

Proof. Fix $T > 0$, $x \in X_e$, and let $y = Mx$, i.e.,

$$y_i = \sum_{j=1}^{l} M_{ij} x_j,$$

$i = 1, ..., l$. Then

$$\|y_i\|_T \leq \sum_{j=1}^{l} \|M_{ij} x_j\|_T \leq \sum_{j=1}^{l} g(M_{ij}) \|x_j\|_T,$$

$i = 1, ..., l$. Now let $X_T = (\|x_1\|_T, ..., \|x_l\|_T)^\mathsf{T}$, $Y_T = (\|y_1\|_T, ..., \|y_l\|_T)^\mathsf{T}$. Then the above inequality assumes the form $Y_T \leq G(M) X_T$. We now have

$$(Y_T)^\mathsf{T} Y_T \leq (X_T)^\mathsf{T} G(M)^\mathsf{T} G(M) X_T.$$

But since $(Y_T)^T Y_T = (\|y\|_T)^2$ and $(X_T)^T X_T = (\|x\|_T)^2$, it follows that

$$(\|y\|_T)^2 \leq (X_T)^T G(M)^T G(M) X_T \leq \lambda_M [G(M)^T G(M)](\|x\|_T)^2. \quad \blacksquare$$

Let us now consider a **multiple input–multiple output feedback system** governed by the set of functional equations

$$y_i = \sum_{j=1}^{l} H_{ij} e_j, \qquad e_i = x_i + w_i + z_i$$

$$\tag{6.2.2}$$

$$z_i = \sum_{j=1}^{l} B_{ij} f_j, \qquad f_i = u_i + v_i + y_i$$

for $i = 1, \ldots, l$. Here w_i and v_i are fixed reference signals in X_i, x_i, $u_i \in X_{ie}$ are inputs, and B_{ij} and H_{ij} are relations on $X_{je} \times X_{ie}$. The error signals e_i and f_i and the outputs y_i and z_i are assumed to exist and belong to X_{ie}.

Let $x^T = (x_1^T, \ldots, x_l^T)$ with u, w, v, z, y, e, and f defined similarly. Let $X = \mathsf{X}_{i=1}^l X_i$ with X_e defined similarly, and let $H = [H_{ij}]$ and $B = [B_{ij}]$ denote $l \times l$ matrix relations on X_e. Then system (6.2.2) can be expressed equivalently as

$$y = He, \qquad e = x + w + z$$

$$z = Bf, \qquad f = u + v + y,$$

which is clearly of the same form as system (6.1.30). Finally, let E_i, F_i, Y_i, and Z_i denote the relations associating given inputs x, $u \in X_e$ with errors e_i, f_i and outputs y_i, z_i, respectively, satisfying Eq. (6.2.2).

According to the small gain theorem (Theorem 6.1.31) and the estimate in Theorem 6.2.1, system (6.2.2) is input–output stable (i.e., the relations E_i, F_i, Y_i, Z_i, $i = 1, \ldots, l$, are all bounded) if $g(B)g(H) < 1$, i.e., if

$$\lambda_M [G(H)^T G(B)^T G(B) G(H)] < 1.$$

Similarly, if

$$\lambda_M [\hat{G}(H)^T \hat{G}(B)^T \hat{G}(B) \hat{G}(H)] < 1,$$

then system (6.2.2) is continuous. In the present section we obtain different and hopefully more readily applicable boundedness conditions and continuity conditions.

Our first result is as follows.

6.2.3. Theorem. If all the gains $g(H_{ij})$ and $g(B_{ij})$ are finite and if the successive principal minors of the $l \times l$ test matrix

$$R = I - G(B) G(H)$$

are positive (I denotes the $l \times l$ identity matrix), then all the relations E_i, F_i, Y_i, and Z_i associated with system (6.2.2) are bounded.

Proof. From Eq. (6.2.2) it follows immediately that

$$e_i = x_i + w_i + \sum_{j=1}^{l} B_{ij}\left(u_j + v_j + \sum_{k=1}^{l} H_{jk} e_k\right).$$ (6.2.4)

Now for any $T > 0$ we have

$$\|e_i\|_T \le \|x_i\|_T + \|w_i\|_T + \sum_{j=1}^{l} g(B_{ij})\left[\|u_j\|_T + \|v_j\|_T + \sum_{k=1}^{l} g(H_{jk})\|e_k\|_T\right]$$

so that

$$\|e_i\|_T - \sum_{k=1}^{l}\left[\sum_{j=1}^{l} g(B_{ij}) g(H_{jk})\right]\|e_k\|_T \le \|x_i\|_T + \|w_i\|_T$$

$$+ \sum_{j=1}^{l} g(B_{ij})[\|u_j\|_T + \|v_j\|_T].$$

Define $E_T = (\|e_1\|_T, ..., \|e_l\|_T)^\mathsf{T}$ and define X_T, W_T, U_T, and V_T analogously. Then the above inequality can be expressed as

$$\{I - G(B) G(H)\} E_T \le X_T + W_T + G(B)[U_T + V_T].$$

Since the off-diagonal terms of matrix $R = I - G(B) G(H)$ are all nonpositive and since by assumption, all the principal minors of R are positive, it follows that R is an M-matrix, and furthermore, it follows from Theorem 2.5.2, that $R^{-1} = \Delta$ exists and that all elements of R^{-1} are nonnegative. Thus,

$$E_T \le \Delta[X_T + W_T + G(B)(U_T + V_T)].$$

Recalling that X is a completely compatible subspace and letting $T \to \infty$, we have

$$\|e\| \le \|\Delta\| \, [\|x\| + \|w\| + \|G(B)\| (\|u\| + \|v\|)] < \infty.$$

The last inequality and Eq. (6.2.2) imply that $\|y\| \le g(H)\|e\| < \infty$, $\|f\| \le \|u\| + \|v\| + \|y\| < \infty$, and $\|z\| \le g(B)\|f\| < \infty$. ∎

By interchanging the roles of B and H in the previous theorem we can prove the following result.

6.2.5. Theorem. If all of the gains $g(B_{ij})$ and $g(H_{ij})$ are finite and if the successive principal minors of the test matrix $R = I - G(H) G(B)$ are positive, then all the relations E_i, F_i, Y_i, and Z_i associated with system (6.2.2) are bounded.

Our next result has essentially the same proof.

6.2.6. Theorem. If all gains $g(B_{ij})$ and $g(H_{ij})$ are finite, if either $u_i + v_i = 0$ for $i = 1, ..., l$ or B_{ij} is linear for $i, j = 1, ..., l$, and if the successive principal minors of the test matrix

$$S = I - G(BH)$$

are positive, then all the relations E_i, F_i, Y_i, and Z_i associated with system (6.2.2) are bounded.

Proof. From the hypotheses and from Eq. (6.2.4) it follows that

$$e_i = x_i + w_i + \sum_{j=1}^{l} B_{ij}(u_j + v_j) + \sum_{k=1}^{l}\left(\sum_{j=1}^{l} B_{ij} H_{jk} e_k\right)$$

for $i = 1, ..., l$. Thus, for any $T > 0$ we have

$$[I - G(BH)]\,E_T \leq X_T + W_T + G(B)[U_T + V_T],$$

as in the proof of Theorem 6.2.3. Since all the off-diagonal terms of matrix S are nonpositive and since all principal minors of matrix S are positive, it follows that S is an M-matrix. Therefore, from Theorem 2.5.2 it follows that $S^{-1} \triangleq \Delta$ exists and that $\Delta \geq 0$. Therefore

$$E_T \leq \Delta(X_T + W_T) + \Delta G(B)(U_T + V_T).$$

To complete the proof we let $T \to \infty$ and proceed as in the proof of Theorem 6.2.3. ∎

Interchanging the roles of B and H in Theorem 6.2.6, we obtain the following result.

6.2.7. Theorem. If all gains $g(B_{ij})$ and $g(H_{ij})$ are finite, if either $x_i + w_i = 0$ for $i = 1, ..., l$, or if H_{ij} is linear for $i, j = 1, ..., l$, and if the successive principal minors of the test matrix $S = I - G(HB)$ are positive, then all of the relations E_i, F_i, Y_i, and Z_i associated with system (6.2.2) are bounded.

We now show that the hypotheses in Theorem 6.2.3 imply that all eigenvalues λ of $G(B)G(H)$ satisfy the condition $|\lambda| < 1$.

6.2.8. Lemma. If the hypotheses of Theorem 6.2.3 are true, then any (possibly complex) eigenvalue λ of the matrix $M = G(B)G(H)$ must satisfy the condition $|\lambda| < 1$.

Proof. Since $M \geq 0$, it is known (see, e.g., Gantmacher [2, p. 66]) that M has a dominant eigenvalue λ_0 which is real, positive, and simple with associated eigenvector $v \geq 0$ such that $|v| = 1$. Now since λ_0 is dominant, $\lambda_0 > |\lambda|$ for any other eigenvalue λ of M. Also, since $\Delta = (I - M)^{-1}$ exists, $\lambda_0 \neq 1$. Let $r = \lambda_0 - 1 \neq 0$. Then $(\lambda_0 I - M)v = 0$, $v \geq 0$, and $v \neq 0$ imply that $(I - M)v + rv = 0$, or $v = -r\Delta v \geq 0$. Thus, r must be negative. ∎

Thus, the hypotheses of Theorem 6.2.3 and the hypothesis

$$\lambda_M[G(H)^T G(B)^T G(B) G(H)] < 1$$

needed to apply the small gain theorem are evidently similar. However, these two stability conditions are apparently not equivalent, nor is one a special case of the other.

The type of condition assumed in Theorem 6.2.3 appears to offer several advantages over other existing results. For example, this condition leads to a linear constraint on the margin of boundedness parameter (which plays a similar role as the degree of stability in previous chapters) to be defined in Section 6.4, Part C, while the small gain condition

$$\lambda_M[G(H)^\mathrm{T} G(B)^\mathrm{T} G(B) G(H)] < 1$$

leads to a corresponding quadratic constraint. The above remarks are also applicable to the matrices $G(H)G(B)$, $G(BH)$, or $G(HB)$, encountered in Theorems 6.2.5–6.2.7.

The stability results considered thus far (Theorems 6.2.3, 6.2.5, 6.2.6, and 6.2.7) have incremental analogs.

6.2.9. Theorem. Let all the incremental gains $\hat{g}(B_{ij})$ and $\hat{g}(H_{ij})$ be finite, let the successive principal minors of the test matrix R be positive, and assume one of the following conditions:

    (i) $R = I - \hat{G}(B)\hat{G}(H)$,
   (ii) $R = I - \hat{G}(H)\hat{G}(B)$,
  (iii) either $u_i + v_i = 0$ for all i or $B = [B_{ij}]$ is linear and $R = I - \hat{g}(BH)$,
  (iv) either $x_i + w_i = 0$ for all i or $H = [H_{ij}]$ is linear and $R = I - \hat{g}(HB)$.

Then all of the input–output relations E_i, F_i, Y_i, and Z_i associated with Eq. (6.2.2) are continuous.

Proof. We outline the proof of part (i). The proofs of the remaining parts are similar. Consider Eq. (6.2.4) for two sets of inputs x, $u \in X_e$ and $\bar{x}$, $\bar{u} \in X_e$ with $\delta x \triangleq x - \bar{x} \in X$ and $\delta u = u - \bar{u} \in X$. Let e and $\bar{e}$ be solutions of Eq. (6.2.4) corresponding to those inputs and let $\delta e \triangleq e - \bar{e}$. In order to estimate δe we subtract the two copies of Eq. (6.2.4) and take seminorms to obtain

$$\|\delta e_i\|_T \leq \|\delta x_i\|_T + \sum_{j=1}^{l} \hat{g}(B_{ij}) \left[\|\delta u_j\|_T + \sum_{k=1}^{l} \hat{g}(H_{jk}) \|\delta e_k\|_T \right].$$

The rest of the proof follows the proof of Theorem 6.2.3, and we obtain

$$\|\delta e\| \leq \|\Delta\| (\|\delta x\| + \|\hat{G}(B)\| \|\delta u\|) < \infty,$$

$$\|\delta y\| \leq \hat{g}(H) \|\delta e\| < \infty,$$

$$\|\delta f\| \leq \|\delta u\| + \|\delta y\| < \infty,$$

and

$$\|\delta z\| \leq \hat{g}(B) \|\delta f\| < \infty. \quad \blacksquare$$

Next, we consider the special case of system (6.2.2) where $H_{ij} = 0$ for all

$i \neq j$. Relabeling $H_{ii} = H_i$, we can rewrite Eq. (6.2.2) as

$$z_i = \sum_{j=1}^{l} B_{ij} f_j, \qquad f_i = u_i + v_i + y_i$$

$$y_i = H_i e_i, \qquad e_i = x_i + w_i + z_i \qquad (\mathscr{S})$$

for $i = 1, \ldots, l$.

System $(\mathscr{S})$ may be viewed as an interconnection of l **isolated subsystems** described by equations of the form

$$z_i = B_{ii} f_i, \qquad f_i = u_i + v_i + y_i$$

$$y_i = H_i e_i, \qquad e_i = x_i + w_i + z_i, \qquad (\mathscr{S}_i)$$

$i = 1, \ldots, l$, with **interconnecting structure** specified by relations B_{ij} for all $i \neq j$. The isolated feedback system $(\mathscr{S}_i)$ and the **interconnected system** $(\mathscr{S})$ are depicted in block diagram form in Figs. 6.2a and 6.2b, respectively.

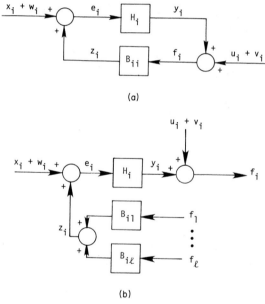

(a)

(b)

Figure 6.2 (a) Block diagram of isolated feedback system $(\mathscr{S}_i)$. (b) Block diagram of interconnected system $(\mathscr{S})$ where $i = 1, \ldots, l$.

Alternatively, we may regard $(\mathscr{S})$ as an interconnection of l isolated subsystems described by equations of the form

$$y_i = H_i e_i, \qquad (\mathrm{H}_i)$$

$i = 1, ..., l$, with interconnecting structure specified by all of the B_{ij}, and speak of **isolated subsystem** (H_i). For either case, the following additional comments are appropriate in the present context. In general, the term $u_i + v_i$ may include external inputs as well as disturbances. If in particular $u_i + v_i$ does not contain external input terms, then f_i represents the *available* output of subsystem H_i under additive disturbances. Furthermore, in applications, the assumption $u_i = v_i = 0$ is frequently valid. In many of our subsequent results, we will address ourselves to this special case.

Since all of the Theorems 6.2.3 and 6.2.5–6.2.7 can directly be applied to composite system ($\mathscr{S}$), the following results are seen to be true.

6.2.10. Theorem. If the gains $g(H_i)$ and $g(B_{ij})$ are finite for $i, j = 1, ..., l$ and if the successive principal minors of the test matrix $R = [r_{ij}]$ defined by

$$r_{ij} = \begin{cases} 1 - g(B_{ii}) g(H_i), & \text{if } i = j \\ -g(B_{ij}) g(H_j), & \text{if } i \neq j \end{cases}$$

are all positive, then all of the input–output relations E_i, F_i, Y_i, and Z_i associated with system ($\mathscr{S}$) are bounded.

6.2.11. Theorem. If all the gains $g(H_i)$ and $g(B_{ij})$ are finite, if either $u_i + v_i = 0$ for all i or $B = [B_{ij}]$ is linear, and if the successive principal minors of the test matrix R defined by

$$r_{ij} = \begin{cases} 1 - g(B_{ii} H_i), & \text{if } i = j \\ -g(B_{ij} H_j), & \text{if } i \neq j \end{cases}$$

are all positive, then all of the input–output relations E_i, F_i, Y_i, and Z_i associated with system ($\mathscr{S}$) are bounded.

In view of Theorem 6.2.9 the following result is also true for system ($\mathscr{S}$).

6.2.12. Theorem. Suppose that all of the incremental gains $\hat{g}(H_i)$ and $\hat{g}(B_{ij})$ are finite, suppose the successive principal minors of the test matrix $R = [r_{ij}]$ are all positive, and suppose that one of the following conditions hold:

(i)
$$r_{ij} = \begin{cases} 1 - \hat{g}(B_{ii}) \hat{g}(H_i), & i = j \\ -\hat{g}(B_{ij}) \hat{g}(H_j), & i \neq j \end{cases}$$

(ii)   either $u + v = (u_1^T, ..., u_l^T)^T + (v_1^T, ..., v_l^T)^T = 0$ or $B = [B_{ij}]$ is linear, and

$$r_{ij} = \begin{cases} 1 - \hat{g}(B_{ii} H_i), & i = j \\ -\hat{g}(B_{ij} H_j), & i \neq j \end{cases}$$

(iii)   either $x + w = (x_1^T, ..., x_l^T)^T + (w_1^T, ..., w_l^T)^T = 0$ or $H = \text{diag}[H_i]$ is linear, and

$$r_{ij} = \begin{cases} 1 - \hat{g}(H_i B_{ii}), & i = j \\ -\hat{g}(H_i B_{ij}), & i \neq j. \end{cases}$$

Then all the input–output relations E_i, F_i, Y_i, and Z_i associated with system $(\mathscr{S})$ are continuous.

6.2.13. Remark. If in Theorems 6.2.3, 6.2.5–6.2.7, and 6.2.9–6.2.12, the gains (incremental gains) are replaced by upper bounds for the corresponding gains (incremental gains), and if the altered test matrix still has positive successive principal minors, then all of the above theorems remain true.

Next, we observe that Theorem 6.2.10 can be written in the following altered form.

6.2.14. Theorem. If all gains $g(H_i)$ and $g(B_{ij})$ are finite, if $g(H_i) > 0$ for all i, and if the successive principal minors of the test matrix $S = [s_{ij}]$ defined by

$$s_{ij} = \begin{cases} g(H_i)^{-1} - g(B_{ii}), & \text{if } i = j \\ -g(B_{ij}), & \text{if } i \neq j \end{cases}$$

are positive, then all of the input–output relations E_i, F_i, Y_i, and Z_i of system $(\mathscr{S})$ are bounded. If all gains are replaced by incremental gains, then the above relations are continuous.

Proof. The crucial point in the proof of Theorem 6.2.3 is to show that $\Delta = R^{-1} = (I - G(B)G(H))^{-1}$ exists and is nonnegative. Since

$$R = I - G(B)G(H) = [G(H)^{-1} - G(B)]G(H) \triangleq SG(H)$$

then $\Delta = R^{-1} = G(H)^{-1}S^{-1}$. Now

$$G(H)^{-1} = \text{diag}[g(H_1)^{-1}, ..., g(H_l)^{-1}] > 0$$

and the hypotheses of this theorem ensure (see Theorem 2.5.2) that $S^{-1} \geq 0$. Thus, $\Delta \geq 0$ and the proof of Theorem 6.2.3 applies.

The proof of the incremental form is essentially the same.   ■

The results and proofs given in the present section are intended to serve as models. More important than these specific theorems is the method of analysis developed. Thus, in specific applications it may be necessary to establish new results, using the present ones as motivation. Frequently, preliminary work (e.g., appropriate transformations of the system in question) can be used to great advantage before applying the present method. These remarks are also applicable to the instability results given in the next section. In subsequent sections of this chapter, the thrust of the above remarks will become further evident.

6.3 Instability of Large Scale Systems

In the present section we consider once more composite system $(\mathscr{S})$ and isolated subsystems (H_i), $i = 1, ..., l$, encountered in Section 6.2. Throughout this section X is assumed to be a completely compatible subset of an extended space X_e, and X_e is assumed to a possess Hilbert space structure. Then for any $T > 0$, the seminorm $\|\cdot\|_T$ is defined by the inner product $\langle \,,\, \rangle_T$ and

$$\langle x, y \rangle = \lim_{T \to \infty} \langle x, y \rangle_T$$

determines an inner product on X. Given any subspace N of X, we define as usual the orthogonal complement of N as

$$N^{\perp} = \{x \in X : \langle x, y \rangle = 0 \text{ for all } y \in N\}.$$

Let $\bar{N}$ denote the closure of N. Then for any $x \in X$ we can find $x_1 \in \bar{N}$ and $x_2 \in N^{\perp} = (\bar{N})^{\perp}$ such that $x = x_1 + x_2$ and such that

$$\|x\| = \|x_1\| + \|x_2\|$$

(see, e.g., Dunford and Schwartz [1, p. 249, Lemma 4]). Given a square matrix M, we define the **spectral radius** $r(M)$ by

$$r(M) = \max\{|\lambda| : \lambda \text{ is an eigenvalue of } M\}.$$

Let us now assume that at least one subsystem (H_i) is unstable for $i = i_0$ and let us consider composite system $(\mathscr{S})$.

6.3.1. Theorem. Suppose that the gains $g(B_{ij})$ are finite for $i, j = 1, ..., l$ and suppose that on X_{ie} the H_i are continuous operators for $i = 1, ..., l$. Let $N_i = S(H_i)$ be the stable manifold of H_i with $g_c(H_i) < \infty$ for all i and suppose that $N_i^{\perp} \neq \{0\}$ for at least one value $i = i_0$. For $R = [r_{ij}] = [g(B_{ij})g_c(H_j)]$ assume either $r(R) < 1$ or $r(R) = 1$ and $r_{ij} > 0$ for all $i, j = 1, ..., l$. If $u = v = 0$, if $x_i + w_i \in (N_i)^{\perp}$ for all i, and if $x_i + w_i \neq 0$ for at least the one value $i = i_0$, then $y \notin X$. (In this case system $(\mathscr{S})$ is said to be **unstable**, or **input–output unstable**.)

Proof. For purposes of contradiction, assume that $y \in X$, i.e., $y_i \in X_i$ for $i = 1, ..., l$. Since y solves $(\mathscr{S})$, then $y_i = H_i e_i \in X_i$ if and only if $e_i = x_i + w_i + z_i \in N_i$. Then $z_i = e_i - (x_i + w_i)$ where $e_i \in N_i$ and $x_i + w_i \in N_i^{\perp}$. It follows now that $\|z_i\| = \|e_i\| + \|x_i + w_i\|$ and

$$\|e_i\| \leq \|z_i\| = \left\| \sum_{j=1}^{l} B_{ij}(H_j e_j) \right\| \leq \sum_{j=1}^{l} g(B_{ij}) g_c(H_j) \|e_j\| = \sum_{j=1}^{l} r_{ij} \|e_j\| \tag{6.3.2}$$

with strict inequality at least for $i = i_0$.

If $r(R) < 1$, replace $g(B_{ij})g_c(H_j)$ by $g(B_{ij})g_c(H_j)+A$ for all i,j. Choose $A > 0$ and so small that for the new matrix R thus formed, we still have $r(R) < 1$. Now for this alternate R, we have $r_{ij} > 0$ for all i,j. So by the Perron theorem (see Gantmacher [2, p. 53]) there are positive numbers α_i such that $\alpha^T = (\alpha_1, ..., \alpha_l) > 0$ is a row eigenvector corresponding to the dominant eigenvalue $r(R)$ of R. Then

$$\alpha^T R = r(R)\alpha^T$$

or

$$\sum_{i=1}^{l} \alpha_i r_{ij} = r(R)\alpha_j, \qquad j = 1, ..., l. \tag{6.3.3}$$

Since all $\alpha_i > 0$ and since $\|x_i + w_i\| > 0$ at least for $i = i_0$, we have

$$\sum_{i=1}^{l} \alpha_i \|e_i\| < \sum_{i=1}^{l} \alpha_i \|z_i\|.$$

This inequality and (6.3.2) and (6.3.3) imply that

$$\sum_{i=1}^{l} \alpha_i \|e_i\| < \sum_{i=1}^{l} \alpha_i \sum_{k=1}^{l} r_{ik} \|e_k\| = \sum_{k=1}^{l} \left(\sum_{i=1}^{l} \alpha_i r_{ik} \right) \|e_k\|$$

$$= \sum_{k=1}^{l} r(R)\alpha_k \|e_k\| = r(R) \sum_{k=1}^{l} \alpha_k \|e_k\|,$$

i.e.,

$$r(R) > 1.$$

However, since $r(R) \leq 1$, we have arrived at a contradiction. This proves the theorem.   ∎

In Theorem 6.3.1 it is assumed that when $i = i_0$, $N_i^{\perp} \neq \{0\}$. We note that for some operators H the assumption that H is unstable does not imply that $[S(H)]^{\perp} \neq \{0\}$. This can be seen from the example $X = L_2$ and

$$Hy(t) = \int_0^t y(\tau)\, d\tau, \qquad t \geq 0, \qquad y \in X.$$

This H is clearly unstable but at the same time its stable manifold contains the set $N_0 = \{x_T : x \in X\}$. Since N_0 is dense in L_2, it follows that $[\overline{S(H)}] = L_2 = X$ and $[S(H)]^{\perp} = \{0\}$. The next result will show when the condition that H is unstable implies $[S(H)]^{\perp} \neq \{0\}$.

6.3.4. Lemma.   If $H: X_e \to X_e$ is a continuous linear mapping, if H is unstable on X and if $g_c(H) < \infty$, then $[S(H)]^{\perp} \neq \{0\}$.

   Proof.   We first show that $N \triangleq S(H)$ is closed. In so doing, pick a sequence $\{x_j\} \subset N$ such that $x_j \to x \in X$ as $j \to \infty$. Now H is continuous on X_e means that for any $T > 0$ there is a number $\mu(T) \geq 0$ such that $\|Hx\|_T \leq \mu(T)\|x\|_T$ for all $x \in X_e$. Thus,

$$\|Hx_j - Hx\|_T = \|H(x_j - x)\|_T \leq \mu(T)\|x - x_j\|_T.$$

Since X is completely compatible, it follows that $\|x - x_j\|_T \leq \|x - x_j\| \to 0$ as $j \to \infty$. In particular, pick j such that $\mu(T)\|x - x_j\| < 1$. For this j, $\|Hx\|_T \leq \|Hx_j\| + 1$. But $g_c(H) < \infty$ implies that $\|Hx_j\| \leq g_c(H)\|x_j\| \leq K$ for some K, which works for all j. Thus, $\|Hx\|_T \leq K + 1$ for all $T > 0$. In the limit, as $T \to \infty$, we have $\|Hx\|_T \to \|Hx\| \leq K + 1 < \infty$. Thus, $Hx \in X$ and $x \in S(H)$. Since the sequence $\{x_j\}$ is an arbitrary convergent sequence in $S(H)$ it follows that $S(H)$ is closed.

Finally, since $S(H)$ is closed and H is unstable, there must be nonzero elements in $[S(H)]^{\perp}$. ∎

Note that Lemma 6.2.8 and Theorem 6.3.1 can be combined to yield the following result.

6.3.5. Theorem. Suppose that the gains $g(B_{ij}) < \infty$ for all $i, j = 1, \dots, l$, suppose that $H_i: X_{ie} \to X_{ie}$ is continuous and linear with stable manifold $N_i = S(H_i)$, $i = 1, \dots, l$, such that $N_i^{\perp} \neq \{0\}$ for at least one $i = i_0$, and suppose $g_c(H_i) < \infty$, $i = 1, \dots, l$. Suppose the successive principal minors of the test matrix

$$R = I - G(B)G_c(H)$$

are positive. Then system $(\mathscr{S})$ is unstable with respect to X.

In the remaining results of this section we require the following concepts.

6.3.6. Definition. An operator $H: X_e \to X_e$ is called

(i)   **passive** if for all $x \in X$ such that $Hx \in X$ we have $\langle Hx, x \rangle \geq 0$;
(ii) **properly passive** if for all $x \in X$ such that $x \neq 0$ and $Hx \in X$ we have $\langle Hx, x \rangle > 0$;
(iii) **strictly passive** (with constant δ) if there is a $\delta > 0$ such that when $x \in X$ and $Hx \in X$, we have $\langle Hx, x \rangle \geq \delta \langle x, x \rangle$.

We now prove the following result.

6.3.7. Theorem. Let $g(B_{ij}) < \infty$ for $i \neq j$, $i, j = 1, \dots, l$. Let $-B_{ii}$ be linear, strictly passive operators with constants $\delta_i > 0$, $i = 1, \dots, l$. Assume that for some $i = i_0$, the stable manifold $N_i = S(B_{ii})$ has nonzero orthogonal complement, $N_i^{\perp} \neq \{0\}$. Let H_i be a bounded, properly passive operator such that $H_i 0 = 0$ for $i = 1, \dots, l$. If $u = v = 0$, if $x_i + w_i \in N_i^{\perp}$, $i = 1, \dots, l$, and if $x_i + w_i \neq 0$ when $i = i_0$, and if the successive principal minors of the test matrix R defined by

$$r_{ij} = \begin{cases} \delta_i, & \text{if } i = j \\ -g(B_{ij}), & \text{if } i \neq j \end{cases}$$

are all positive, then any solution e of $(\mathscr{S})$ cannot be in X (and $(\mathscr{S})$ is input–output unstable).

Proof. For purposes of contradiction, assume that $e \in X$, i.e., assume that $e_i \in X_i$, $i = 1, ..., l$. Using the equations describing $(\mathscr{S})$ we have

$$e_i = (x_i + w_i) + \sum_{j=1}^{l} B_{ij} H_j e_j, \qquad i = 1, ..., l. \tag{6.3.8}$$

Since all of the elements in this equation belong to X and since $H_i e_i \in X$, it follows that $H_i e_i \in N_i$. Taking the inner product of Eq. (6.3.8) with $H_i e_i$, we have

$$\langle H_i e_i, e_i \rangle = \langle H_i e_i, x_i + w_i \rangle + \sum_{j=1}^{l} \langle H_i e_i, B_{ij} H_j e_j \rangle.$$

Since $H_i e_i \in N_i$ and since $x_i + w_i \in N_i^{\perp}$, we have $\langle H_i e_i, x_i + w_i \rangle = 0$ and

$$\langle H_i e_i, -B_{ii} H_i e_i \rangle = -\langle H_i e_i, e_i \rangle + \sum_{j=1, i \neq j}^{l} \langle H_i e_i, B_{ij} H_j e_j \rangle.$$

Since H_i is passive and since $-B_{ii}$ is strictly passive, we can estimate

$$\delta_i \| H_i e_i \|^2 \leq -\langle H_i e_i, e_i \rangle + \sum_{j=1, i \neq j}^{l} \langle H_i e_i, B_{ij} H_j e_j \rangle$$

$$\leq \sum_{j=1, i \neq j}^{l} g(B_{ij}) \| H_i e_i \| \| H_j e_j \|$$

or

$$\delta_i \| H_i e_i \| - \sum_{j=1, i \neq j}^{l} g(B_{ij}) \| H_j e_j \| \leq 0,$$

$i = 1, ..., l$. These l inequalities can be written as $RA \leq 0$, where $A^T = [\| H_1 e_1 \|, ..., \| H_l e_l \|] \geq 0$ and $R = [r_{ij}]$ is the given test matrix. The hypotheses of the theorem imply that R^{-1} exists and that $R^{-1} \geq 0$ so that $A \leq 0$. This means $H_i e_i = 0$ for $i = 1, ..., l$ and $\langle H_i e_i, e_i \rangle = 0$. Since H_i is properly passive, we have $e_i = 0$ for $i = 1, ..., l$. This and Eq. (6.3.8) imply that $x_i + w_i = 0$ for $i = 1, ..., l$, a contradiction. This completes the proof.   ∎

The above theorem states that under suitable conditions, if at least one subsystem $(\mathscr{S}_i)$ is unstable, then the composite system $(\mathscr{S})$ is also unstable. In our next result, the instability can occur in a subsystem, in an interconnection term, or in both the subsystem and interconnecting structure.

6.3.9. Theorem. Let $-B = [-B_{ij}]$ be a passive linear operator on X_e. If $N = S(B)$ is the stable manifold of B, let $N^\perp \neq \{0\}$. Let H_i be a bounded, properly passive operator with $H_i 0 = 0$ for $i = 1, \ldots, l$. If $u = v = 0, x + w \in N^\perp$ and $x + w \neq 0$, then $e \notin X$.

Proof. System $(\mathscr{S})$ can be rewritten equivalently as

$$e = x + w + BHe \qquad (6.3.10)$$

where $(He)^T = [H_1 e_1, \ldots, H_l e_l]^T$. For purposes of contradiction we assume that $e \in X$. Then He must belong to N. Thus,

$$\langle He, e \rangle = \langle He, x + w \rangle + \langle He, BHe \rangle = \langle He, BHe \rangle.$$

If $e \neq 0$, then

$$0 < \langle He, e \rangle = \langle He, BHe \rangle \leq 0.$$

Therefore $e = 0$. Then $He = 0$ and it follows from Eq. (6.3.10) that $x + w = 0$, which is a contradiction. This concludes the proof. ∎

Finally, we note some of the conditions which imply that B (in Theorem 6.3.9) is unstable.

(a) For some pair i, j, the entry B_{ij} is unstable.
(b) There are constants α_i such that the column sum $\sum_{i=1}^l \alpha_i B_{ij}$ is unstable.
(c) There are constants β_j such that the row sum $\sum_{j=1}^l B_{ij} \beta_j$ is unstable.
(d) There exist constant matrices C and D such that CBD is unstable.

Note that statement (d) includes the other statements. Moreover, (d) can be verified by noting that if B is stable, then CBD is also stable. Any of the preceding conditions can be used, together with Lemma 6.3.4, to see that $N^\perp \neq \{0\}$.

6.4 Stability of Large Scale Systems: Results Involving Sector and Conicity Conditions

The present section consists of four parts. In the first of these we discuss a special transformation with the property that the stability (i.e., boundedness and continuity) of the transformed system implies the stability of the original system. It will be seen that this transformation widens the applicability of the results of Section 6.2 considerably. In particular, we will be able to formulate stability results for interconnected systems involving sector conditions and conicity conditions, to be introduced in the second part of this section. In the third part we present stability results for composite systems phrased in terms of some useful design parameters, called margin of boundedness and gain factor. The present results are applied to two specific examples in the last part of this section.

A. Useful Transformation

Let us restate composite system ($\mathscr{S}$) first introduced in Section 6.2,

$$z_i = \sum_{j=1}^{l} B_{ij} f_j, \qquad f_i = u_i + v_i + y_i$$

$$\text{(}\mathscr{S}\text{)}$$

$$y_i = H_i e_i, \qquad e_i = x_i + w_i + z_i$$

$i = 1, ..., l$. We now rewrite system ($\mathscr{S}$) in vector-matrix form. Let $B = [B_{ij}]$, $H = [H_i \delta_{ij}]$ where $\delta_{ij} = 0$ if $i \neq j$ and $\delta_{ii} = 1$, $z^\mathrm{T} = (z_1{}^\mathrm{T}, ..., z_l{}^\mathrm{T})$, and $f^\mathrm{T} = (f_1{}^\mathrm{T}, ..., f_l{}^\mathrm{T})$ with u, v, e, y, x, and w similarly defined. Then ($\mathscr{S}$) can be rewritten as

$$z = Bf, \qquad f = u + v + y$$

$$(6.4.1)$$

$$y = He, \qquad e = x + w + z.$$

System (6.4.1) in turn can be rewritten as

$$e = (x+w) + Bf$$

$$\text{(}\mathscr{F}\text{)}$$

$$f = (u+v) + He.$$

6.4.2. Theorem. Let C and D be linear operators and define $E \triangleq I + (B+C)D$. Assume that either B is linear or $D = 0$. If $D \neq 0$, assume that E is a one-to-one operator on X_e with inverse E^{-1}. Then the transformation

$$e' = e + Cf - C(u+v), \qquad\qquad f' = De + (I+DC)f$$

or

$$\text{(T)}$$

$$e = (I+CD)[e' + C(u+v)] - Cf', \qquad f = f' - D[e' + C(u+v)]$$

takes system ($\mathscr{F}$) to the new system

$$e' = E^{-1}(x+w) - C(u+v) + B'f'$$

$$\text{(}\mathscr{S}'\text{)}$$

$$f' = (I+DC)(u+v) + H'e'$$

where $B' = E^{-1}(B+C)$ and $H' = (H^{-1}+C)^{-1} + D$.

Proof. From $f - u - v = He$ we get

$$H^{-1}(f - u - v) = e. \tag{6.4.3}$$

Adding $C(f - u - v)$ to both sides of Eq. (6.4.3) and recalling the definition of e' we see that

$$(H^{-1}+C)(f - u - v) = e + C(f - u - v) = e'.$$

Thus we have $(H^{-1}+C)He = e'$ and

$$He = (H^{-1}+C)^{-1}e'. \tag{6.4.4}$$

Using the definitions of e' and f we obtain

$$
\begin{aligned}
e' &= e + Cf - C(u+v) = (x+w+Bf) + Cf - C(u+v) \\
&= x + w - C(u+v) + (B+C)f \\
&= x + w - C(u+v) + (B+C)[f' - De' - DC(u+v)].
\end{aligned}
$$

If B is nonlinear, then $D = 0$ and we are done with e'. If B is linear then we can write

$$
e' + (B+C)De' = x + w - [I + (B+C)D]C(u+v) + (B+C)f'.
$$

Recalling the definition of E we obtain

$$
Ee' = x + w - EC(u+v) + (B+C)f'
$$

or

$$
e' = E^{-1}(x+w) - C(u+v) + E^{-1}(B+C)f'.
$$

Thus, for the case when B is linear and $D \neq 0$, we are also done with e'.

Using the definition of f' and Eq. $(\mathscr{S})$ we obtain after obvious manipulations,

$$
\begin{aligned}
f' &= f + De' + DC(u+v) = u + v + He + De' + DC(u+v) \\
&= (I+DC)(u+v) + He + De'.
\end{aligned}
$$

Applying Eq. (6.4.4) to the above, we see that

$$
\begin{aligned}
f' &= (I+DC)(u+v) + (H^{-1}+C)^{-1}e' + De' \\
&= (I+DC)(u+v) + H'e'.
\end{aligned}
$$

This completes the proof. ∎

The following result follows now immediately.

6.4.5. Corollary. Assume that the hypotheses of Theorem 6.4.2 are satisfied. If C and D are bounded with respect to X and if $u+v \in X$, then e and f belong to X if and only if e' and f' are in X.

The above corollary states that the stability properties of the original system $(\mathscr{S})$ and those of the transformed system $(\mathscr{S}')$ are equivalent. This allows us to deduce the stability properties of system $(\mathscr{S})$ from those of system $(\mathscr{S}')$.

B. Sector and Conicity Conditions

Let us now apply the transformation (T) to a more specific model. Assume that B is linear and that X_e has Hilbert space structure. Let C_0 and D_0 be disjoint subsets of integers such that $C_0 \cup D_0 = \{1, ..., l\}$. Then for each

$i = 1, 2, ..., l$, pick constants d_i and c_i such that $d_i = 0$ if $i \notin D_0$ and $c_i = 0$ if $i \notin C_0$. Define $C = \mathrm{diag}[c_1, ..., c_l]$ and $D = \mathrm{diag}[d_1, ..., d_l]$. Under the transformation (T), H' is determined in this case by $H' = \mathrm{diag}[H_1', ..., H_l']$ where

$$H_i' = \begin{cases} H_i + d_i I, & \text{if } i \in D_0 \\ (H_i^{-1} + c_i I)^{-1}, & \text{if } i \in C_0. \end{cases}$$

This means that a given subsystem (H_i) has feedback c_i or feedforward d_i associated with it, but not both.

6.4.6. Definition. A relation H_i is called **conic (incrementally conic)** with constants (a, b) if for $a \leq b$ the relation is inside (incrementally inside) the sector $\{a, b\}$ and for $a > b$ it is outside (incrementally outside) the sector $\{b, a\}$.

Our next result yields boundedness conditions for a multi-loop system $(\mathcal{S})$ when certain sector conditions are met.

6.4.7. Theorem. Let B be linear and let X_e have Hilbert space structure. Let C_0 and D_0 be the sets of integers defined above. Assume that H_i is conic with constants (a_i, b_i) where $b_i > a_i$ if $i \in D_0$ and $a_i b_i (b_i - a_i) < 0$ if $i \in C_0$. Define

$$d_i = \begin{cases} \dfrac{-(b_i + a_i)}{2}, & \text{if } i \in D_0 \\ 0, & \text{if } i \in C_0 \end{cases}$$

and

$$c_i = \begin{cases} 0, & \text{if } i \in D_0 \\ \dfrac{-(b_i + a_i)}{2 a_i b_i}, & \text{if } i \in C_0. \end{cases}$$

Assume that $I + BD$ has an inverse and define

$$B' = [I + (B + C) D]^{-1} (B + C) = (I + BD)^{-1} (B + C).$$

Define

$$\eta_i = \begin{cases} \dfrac{b_i - a_i}{2}, & \text{if } i \in D_0 \\ \dfrac{-2 a_i b_i}{b_i - a_i}, & \text{if } i \in C_0. \end{cases}$$

If $B' = [B_{ij}']$ has finite gain (i.e., $g(B_{ij}') < \infty$) and if the successive principal minors of the test matrix

$$R = I - G(B') \, \mathrm{diag}[\eta_1, \eta_2, ..., \eta_l]$$

are all positive, then all of the input–output relations Z_i, Y_i, F_i, and E_i associated with system $(\mathcal{S})$ are bounded with respect to X.

Proof. Given any relation M on X_e note that the following results are true.

(i)   If M is conic with constants (a, b), then for any real number k, $M + kI$ is conic with constants $(a + k, b + k)$;

(ii)   if M is conic with constants (a, b) and $k > 0$, then kM is conic with constants (ka, kb) and if $k < 0$, then kM is conic with constants (kb, ka);

(iii)   if M is conic with constants (a, b) and if $ab \neq 0$, then M^{-1} is conic with constants (b^{-1}, a^{-1});

(iv)   if $M + aI$ is positive and if $a \neq 0$, then M^{-1} is conic with constants $(0, a^{-1})$; and

(v)   if $-M + bI$ is positive and $b \neq 0$, then M^{-1} is conic with constants $(b^{-1}, 0)$.

The proofs of these statements are straightforward. For example, for item (i) we note that $Mx - rx = (M + k)x - (r + k)x$ for $r = a$ or $r = b$. It follows that

$$\langle Mx - ax, Mx - bx \rangle_T \leq 0 \quad (\text{or} \geq 0)$$

is equivalent to

$$\langle (M + k)x - (a + k)x, (M + k)x - (b + k)x \rangle_T \leq 0 \quad (\text{or} \geq 0).$$

To verify part (ii) it is only necessary to factor out the constant $(\pm k)^2$, i.e.,

$$\langle kMx - akx, kMx - bkx \rangle_T = k^2 \langle Mx - ax, Mx - bx \rangle_T.$$

In part (iii) we note that if $(x, y) \in M$ then $(y, x) \in M^{-1}$. If $ab \neq 0$ and if

$$\langle y - ax, y - bx \rangle_T \geq 0$$

then

$$ab\langle a^{-1}y - M^{-1}y, b^{-1}y - M^{-1}y \rangle_T = ab\langle a^{-1}y - x, b^{-1}y - x \rangle_T$$
$$= ab\langle x - a^{-1}y, x - b^{-1}y \rangle_T \geq 0.$$

Parts (iv) and (v) are similarly proved.

We now use these statements with $M = H_i$. If $i \in D_0$ then $H_i' = H_i + dI$. Since H_i is conic with constants (a_i, b_i), it follows that H_i' is conic with constants $(a_i + d_i, b_i + d_i)$. But $d_i = -(b_i + a_i)/2$. So $a_i + d_i = (a_i - b_i)/2 = -\eta_i$ and $b_i + d_i = (b_i - a_i)/2 = \eta_i$. Thus, H_i' is interior conic $(0, \eta_i)$ (i.e., H_i' is conic with center 0 and radius η_i).

Now assume that $i \in C_0$ so that $H_i' = (H_i^{-1} + c_i I)^{-1}$. Since H_i is conic with constants (a_i, b_i) and $a_i b_i \neq 0$, it follows that H_i^{-1} is conic with constants (b_i^{-1}, a_i^{-1}), $H_i^{-1} + c_i I$ is conic with constants $(b_i^{-1} + c_i, a_i^{-1} + c_i)$ and

$$b_i^{-1} + c_i = -(b_i - a_i)/(2a_i b_i) = -(a_i^{-1} + c_i) = \eta_i^{-1}.$$

Using item (iii) above again we see that $H_i' = (H_i^{-1} + c_i I)^{-1}$ is conic with constants $(-\eta_i, \eta_i)$. Thus, H_i' is interior conic $(0, \eta_i)$ and $g(H_i') \leq \eta_i$.

The stability of $(\mathscr{S})$ now follows by applying Theorem 6.2.10 and Corollary 6.4.5.   ∎

6.4.8. Remark. In Theorem 6.4.7 the conditions on a_i and b_i mean the following. If $i \in D_0$, we must have $b_i > a_i$ and H_i must be inside of the sector $\{a_i, b_i\}$. If $i \in C_0$, there are three possibilities.

(i) $b_i > 0 > a_i$ and H_i is inside the sector $\{a_i, b_i\}$,

(ii) $a_i > b_i > 0$ and H_i is outside of the sector $\{a_i, b_i\}$, and

(iii) $b_i < a_i < 0$ and H_i is outside of the sector $\{a_i, b_i\}$.

In (i) the limiting case $b_i \to \infty$ can be used. In this case $c_i = (-2a_i)^{-1} = (\eta_i)^{-1}$. In (ii) $a_i \to \infty$ can be used with $-H_i + b_i I$ positive and $c_i = (-2b_i)^{-1} = (\eta_i)^{-1}$. In (iii) the case $b_i \to -\infty$ can be used with $H_i - a_i I$ positive and $\eta_i = -2a_i = (c_i)^{-1}$.

When B is not linear, we can prove the following alternate form of Theorem 6.4.7.

6.4.9. Theorem. Assume that X_e has Hilbert space structure, that D_0 is the empty set, and that $C_0 = \{1, 2, \ldots, l\}$. Assume that H_i is conic with constants (a_i, b_i) where $a_i b_i (b_i - a_i) < 0$. Define

$$c_i = -\frac{(b_i + a_i)}{2a_i b_i} \quad \text{and} \quad \eta_i = \frac{-2a_i b_i}{(b_i - a_i)}.$$

If $g(B_{ij}) < \infty$ for all $i, j = 1, \ldots, l$ and if the successive principal minors of the test matrix

$$I - G(B') \operatorname{diag}[\eta_1, \ldots, \eta_l]$$

are all positive, then all the input–output relations E_i, F_i, Y_i, and Z_i associated with system $(\mathscr{S})$ are bounded with respect to X.

Proof. Since in the present case the transformation (T) transforms system $(\mathscr{S})$ to system $(\mathscr{S}')$ with $D = 0$, the proof of Theorem 6.4.7 still works. ∎

Remark 6.4.8 applies also to Theorem 6.4.9, with the understanding that D_0 is empty so that $i \in D_0$ is impossible.

Corresponding to the preceding two boundedness theorems, we can establish analogous continuity results. For example, we have the following result.

6.4.10. Theorem. Assume that X_e has Hilbert space structure and that either B is linear or that D_0 is empty. Assume that H_i is incrementally conic with constants (a_i, b_i) where $a_i b_i (b_i - a_i) < 0$ if $i \in C_0$ and $b_i > a_i$ if $i \in D_0$. Define

$$d_i = \begin{cases} -(b_i + a_i)/2, & \text{if } i \in D_0 \\ 0, & \text{if } i \in C_0 \end{cases}$$

and

$$c_i = \begin{cases} 0, & \text{if } i \in D_0 \\ -(b_i + a_i)/(2a_i b_i), & \text{if } i \in C_0. \end{cases}$$

Suppose that $I + BD$ has an inverse. Define $B' = \{I + (B + C)D\}^{-1}(B + C) = (I + BD)^{-1}(B + C)$, and

$$\eta_i = \begin{cases} (b_i - a_i)/2, & \text{if } i \in D_0 \\ (-2a_i b_i)/(b_i - a_i), & \text{if } i \in C_0. \end{cases}$$

If all incremental gains $\hat{g}(B'_{ij})$ are finite and if the successive principal minors of the test matrix

$$I - \hat{G}(B') \operatorname{diag}[\eta_1, \ldots, \eta_l]$$

are all positive, then all input–output relations E_i, F_i, Y_i, and Z_i associated with system $(\mathscr{S})$ are continuous with respect to X.

The proof of Theorem 6.4.10 is similar to the proof of Theorems 6.4.7 and 6.4.9.

C. Margin of Boundedness and Gain Factor

We now present stability results which are phrased in terms of some useful parameters. One of these, called margin of boundedness, is somewhat analogous to the degree of stability (of the entire composite system or of individual subsystems) considered in the previous chapters. The second parameter, called gain factor, provides a measure of gain of the overall system (i.e., of the overall composite system or of individual subsystems). As in the preceding chapters, these parameters are especially well suited in providing qualitative trade off information between the various isolated subsystems and the interconnecting structure. Although we define these parameters only for system $(\mathscr{S})$, these definitions are of course also applicable to isolated subsystems $(\mathscr{S}_i)$ (using obvious changes in notation, if necessary), as well as to the more general multiple input–multiple output system described by Eq. (6.2.2).

6.4.11. Definition. System $(\mathscr{S})$, viewed as a single loop system, is said to possess **margin of boundedness** δ or δ', where $0 < \delta < 1$ and $0 < \delta' < 1$, if $(\mathscr{S})$ is stable with respect to X and

  (i) for $u = v = 0$, $\|e\| \leq \delta^{-1}\|x + w\|$ for all $x, w \in X$, or
  (ii) for $x = w = 0$, $\|f\| \leq (\delta')^{-1}\|u + v\|$ for all $u, v \in X$.

6.4.12. Definition. System $(\mathscr{S})$, viewed as a single loop system, is said to possess **gain factor** μ or μ' if $(\mathscr{S})$ is stable with respect to X and

  (i) for $u = v = 0$, $\|y\| \leq \mu\|x + w\|$ for all $x, w \in X$, or
  (ii) for $x = w = 0$, $\|z\| \leq \mu'\|u + v\|$ for all $u, v \in X$.

In the corresponding incremental definitions, $\hat{\delta}, \hat{\delta}'$ or $\hat{\mu}, \hat{\mu}'$ are called a **margin of continuity** and an **incremental gain factor**, respectively.

Theorems 6.4.7, 6.4.9, and 6.4.10 can be used to estimate the parameters δ and μ, and by symmetric arguments, δ' and μ'.

6.4.13. Theorem. Let X_e have Hilbert space structure. Assume either

(i) H is inside the sector $\{a, b\}$, $b > a$, $d = -(a+b)/2$, $\eta_1 = (b-a)/2$, B is linear, $I + Bd$ is one-to-one, and $B' \triangleq (I + Bd)^{-1}B$ has gain $g(B') \leq \eta_2$, or

(ii) H is conic with constants (a, b), $ab(b-a) < 0$, $\eta_1 = -2ab/(b-a)$, $c = -(a+b)/(2ab)$, $B' \triangleq B + cI$, and $g(B') \leq \eta_2$.

If $\delta = 1 - \eta_1 \eta_2 > 0$, then δ is a margin of boundedness for $(\mathscr{S}')$ and $\mu = \eta_1/\delta$ is a gain factor for $(\mathscr{S})$.

The incremental form of Theorem 6.4.13 is also true. It follows directly from Theorem 6.4.10.

In case (ii) of Theorem 6.4.13, the gain $g(B')$ can be estimated in several cases as follows.

(a)   If $ab(b-a) < 0$, then $\delta = 1 - \eta_1 \eta_2 > 0$ provided that $-B$ is inside the sector $\{-b^{-1} - \delta(b-a)/(2ab), a^{-1} + \delta(b-a)/(2ab)\}$. Indeed, this statement is equivalent to saying that B is inside of the sector $\{c - ((1+\delta)/\eta_1), c + ((1-\delta)/\eta_1)\}$.

(b)   If $a < 0$, $b = +\infty$, and $H_1 - aI$ is positive, then $\delta = 1 - \eta_1 \eta_2 > 0$ provided that $-B$ is inside the sector $\{-\delta(\tfrac{1}{2}a^{-1}), -a^{-1} + \delta(\tfrac{1}{2}a^{-1})\}$.

The parameters δ, δ' and μ, μ' can also be estimated, using conicity conditions. We have the following result.

6.4.14. Theorem. Suppose that for some real constants $r \geq 0$ and c we know that B is interior conic $(c, (1-\delta)r)$, and suppose that H satisfies one of the following conditions.

(i) $c^2 > r^2$ and H is exterior conic $(-c/(c^2 - r^2), r/(c^2 - r^2))$,

(ii) $r^2 > c^2$ and H is interior conic $(c/(r^2 - c^2), r/(r^2 - c^2))$, or

(iii) $r^2 = c^2$ and $2cH + I$ is positive.

Then δ and $\delta' = \delta/(1 + |c|/r)$ are margin of boundedness parameters for $(\mathscr{S}')$ and $\mu = (\delta r)^{-1}, \mu' = (\delta')^{-1}(|c| + r^{-1})$ are gain factor parameters for $(\mathscr{S})$.

Proof. Let $H' = (H^{-1} + cI)^{-1}$ and $B' = B + cI$ as above. Then the method of proof used in Theorem 6.4.7 shows that H' is interior conic $(0, r^{-1})$ and B' is interior conic $(0, (1-\delta)r)$. Their respective gains are $g(H') \leq r^{-1}$ and $g(B') \leq (1-\delta)r$. Thus, $1 - g(H')g(B') = 1 - (1-\delta) = \delta$. The small gain theorem applied to $(\mathscr{S}')$ shows that if $u = v = 0$, then

$$\|e'\| \leq \delta^{-1} \|x' + w'\| = \delta^{-1} \|x + w\|.$$

Since $y = y'$, $x' = x$, and $w = w'$, we have

$$\|y\| \leq g(H')\|e'\| \leq [1/(r\delta)]\|x+w\|.$$

If $x = w = 0$, then by the small gain theorem, system $(\mathscr{S})$ is stable with respect to X. Also,

$$\|f\| = \|u+v+H'[-c(u+v)+B'f]\|$$
$$\leq \|u+v\| + g(H')[|c|\,\|u+v\|+g(B')\|f\|]$$
$$= \|u+v\|(1+r^{-1}|c|) + r^{-1}(1-\delta)r\|f\|.$$

Therefore,

$$\|f\| \leq \delta^{-1}(1+r^{-1}|c|)\|u+v\|$$

and

$$\|z\| = \|z'-cf\| = \|B'f-cf\| \leq [|c|+g(B')]\|f\| \leq (|c|+r^{-1})\|f\|.$$

This completes the proof. ■

The sector conditions and the conicity conditions are the same conditions in different forms. To see this, take

$$c = -\frac{1}{2}\left(\frac{1}{a}+\frac{1}{b}\right), \qquad r = \frac{1}{2}\left(\frac{1}{a}-\frac{1}{b}\right)$$

or equivalently,

$$a = \frac{1}{r-c} \qquad \text{and} \qquad b = \frac{-1}{r+c}.$$

We now state and prove a boundedness result for a special case of $(\mathscr{S})$ using gain factors. Let us consider interconnected feedback systems of the type depicted in Fig. 6.3, which are described by functional equations of the form

$$e_i = u_i + B_i y_i, \qquad y_i = H_i e_i, \qquad u_i = x_i + w_i + \sum_{j=1}^{l} C_{ij} y_j, \qquad (6.4.15)$$

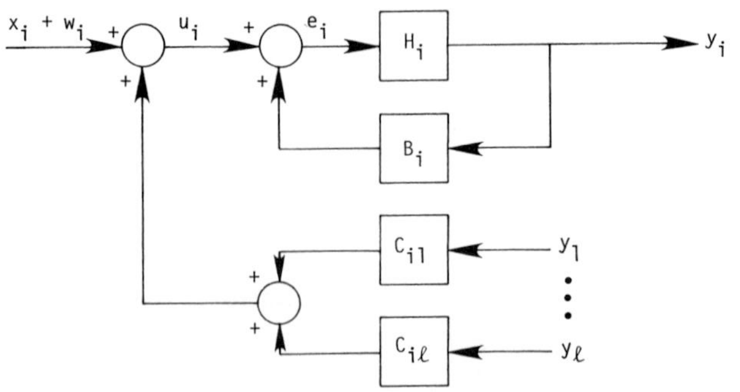

Figure 6.3   Interconnected feedback system described by Eq. (6.4.15) where $i = 1, ..., l$.

$i = 1, ..., l$, where w_i denotes a reference signal, x_i is an input, and y_i and e_i are output signals. Assume that $x_i \in X_{ie}$, $w_i \in X_i$, $e_i \in X_{ie}$, and $y_i \in X_{ie}$. Here, H_i and B_i are relations on X_{ie} while C_{ij} is a relation on $X_{je} \times X_{ie}$.

System (6.4.15) may be viewed as an interconnection of l isolated subsystems described by the equations

$$e_i = x_i + w_i + B_i y_i, \qquad y_i = H_i e_i, \qquad (6.4.16)$$

$i = 1, ..., l$.

6.4.17. Theorem. Suppose that each isolated subsystem described by Eq. (6.4.16) is bounded with respect to X_i and has gain factor μ_i. If the gains $g(C_{ij})$, $i, j = 1, ..., l$, are finite and if the successive principal minors of the test matrix

$$r_{ij} = \begin{cases} 1 - g(C_{jj})\mu_j, & i = j \\ -g(C_{ij})\mu_j, & i \neq j \end{cases}$$

are all positive, then the input–output relations E_i, U_i, and Y_i associated with system (6.4.15) are all bounded with respect to $X = \chi^l_{i=1} X_i$. (U_i denotes the relation associating a given input x with the variable u_i.)

Proof. Since each subsystem has gain factor μ_i, we have for any time T,

$$\|y_j\|_T \leq \mu_j \|u_j\|_T$$

for all $j = 1, ..., l$, so that

$$\|u_i\|_T \leq \|x_i\|_T + \|w_i\|_T + \sum_{j=1}^{l} \|C_{ij} y_j\|_T \leq \|x_i\|_T + \|w_i\|_T + \sum_{j=1}^{l} g(C_{ij}) \mu_j \|u_j\|_T.$$

Define $U_T = (\|u_1\|_T, ..., \|u_l\|_T)^T$ and define X_T and W_T similarly. Then the above inequality can be written as

$$R U_T \leq X_T + W_T$$

where R is the test matrix given in the hypothesis. Since all off-diagonal terms of matrix R are nonpositive and since all successive principal minors of R are by assumption positive, R is an M-matrix. It follows from Theorem 2.5.2 that $\Delta = R^{-1}$ exists and that $R^{-1} \geq 0$. Therefore,

$$U_T \leq \Delta(X_T + W_T).$$

Letting $T \to \infty$ we see that $U \leq \Delta(X + W) < \infty$. Thus, each $u_i \in X_i$. For $i = 1, ..., l$ we now have

$$\|y_i\| \leq \mu_i \|u_i\| < \infty \qquad \text{and} \qquad \|e_i\| \leq \|u_i\| + g(B_i) \|y_i\| < \infty.$$

This concludes the proof. ■

Note that in proving Theorem 6.4.17 the method of Section 6.2 was used, rather than an application of any specific result from that section. Furthermore, note that *a continuity theorem corresponding to Theorem 6.4.17 is also true.*

In the subsequent specific examples we shall see that in applications the margin of boundedness parameter δ_i of the isolated subsystems is actually of more importance than the gain factor μ_i.

D. Examples

We now consider two specific examples to indicate the types of elements that can be used as subsystems and interconnections and to illustrate application of the theory developed. In both examples the underlying space is $X_e = L_{2e}$ while $X = L_2$ and boundedness and continuity are interpreted in the sense of the L_2-norm.

6.4.18. Example. Consider the multiple-loop system depicted in Fig. 6.4. This system may be viewed as an interconnection of three linear time-invariant systems (blocks H_1, H_2, H_3), a time-varying gain (block H_5), a piecewise nonlinearity (block H_4), and a hysteresis nonlinearity (block H_6). This illustrates the variety of types of elements allowable. Elements with time delays can also be handled.

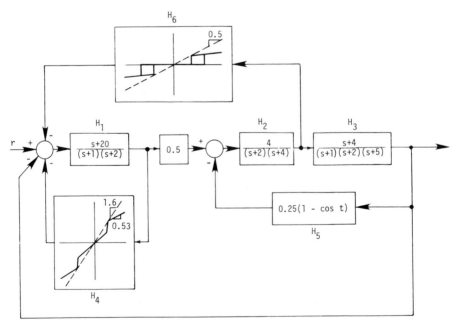

Figure 6.4 Block diagram of Example 6.4.18.

Now let $h_1(t)$ be the inverse Laplace transform of $(s+20)/[(s+1)(s+2)]$. Then the block labeled H_1 represents a system having input x and output

$$v(t) = \int_0^t h_1(t-\tau)x(\tau)\,d\tau + z(t)$$

for $t \geq 0$, where $z(t)$ accounts for initial conditions. Similar statements apply to the blocks labeled H_2 and H_3.

We first analyze the more general system specified in Fig. 6.5 and then return to the more special case of Fig. 6.4. The blocks labeled H_1, H_2, and H_3 are assumed to be linear and time-invariant operators belonging to class $\mathscr{L}_0$.

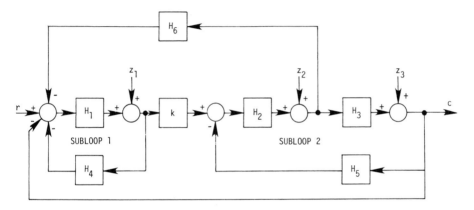

Figure 6.5   A multi-loop system.

Zero input responses of these linear blocks are accounted for by the functions z_1, z_2, and z_3 which are assumed to belong to $L_2[0, \infty)$. The blocks labeled H_4, H_5, and H_6 represent time-varying nonlinearities which are modeled by relations on $L_{2e}[0, \infty)$. The constant k is assumed to be nonnegative.

We now use Theorem 6.4.7 to obtain boundedness conditions for the system of Fig. 6.5. Assume that the single loop system containing the two relations H_1 and $-H_4$, identified as subloop 1, is stable with margin of boundedness δ. This is ensured by assuming that H_1 is conic with constants (a_1, b_1) and by assuming that H_4 is inside of the sector

$$\{-b_1^{-1} - \delta(b_1-a_1)(2a_1b_1)^{-1}, -a_1^{-1} + \delta(b_1-a_1)(2a_1b_1)^{-1}\},$$

where $b_1 < a_1 < 0$ or $a_1 < 0 < b_1$. Further, assume that H_2 is inside the sector $\{-b_2, b_2\}$, H_3 is inside the sector $\{-b_3, b_3\}$, H_5 is inside the sector $\{0, b_5\}$, and H_6 is inside the sector $\{0, b_6\}$, where b_2, b_3, b_5, and b_6 are all positive.

We now show that the system of Fig. 6.5 is bounded if

$$\delta(1-b_2 b_3 b_5) + kb_2(b_3+b_6)(2b_1 a_1/(b_1-a_1)) > 0. \qquad (6.4.19)$$

First we see that each H_i is conic with constants (a_i, b_i), where $a_4 = -b_1^{-1} - \delta(b_1-a_1)/(2a_1 b_1)$, $b_4 = -a_1^{-1} + \delta(b_1-a_1)/(2b_1 a_1)$, $a_2 = -b_2$, $a_3 = -b_3$, and $a_5 = a_6 = 0$. In Theorem 6.4.7 select $D_0 = \{2,3,4,5,6\}$ and $C_0 = \{1\}$. Then $d_2 = d_3 = 0$, $d_5 = -b_5/2$, $d_6 = -b_6/2$, $\eta_1 = -(2b_1 a_1)/(b_1-a_1)$, $\eta_2 = b_2$, $\eta_3 = b_3$, $\eta_5 = b_5/2$, and $\eta_6 = b_6/2$. From the block diagram of Fig. 6.5 it is clear that matrix B assumes the form

$$B = \begin{bmatrix} 0 & 0 & -1 & -1 & 0 & -1 \\ k & 0 & 0 & 0 & -1 & 0 \\ 0 & 1 & 0 & 0 & 0 & 0 \\ 1 & 0 & 0 & 0 & 0 & 0 \\ 0 & 0 & 1 & 0 & 0 & 0 \\ 0 & 1 & 0 & 0 & 0 & 0 \end{bmatrix}$$

so that

$$(I+BD)^{-1} = \begin{bmatrix} 1 & 0 & 0 & d_4 & 0 & d_6 \\ 0 & 1 & 0 & 0 & d_5 & 0 \\ 0 & 0 & 1 & 0 & 0 & 0 \\ 0 & 0 & 0 & 1 & 0 & 0 \\ 0 & 0 & 0 & 0 & 1 & 0 \\ 0 & 0 & 0 & 0 & 0 & 1 \end{bmatrix}$$

and $B' = (I+BD)^{-1}(B+C)$ is computed as

$$B' = \begin{bmatrix} c_1+d_4 & d_6 & -1 & -1 & 0 & -1 \\ k & 0 & d_5 & 0 & -1 & 0 \\ 0 & 1 & 0 & 0 & 0 & 0 \\ 1 & 0 & 0 & 0 & 0 & 0 \\ 0 & 0 & 1 & 0 & 0 & 0 \\ 0 & 1 & 0 & 0 & 0 & 0 \end{bmatrix}$$

Since subloop 1 has margin of boundedness δ we have $1-\eta_1\eta_4 = \delta$. Also note that $c_1 + d_4 = 0$. We now finally have

$$R = I - [\,|b'_{ij}|\,\eta_j] = \begin{bmatrix} 1 & \eta_2 d_6 & -\eta_3 & -\eta_4 & 0 & -\eta_6 \\ -k\eta_1 & 1 & \eta_3 d_5 & 0 & -\eta_5 & 0 \\ 0 & -\eta_2 & 1 & 0 & 0 & 0 \\ -\eta_1 & 0 & 0 & 1 & 0 & 0 \\ 0 & 0 & -\eta_3 & 0 & 1 & 0 \\ 0 & -\eta_2 & 0 & 0 & 0 & 1 \end{bmatrix}.$$

The successive principal minors of R are all positive if

$$1 + k\eta_1\eta_2 d_6 > 0$$

$$(1 + \eta_2\eta_3 d_5) + k\eta_1\eta_2(d_6 - \eta_3) > 0$$

$$(1 - \eta_4\eta_1)(1 + \eta_2\eta_3 d_5) + k\eta_1\eta_2(d_6 - \eta_3) > 0$$

$$(1 - \eta_4\eta_1)(1 + 2\eta_2\eta_3 d_5) + k\eta_1\eta_2(d_6 - \eta_3) > 0$$

$$(1 - \eta_4\eta_1)(1 + 2\eta_2\eta_3 d_5) + k\eta_1\eta_2(2d_6 - \eta_3) > 0.$$

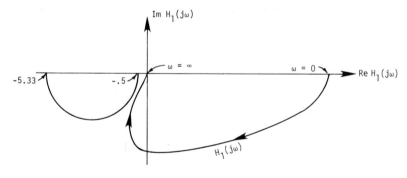

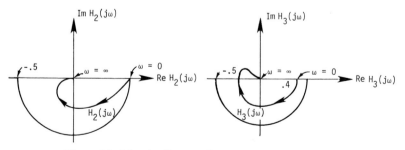

Figure 6.6   Nyquist diagrams for operators H_1, H_2, and H_3.

Since $d_5 < 0$ and $d_6 < 0$, the last inequality implies the previous four. Using the expressions for η_i and d_i and $1 - \eta_1\eta_4 = \delta$, the last inequality becomes inequality (6.4.19).

We now show that in particular, the system of Fig. 6.4 satisfies inequality (6.4.19), and thus, this system is bounded. First we note that the linear operators H_1, H_2, and H_3 have impulse response functions belonging to $L_1[0, \infty)$. Their Nyquist diagrams are shown in Fig. 6.6. From these diagrams it is clear that H_1 is outside the sector $\{-5.33, -0.5\}$ and H_2 and H_3 are inside the sector $\{-0.5, 0.5\}$. Note also that H_5 has input–output relation

$$v(t) = 0.25(1 - \cos t)x(t) = N_5(x(t), t)$$

and that $0 \le \partial N_5/\partial x \le 0.5$. Thus, H_5 is inside of the sector $\{0, 0.5\}$. Similarly, H_4 is inside of the sector $\{0.53, 1.6\}$ and H_6 is inside of the sector $\{0, 0.5\}$.

Recalling that $a_1 = -0.5$, $b_1 = -5.33$, $b_2 = b_3 = b_5 = b_6 = 0.5$, and $k = 0.5$ we compute $\delta = 0.375$,

$$-b_1^{-1} - \delta\frac{(b_1 - a_1)}{2a_1 b_1} = 0.526, \qquad -a_1^{-1} + \delta\frac{(b_1 - a_1)}{2a_1 b_1} = 1.66$$

and see that inequality (6.4.19) is satisfied.

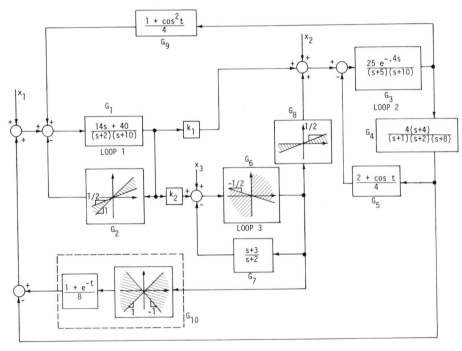

Figure 6.7   Block diagram of Example 6.4.20.

6.4.20. Example. Consider now the interconnected system depicted in Fig. 6.7. We will use Theorem 6.4.17 to analyze this system. We make the identifications $l = 3$, $H_1 = G_1$, $B_1 = -G_2$, $C_{21} = k_1$, $C_{31} = k_2$, $H_2 = G_3$, $B_2 = -G_5 G_4$, $C_{32} = 0$, $C_{12} = G_9 - G_4$, $H_3 = G_6$, $B_3 = -G_7$, $C_{13} = G_{10}$, $C_{23} = G_8$, $C_{11} = C_{22} = C_{33} = 0$. The operators G_1, G_3, G_4, and G_7 are in class $\mathscr{L}$ (see Section 6.1, Part D) and are characterized by their Laplace transforms. Also, G_2, G_6, G_8 belong to class $\mathscr{N}$ (see Section 6.1, Part C) and have graphs which lie in the indicated shaded regions. Furthermore, G_5 and G_9 are time-varying linear gains and G_{10} is an operator of class $\mathscr{N}$ in cascade with a time-varying linear gain. The interconnections k_1 and k_2 are constant multipliers.

For the sake of illustration we shall at first regard the linear elements G_1, G_3, G_4, G_7 as being adjustable. At the end of our discussion we will make the additional assumption that these linear elements are those indicated in the figure. First we need to determine margins of boundedness and gain factors for each isolated subsystem (i.e., for Loops 1, 2, 3, when disconnected from the system).

Loop 1. G_2 is interior conic $(\tfrac{3}{4}, \tfrac{1}{4})$. Setting $c_1 = -\tfrac{3}{4}$ and $(1 - \delta_1) r_1 = \tfrac{1}{4}$, we note that $c_1^2 > r_1^2$, provided that $\tfrac{2}{3} > \delta_1 > 0$ and $r_1^2 > c_1^2$ if $1 > \delta_1 > \tfrac{2}{3}$. Therefore the isolated subsystem determined by Loop 1 has margin of boundedness δ_1 with $\tfrac{2}{3} > \delta_1 > 0$ if $-G_1$ lies outside of the sector $\{4(1-\delta_1)/(4-3\delta_1), 4(1-\delta_1)/(2-3\delta_1)\}$ and $1 > \delta_1 > \tfrac{2}{3}$ if $-G_1$ lies in the same sector (see Theorem 6.4.14). The gain factor $\mu_1 = r_1^{-1} \delta_1^{-1} = 4(1-\delta_1)\delta_1^{-1}$.

Loop 2. The operator G_5 has gain 0.75. If we require G_3 and G_4 to be interior conic with constants $(0, b_3)$ and $(0, b_4)$, respectively, then this loop has margin of boundedness $\delta_2 = 1 - 0.75 b_3 b_4$ if δ_2 is in the interval $0 < \delta_2 < 1$. This can be checked by using $c_2 = 0$, $(1 - \delta_2) r_2 = 0.75 b_4$, and $b_3 = r_2^{-1}$ in part (ii) of Theorem 6.4.14. The corresponding gain factor is $\mu_2 = r_2^{-1} \delta_2^{-1} = b_3(1 - 0.75 b_3 b_4)$.

Loop 3. The operator $2G_6 + I$ is positive. If $c_3 = r_3 = 1$ in part (iii) of Theorem 6.4.14 and if $-G_7$ is interior conic $(-1, 1 - \delta_3)$, then this subsystem has margin of boundedness δ_3 and gain factor $\mu_3 = \delta_3^{-1}$.

Next, we need to form the test matrix R,

$$R = \begin{bmatrix} 1 & -(0.5 + b_4)\mu_2 & -0.25\delta_3^{-1} \\ -|k_1|\mu_1 & 1 & -0.5\delta_3^{-1} \\ -|k_2|\mu_1 & 0 & 1 \end{bmatrix}.$$

This matrix has positive successive principal minors if and only if

$$1 - (0.5 + b_4)|k_1|\mu_1\mu_2 - |k_2|\mu_1\{0.5(0.5 + b_4)\mu_2\delta_3^{-1} + 0.25\delta_3^{-1}\} > 0.$$

Recalling that $\mu_1 = 4(1-\delta_1)\delta_1^{-1}$, $\delta_2 = 1-0.75b_3b_4$, and $\mu_2 = b_3\delta_2^{-1}$, we get the inequality

$$\delta_1\delta_2\delta_3 - (1-\delta_1)\{[b_3 + \tfrac{8}{3}(1-\delta_2)][2|k_1|\delta_3 + |k_2|] + |k_2|\delta_2\} > 0. \tag{6.4.21}$$

For each choice of δ_1, δ_2, δ_3, and b_3 there is a corresponding set of k_1 and k_2 for which the stability condition (6.4.21) holds. In particular, this condition is satisfied for all k_1 and k_2 for which $|k_1|$ and $|k_2|$ are sufficiently small.

Next, consider the Laplace transforms indicated in Fig. 6.7. From the Nyquist plot of G_1 given in Fig. 6.8 it is seen that G_1 is inside of the sector $\{-0.5, 2\}$. Therefore it is easy to calculate that $\delta_1 = 0.8$. Similarly, $G_3(j\omega)$ and $G_4(j\omega)$ lie inside circles in the complex plane centered at the origin with radii 0.5 and 1, respectively (see Fig. 6.8). Therefore, $b_3 = 0.5$, $b_4 = 1$, and $\delta_2 = \tfrac{5}{8}$. The Nyquist plot of G_7, shown in Fig. 6.8, indicates that G_7 is inside the sector $\{0.5, 1.5\}$ so that $\delta_3 = 0.5$. Substituting these numbers into inequality (6.4.21) we find that a sufficient condition for boundedness of the system given in Figure 6.7 is given by

$$0 < 1.2|k_1| + 1.7|k_2| < 1. \tag{6.4.22}$$

Examples 6.4.18 and 6.4.20 show that analysis of rather complicated interconnected systems by the present methods is indeed possible. These results

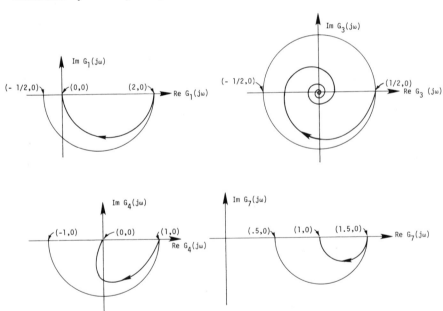

Figure 6.8 Nyquist plots for Example 6.4.20.

may be conservative, but not excruciatingly so. In particular, these examples illustrate the variety of system elements that can be treated by the present approach. In addition, an indication is given of some of the methods which can be employed to verify sector or conicity conditions. We emphasize that many other types of relations may be analyzed by the present results. Note that in Example 6.4.20, condition (6.4.22) is sufficient for stability for *any* nonlinear memoryless elements with graphs in the indicated shaded regions. Terms with memory or delays can be treated just as easily. Thus, in Example 6.4.20, G_2 may be replaced by any relation which is conic $(0.75, 0.25)$. This is the case if, for example,

$$(G_2 x)(t) = 0.75x(t) + a_1 x(t-\tau_1) + a_2 x(t-\tau_2)$$

where $|a_1| + |a_2| \le 0.25$, $0 < \tau_1 < \infty$, and $0 < \tau_2 < \infty$. This is also the case if, for example,

$$(G_2 x)(t) = 0.75x(t) + a \int_0^t e^{-b(t-\tau)} x(\tau) \, d\tau$$

where $b > 0$ and $|a|/b < 0.25$.

6.5 Stability of Large Scale Systems: Popov-Like Conditions

The systems considered in this section are special cases of composite system (6.4.15). In particular, we shall treat those systems which may be viewed as an interconnection of scalar input–scalar output subsystems of the Popov type. We will consider two types of system configurations.

First we consider systems with configuration as shown in Fig. 6.9a, which are described by functional equations of the form

$$e_i = x_i + w_i - L_i y_i + \sum_{j=1}^l C_{ij} y_j, \qquad y_i = N_i e_i, \tag{6.5.1}$$

$i = 1, \ldots, l$. Here x_i, e_i, $y_i \in L_{2e}(R^+, R) = X_e$ and $w_i \in L_2(R^+, R) = X$. The isolated subsystems of system (6.5.1) are given by

$$e_i = x_i + w_i - L_i y_i, \qquad y_i = N_i e_i, \tag{6.5.2}$$

$i = 1, \ldots, l$.

6.5.3. Definition. Isolated subsystem (6.5.2) is said to have **Property A** if (i) $N_i \in \mathcal{N}$ and N_i is interior conic with center $b_i/2$ and radius $b_i/2$ for some $b_i \in (0, \infty)$; (ii) $L_i \in \mathcal{L}$ with Fourier transform $L_i^*(j\omega)$; (iii) $L_i = L_{i2} L_{i1}$

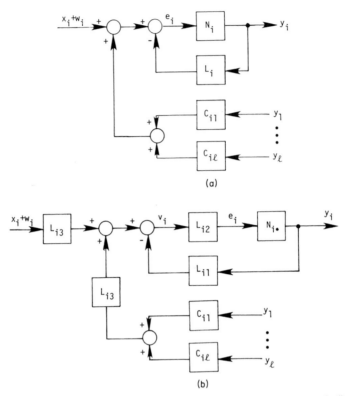

Figure 6.9 (a) Interconnected feedback system (6.5.1) where $i = 1, ..., l$. (b) Transformed interconnected feedback system (6.5.1) where $i = 1, ..., l$.

where L_{i2} is a time-invariant linear operator with Fourier transform $L_{i2}^*(j\omega) = 1/(1 + j\omega q_i)$ for some positive constant q_i; (iv) $|L_i^*(j\omega)|$ and $|j\omega L_i^*(j\omega)|$ are bounded for all $\omega \geq 0$; and (v) the operator $L_{i1} + (b_i^{-1} - \Delta_i)I$ is positive for some constant $\Delta_i > 0$.

Let X_i be the range of L_{i2}, i.e., $L_{i2}: L_2(R^+, R) \to X_i \subset L_2(R^+, R)$ and define $X = \chi_{i=1}^l X_i$. Thus, X_i is the set of functions in $L_2(R^+, R)$ which are absolutely continuous on finite subintervals of R^+ and whose derivatives are in L_2. Define a norm $\| \cdot \|_i$ on X_i by

$$\|f\|_i = \left\{ \int_0^\infty |f(t)|^2 \, dt \right\}^{1/2} + \left(\int_0^\infty \left| \frac{df(t)}{dt} \right|^2 \, dt \right)^{1/2}.$$

Then X is a compatible subspace of $X_e = L_{2e}(R^+, R^l)$. Now let $L_{i3}: X_i \to L_2$ be the right and left inverse of L_{i2}. Then L_{i3} involves differentiation. Note that in this case boundedness with respect to (X, L_2) will mean that whenever

x_i, dx_i/dt, w_i, $dw_i/dt \in L_2$, then the outputs $y_i \in L_2$. Since in our first stability result for composite system (6.5.1) (Theorem 6.5.7) the e_i will be shown to be in X_i, we have e, $de/dt \in L_2(R^+, R^l)$ so that $e \in C_0$. Moreover, $y_i(t) = N_i e_i(t)$, so that $y \in C_0$ whenever the N_i are all time invariant. This means that our proof (of Theorem 6.5.7) will not only show boundedness with respect to (X, L_2) but also with respect to (X, C_0).

6.5.4. Definition. The interconnections C_{ij}, $i,j = 1, ..., l$, in Eq. (6.5.1) are said to possess **Property B** if they are time-invariant linear operators mapping $L_2(R^+, R) \to X_i$ with finite gains.

We note that if Property A and Property B are both true, then $L_{i1} = L_{i3}L_i$ maps L_{2e} into itself, L_{i1} is bounded with respect to L_2, and has gain

$$g(L_{i1}) = \operatorname*{ess\,sup}_{\omega \geq 0} |L_{i1}^*(j\omega)| < \infty.$$

Similarly, $L_{i3}C_{ik}$ is bounded with respect to L_2 and

$$g(L_{i3}C_{ik}) = \operatorname*{ess\,sup}_{\omega \geq 0} |(1+j\omega q_i)C_{ik}^*(j\omega)| < \infty.$$

Note that condition (v) of Definition 6.5.3 is equivalent to the requirement that

$$\operatorname{Re}\{(1+j\omega q_i)L_i^*(j\omega)\} + b_i^{-1} \geq \Delta_i > 0 \qquad \text{a.e. for} \quad \omega \geq 0,$$

the familiar **Popov condition**. Here Δ_i is the minimum distance, parallel to the real axis, between the graph of the modified frequency response of L_i and the Popov line with intercept $-(b_i)^{-1}$ and slope $(q_i)^{-1}$. We note that the parameter Δ_i will play a role analogous to that of δ_i, the margin of boundedness, considered in Section 6.4.

Before considering the main results of this section, we recall a result from the Popov theory.

6.5.5. Lemma. Let $\lambda > 0$ and define Kx by

$$Kx(t) = \lambda \int_0^t e^{-\lambda(t-\tau)} x(\tau)\, d\tau, \qquad \text{for} \quad t \geq 0,$$

where $x \in L_{2e}$. Suppose that $g \in \mathcal{N}$ and satisfies $|g(x,t) - \gamma x| \leq \gamma |x|$ for all $x \in R$ and $t \in R^+$. If $G(x)(t) \triangleq g(Kx(t), t)$ for all $t \geq 0$, then

$$\|G(x) - \gamma x\|_T \leq \gamma \|x\|_T$$

for all $x \in L_{2e}$ and $T \geq 0$.

Proof. The conclusion of this lemma is equivalent to

$$\langle G(x) - \gamma x, G(x) - \gamma x \rangle_T \leq \langle \gamma x, \gamma x \rangle_T$$

or

$$2\gamma \langle x, G(x) \rangle_T \geq \langle G(x), G(x) \rangle_T.$$

If $y = Kx$, then (letting $g(y) = g(y(\cdot), \cdot)$) this is equivalent to

$$\langle g(y), g(y)\rangle_T \le 2\gamma \langle x, g(y)\rangle_T$$

or since $\lambda x = (dy/dt) + \lambda y$,

$$\langle g(y), g(y)\rangle_T \le \frac{2\gamma}{\lambda}\left\langle \frac{dy}{dt}, g(y)\right\rangle_T + 2\gamma \langle y, g(y)\rangle_T.$$

Since $ug(u) > 0$ for $u \ne 0$ and since $|g(u)| \le 2\gamma |u|$, we have

$$\langle g(y), g(y)\rangle_T = \int_0^T g(y(\tau))^2 \, d\tau \le \int_0^T 2\gamma y(\tau) g(y(\tau)) \, d\tau.$$

Moreover,

$$\left\langle \frac{dy}{dt}, g(y)\right\rangle_T = \int_0^T \frac{dy(\tau)}{d\tau} g(y(\tau)) \, d\tau = \int_{y(0)}^{y(T)} g(u) \, du \ge 0.$$

Therefore,

$$\langle g(y), g(y)\rangle_T \le 2\gamma \int_0^T y(\tau) g(y) \, d\tau = 2\gamma \langle y, g(y)\rangle_T$$

$$\le 2\gamma \langle y, g(y)\rangle_T + \frac{2\gamma}{\lambda}\left\langle \frac{dy}{dt}, g(y)\right\rangle_T.$$

This concludes the proof. ∎

We will also require the following preliminary result.

6.5.6. Lemma. Let $N \in \mathcal{N}$ be interior conic $(b/2, b/2)$ where $0 < b < \infty$, let L be a linear time invariant operator on L_{2e}, let $\Delta > 0$ be a constant such that $L + (b^{-1} - \Delta)I$ is positive, and let e and y be given by

$$e = x - Ly, \qquad y = Ne$$

for some fixed $x \in L_{2e}$. Then $\|y\|_T \le \mu \|x\|_T$ for all $T \ge 0$ where $\mu \le \Delta^{-1}$.

Proof. Since $x = e + Ly$ we have

$$\langle x, y\rangle_T = \langle e, y\rangle_T + \langle Ly, y\rangle_T \ge \langle e, y\rangle_T + (\Delta - b^{-1})\langle y, y\rangle_T$$

for any $T > 0$. Since N is inside the sector $(0, b)$ it follows that $N^{-1} - b^{-1}I$ is positive. Thus,

$$\langle e, y\rangle_T = \langle N^{-1}y, y\rangle_T \ge b^{-1}\langle y, y\rangle_T$$

and

$$\|x\|_T \|y\|_T \ge \langle x, y\rangle_T \ge (\Delta - b^{-1})\langle y, y\rangle_T + b^{-1}\langle y, y\rangle_T$$

so that $\Delta \|y\|_T \le \|x\|_T$ for all $T \ge 0$. ∎

We now state and prove the following stability result.

6.5.7. Theorem. Assume that all isolated subsystems (6.5.2) possess Property A and that all interconnections of composite system (6.5.1) possess Property B. If the successive principal minors of the test matrix $R = [r_{ij}]$ defined by

$$r_{ij} = \begin{cases} \Delta_i - g(L_{i3} C_{ii}), & \text{if} \quad i = j \\ -g(L_{i3} C_{ij}), & \text{if} \quad i \neq j \end{cases}$$

are all positive, then the input–output relations Y_i of composite system (6.5.1) are bounded with respect to (X, L_2), i.e., for all $x, e \in X$, the solutions y of Eq. (6.5.1) belong to $L_2(R^+, R^l)$.

Proof. Operating on both sides of Eq. (6.5.1) by L_{i3} we obtain

$$v_i = L_{i3}(x_i + w_i) - L_{i1} y_i + \sum_{j=1}^{l} (L_{i3} C_{ij}) y_j, \qquad y_i = N_i e_i, \qquad e_i = L_{i2} v_i$$

for $i = 1, ..., l$. This *transformed system* (see Fig. 6.9b) has isolated subsystems described by

$$v_i = L_{i3}(x_i + w_i) - L_{i1} y_i, \qquad y_i = N_i L_{i2} v_i. \tag{6.5.8}$$

By Lemma 6.5.5, $N_i L_{i2}$ is interior conic $(b_i/2, b_i/2)$. Thus, by Lemma 6.5.6, all isolated subsystems (6.5.8) have gain margins $\mu_i = \Delta_i^{-1}$ with respect to $L_2(R^+, R)$. Noting that the test matrix R is an M-matrix, we conclude from Theorem 2.5.2 that R^{-1} exists and that $R^{-1} \geq 0$. Let δ_{ij} denote the Kronecker delta and let

$$S = [\delta_{ij} - g(L_{i3} C_{ij}) \mu_j] = R \operatorname{diag}[\mu_1, ..., \mu_l].$$

Then $S^{-1} = \operatorname{diag}[\Delta_1, ..., \Delta_l] R^{-1} \geq 0$. The conclusion of the theorem follows now by applying Theorem 6.4.17. ∎

The second configuration of interconnected systems which we consider is depicted in Fig. 6.10a and is described by functional equations of the form

$$e_i = x_i + w_i - N_i y_i + \sum_{j=1}^{l} C_{ij} y_j, \qquad y_i = L_i e_i, \tag{6.5.9}$$

$i = 1, ..., l$. In Eq. (6.5.9) all symbols have the same meaning as in Eq. (6.5.1). The isolated subsystems are in this case described by the set of equations

$$e_i = x_i + w_i - N_i y_i, \qquad y_i = L_i e_i, \tag{6.5.10}$$

$i = 1, ..., l$. Isolated subsystem (6.5.10) is said to possess **Property A** if L_i and N_i possess all the properties enumerated in Definition 6.5.3.

We now prove the following result.

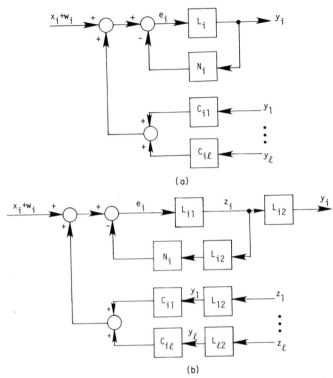

(a)

(b)

Figure 6.10 (a) Interconnected feedback system (6.5.9) where $i = 1, ..., l$. (b) Transformed interconnected feedback system (6.5.12) where $i = 1, ..., l$.

6.5.11. Theorem. Assume that all isolated subsystems (6.5.10) possess Property A and that for all interconnections of composite system (6.5.9) we have $g(C_{ij} L_{j2}) < \infty$. Let

$$D_i = \Delta_i \{ g(L_{i1}) [\Delta_i + g(L_{i1})] \}^{-1}.$$

If the successive principal minors of the test matrix $R = [r_{ij}]$ defined by

$$r_{ij} = \begin{cases} D_i - g(C_{ii} L_{j2}), & \text{for} \quad i = j \\ -g(C_{ij} L_{j2}), & \text{for} \quad i \neq j \end{cases}$$

are all positive, then the input–output relations Y_i of composite system (6.5.9) are bounded with respect to (X, L_2).

Proof. System (6.5.9) is equivalent to the *transformed system* (see Fig. 6.10b) described by the equations

$$e_i = (x_i + w_i) - (N_i L_{i2}) z_i + \sum_{j=1}^{l} C_{ij} L_{j2} z_j, \qquad z_i = L_{i1} e_i \quad (6.5.12)$$

when $y_i = L_{i2} z_i$. Let

$$r_i = x_i + w_i + \sum_{j=1}^{l} C_{ij} L_{j2} z_j, \qquad N_i L_{i2} z_i = u_i$$

so that

$$e_i = r_i - u_i \qquad \text{or} \qquad z_i = L_{i1} r_i - L_{i1} u_i.$$

Taking the inner product at time T with u_i, we see that

$$\langle N_i L_{i2} z_i, z_i \rangle_T + \langle u_i, L_{i1} u_i \rangle_T = \langle u_i, L_{i1} r_i \rangle_T.$$

As before, we have

$$\langle N_i L_{i2} z_i, z_i \rangle_T \geq b_i^{-1} \langle N_i L_{i2} z_i, N_i L_{i2} z_i \rangle_T = b_i^{-1} \|u_i\|_T^2$$

and

$$\langle u_i, L_{i1} u_i \rangle_T \geq (\Delta_i - b_i^{-1}) \langle u_i, u_i \rangle_T = (\Delta_i - b_i^{-1}) \|u_i\|_T^2$$

so that

$$\Delta_i \|u_i\|_T^2 \leq \langle u_i, L_{i1} r_i \rangle_T \leq \|u_i\|_T \|L_{i1} r_i\|_T$$

and

$$\Delta_i \|u_i\|_T \leq g(L_{i1}) \|r_i\|_T.$$

Since $z_i = L_{i1} r_i - L_{i1} u_i$, we have

$$\|z_i\|_T \leq g(L_{i1})(\|r_i\|_T + \|u_i\|_T) \leq g(L_{i1})[1 + g(L_{i1})/\Delta_i] \|r_i\|_T = D_i^{-1} \|r_i\|_T.$$

Recalling the definition of r_i and applying the triangle inequality, we have

$$D_i \|z_i\|_T \leq \|x_i\|_T + \|w_i\|_T + \sum_{j=1}^{l} g(C_{ij} L_{j2}) \|z_j\|_T.$$

Letting $Z_T = (\|z_1\|_T, \dots, \|z_l\|_T)^T$ and defining X_T and W_T similarly, the above inequality can be rewritten as

$$RZ_T \leq X_T + W_T.$$

Noting that R is an M-matrix, it follows that R^{-1} exists and that $R^{-1} \geq 0$ (see Theorem 2.5.2). We thus have

$$Z_T \leq R^{-1}(X_T + W_T) \leq R^{-1}(X + W) < \infty$$

for all $T \geq 0$, where $X + W = (\|x_1\| + \|w_1\|, \dots, \|x_l\| + \|w_l\|)^T$. This concludes the proof. ■

Note that in Theorem 6.5.11, if C_{ik} has Property B, then

$$g(C_{ik} L_{k2}) = \operatorname*{ess\,sup}_{\omega \geq 0} \|C_{ik}^*(j\omega)(1 + j\omega q_i)^{-1}\|.$$

Let us next reconsider composite system (6.5.9), with somewhat different assumptions. Specifically, we assume that $N_i \in \mathcal{N}$ with N_i inside of the sector

$\{a_i, a_i+b_i\}$ for some real constants $a_i < 0$ and b_i such that $a_i+b_i > 0$. As before $L_i = L_{i2}L_{i1}$ where $L_{i2}^*(j\omega) = (1+j\omega q_i)^{-1}$ and $-L_{i1}$ is inside the sector $\{a_i^{-1}, [a_i+b_i(1-b_i\delta_i)^{-1}]^{-1}\}$ for some $\delta_i > 0$. When these conditions are true, we say that isolated subsystem (6.5.10) possesses **Property C**.

Now let us introduce the notation

$$L_{i1}' = (L_{i1}^{-1}+a_i I)^{-1} \quad \text{and} \quad D_i' = \{g(L_{i1}')[1+g(L_{i1}')/\delta_i]\}^{-1}$$

where the indicated gains are with respect to the space $L_2(R^+, R)$.

6.5.13. Theorem. Assume that all gains $g(L_{i1}')$ and $g(C_{ij}L_{j2})$ are finite and that all isolated subsystems (6.5.10) possess Property C. If the successive principal minors of the test matrix $R = [r_{ij}]$ given by

$$r_{ij} = \begin{cases} D_i' - g(C_{ii}L_{i2}), & \text{if } i = j \\ -g(C_{ij}L_{j2}), & \text{if } i \neq j \end{cases}$$

are all positive, then the input–output relations Y_i of composite system (6.5.9) are bounded with respect to (X, L_2).

Proof. Let $A = \text{diag}[a_1, ..., a_l]$. In the present case the transformation (T) of Section 6.4, Part A, reduces to

$$e' = e + Az, \qquad z' = z$$

so that Eq. (6.5.12) transforms to

$$e_i' = (x_i + w_i) - N_i'z_i + \sum_{j=1}^{l} C_{ij}L_{j2}z_j, \qquad z_i = L_{i1}'e_i',$$

where $y_i = L_{i2}z_i$ and $N_i' = N_i L_{i2} - a_i I$. The transformed operator N_i' is in the sector $\{0, b_i\}$ and $L_i' + b_i^{-1} - \delta_i$ is positive. Thus, Theorem 6.5.11 is applicable and the boundedness of z_i follows. Since $y_i = L_{i2}z_i$, it follows that $y \in L_2(R^+, R^l)$. ∎

Before considering a specific example, we point out that the method of the present section does not succeed in proving corresponding continuity conditions for interconnected systems of the Popov type considered herein.

6.5.14. Example. To demonstrate the applicability of the method of the present section, we consider the system shown in Fig. 6.11. In the present case we make the identifications $l = 4$, $C_{11} = C_{22} = C_{33} = C_{44} = 0$, $C_{31} = C_{32} = L_5$, $C_{34} = L_8$, $C_{41} = C_{42} = L_6$, $C_{43} = 0$, $C_{12} = C_{21} = 0$, $C_{13} = C_{23} = L_7$, $C_{14} = C_{24} = L_9$.

Each N_i, $i = 1, 2, 3, 4$, is assumed to be a memoryless nonlinearity characterized by a function n_i with the property $0 \le xn_i(x) \le b_i x^2$ for all $x \in R$ and $b_i > 0$ some constant. Also for each i, $L_i \in \mathscr{L}$ has Laplace transform $L_i^*(s)$ with $|L_i^*(j\omega)|$ and $|j\omega L_i^*(j\omega)|$ bounded for all $\omega \in R^+$.

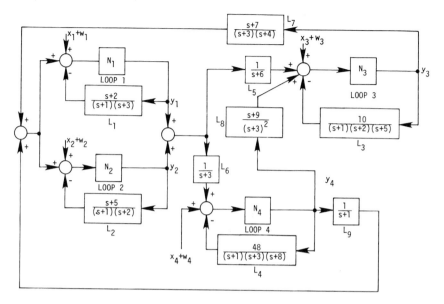

Figure 6.11 Block diagram of Example 6.5.14.

We seek to determine restrictions on the nonlinearities which are sufficient to ensure boundedness of the system of Fig. 6.11. This means that for each isolated subsystem determined by the four indicated loops we must find a Popov line with intercept $-b_i^{-1}$ and slope q_i^{-1} so that the plot of the modified frequency response, $I_i(j\omega)$, of the corresponding linear element L_i lies entirely to the right of this line. Furthermore, the distance from $I_i(j\omega)$ to its Popov line, measured parallel to the real axis, is the parameter Δ_i for the ith isolated subsystem.

Plots of the modified frequency responses of L_i, $i = 1, 2, 3, 4$, are given in Fig. 6.12, together with the Popov lines which are now discussed in the following.

For the time being, let us assume that a suitable set of Popov lines has been found and that the corresponding margins of boundedness, Δ_i, have been identified. The next step is to form the test matrix R of Theorem 6.5.7. Boundedness of the system of Fig. 6.11 is then ascertained if the successive principal minors of R are all positive. Note however that the choice of the Popov slope parameters q_i affects both Δ_i and the various gains which enter matrix R in somewhat subtle ways. Therefore, although the adjustable parameters q_i provide a desirable degree of flexibility, there does not seem to be a straightforward way of selecting the q_i's in an "optimal" manner.

Returning to the example on hand, we must compute the various gains

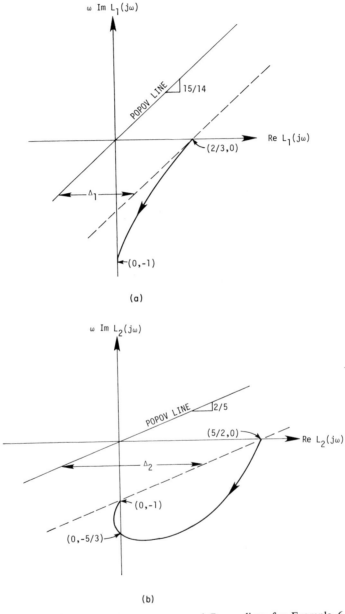

(a)

(b)

Figure 6.12   Modified frequency responses and Popov lines for Example 6.5.14. (a) $\Delta_1 = \frac{2}{3}$; $q_1 = \frac{14}{15}$. (b) $\Delta_2 = \frac{5}{2}$; $q_2 = \frac{5}{2}$.

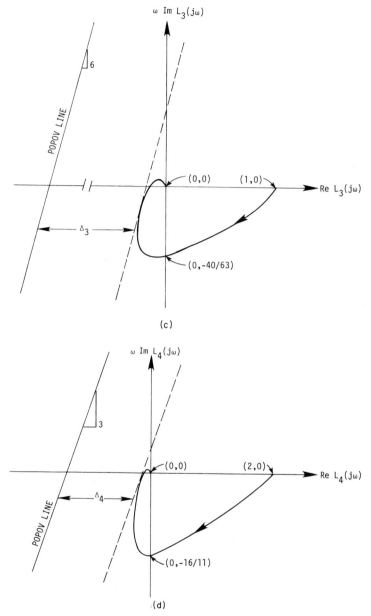

Figure 6.12 Modified frequency responses and Popov lines for Example 6.5.14. (c) $\Delta_3 = 2.0$; $q_3 = \frac{1}{6}$. (d) $\Delta_4 = 1.7$; $q_4 = \frac{1}{3}$.

$g(L_{i3}C_{ij})$ which occur in matrix R. As a sample calculation, we have

$$g(L_{33}C_{31}) = \max_{\omega \geq 0} |(1+j\omega q_3)(6+j\omega)^{-1}| = \max\{\tfrac{1}{6}, q_3\}.$$

In this fashion we obtain

$$R = \begin{bmatrix} \Delta_1 & 0 & -g(L_{13}C_{13}) & -\max\{1, q_1\} \\ 0 & \Delta_2 & -g(L_{23}C_{23}) & -\max\{1, q_2\} \\ -\max\{\tfrac{1}{6}, q_3\} & -\max\{\tfrac{1}{6}, q_3\} & \Delta_3 & -g(L_{33}C_{34}) \\ -\max\{\tfrac{1}{3}, q_4\} & -\max\{\tfrac{1}{3}, q_4\} & 0 & \Delta_4 \end{bmatrix}.$$

The remaining gains in R cannot be specified in simple closed form until more is known about q_1, q_2, and q_3. Roughly speaking, we want to choose the q_i's so that the diagonal terms Δ_i are large and the off-diagonal terms are small in magnitude.

As an illustration of the type of stability criteria which can be obtained, suppose that we allow b_1 and b_2 to be arbitrarily large, i.e., suppose we seek conditions on b_3 and b_4 sufficient to ensure boundedness for any finite, fixed values of b_1 and b_2. In an attempt to make Δ_1 and Δ_2 as large as possible without making the off-diagonal elements of R unnecessarily large, let us try $q_1 = 14/15$ and $q_2 = \tfrac{5}{2}$. Then, as indicated in Fig. 6.12, we have $\Delta_1 \geq \tfrac{2}{3}$ and $\Delta_2 \geq \tfrac{5}{2}$ for any finite b_1 and b_2. For this q_1 we have

$$g(L_{13}C_{13}) \triangleq r \equiv \max_{\omega \geq 0} |(1+j\omega q_1)(7+j\omega)(3+j\omega)^{-1}(4+j\omega)^{-1}|.$$

The maximum is attained at $\omega^2 \cong 5.50$, which yields $r \cong 1.01$. Similarly for this q_2 we have

$$g(L_{23}C_{23}) = \max_{\omega \geq 0} |(1+j\omega q_2)(7+j\omega)(3+j\omega)^{-1}(4+j\omega)^{-1}| \triangleq s \cong 2.6525$$

with maximum attained at $\omega^2 \cong 7.695$.

In order to minimize the remaining off-diagonal terms, we also require $q_3 \leq \tfrac{1}{6}$ and $q_4 \leq \tfrac{1}{3}$. It is then easy to show that

$$g(L_{13}C_{13}) = \max_{\omega \geq 0} |(1+j\omega q_3)(9+j\omega)(3+j\omega)^{-2}| = 1$$

with the maximum attained at $\omega = 0$.

The off-diagonal elements of matrix $R = [r_{ij}]$ are now completely determined, and we have

$$R = \begin{bmatrix} \Delta_1 & 0 & -r & -1 \\ 0 & \Delta_2 & -s & -\tfrac{5}{2} \\ -\tfrac{1}{6} & -\tfrac{1}{6} & \Delta_3 & -1 \\ -\tfrac{1}{3} & -\tfrac{1}{3} & 0 & \Delta_4 \end{bmatrix}.$$

It is now an easy matter to show that all of the successive principal minors of the test matrix R are positive if and only if

$$3\Delta_1 \Delta_2 \Delta_3 \Delta_4 - (s\Delta_1 + r\Delta_2)(1 + \tfrac{1}{2}\Delta_4) - (\tfrac{5}{4}\Delta_1 + \Delta_2)\Delta_3 > 0.$$

If we set Δ_1 and Δ_2 equal to their minimum possible values and substitute for r and s, we obtain the condition

$$\Delta_3 \Delta_4 > 0.833\Delta_3 + 0.429\Delta_4 + 0.854 \tag{6.5.15}$$

as a sufficient condition for boundedness. Therefore, the composite system of Fig. 6.11 will be bounded if we can find Popov lines for the isolated subsystems determined by Loop 3 and Loop 4 with $q_3 \le \tfrac{1}{6}$ and $q_4 \le \tfrac{1}{3}$ and with margins of boundedness satisfying inequality (6.5.15). One suitable choice is $q_3 = \tfrac{1}{6}$, $q_4 = \tfrac{1}{3}$, $\Delta_3 = 2.0$, and $\Delta_4 = 1.7$, which leads to the Popov lines shown in Fig. 6.12. These choices are made most conveniently by graphical rather than analytic techniques. An approximate graphical analysis yields the bounds for the corresponding nonlinearities as

$$b_3 \cong 0.47 \quad \text{and} \quad b_4 \cong 0.55.$$

6.6 L_∞-Stability and l_∞-Stability of Large Scale Systems

In the present section we prove some results for L_∞ boundedness (and l_∞ boundedness) using the shifted L_2-theory mentioned in Section 6.1. This approach is not the only method of proving boundedness with respect to this space. For example, it may be possible to use the theory developed in Section 6.2, together with some space such as L_∞, l_∞, BC, C_1, C_0, and the like. Also, recall from Section 6.5 that in Theorem 6.5.7 the solutions y_i of system (6.5.1) are not only in $L_2(R^+, R)$ but also in $C_0 \subset L_\infty$ (see the remark preceding Definition 6.5.4).

Since shifted L_2-spaces are used repeatedly in this section, we employ the notation

$$X(\sigma, m) = (L_2(R^+, R^m))_g$$

where g is the weighting function $g(t) = \exp(\sigma t)$ and $X_e(\sigma, m) = (L_{2e}^m)_g$ is the extended space. Also, we use the notation

$$\langle f_1, f_2 \rangle_{\sigma, T} = \int_0^T f_2^{\mathrm{T}}(t) f_1(t) \exp(-2\sigma t) \, dt$$

and

$$\|f\|_{\sigma, T} = \langle f, f \rangle_{\sigma, T}^{1/2}.$$

We will require the following result.

6.6.1. Lemma. If $0 > \sigma > \sigma_1 = \sigma - \mathscr{E}$ and if the relation H satisfies any one of the properties

   (i) H is inside the sector $\{a, b\}$,
   (ii) H is outside the sector $\{a, b\}$,
  (iii) $H + aI$ is positive,

with respect to the space $X(\sigma_1, m)$, then H satisfies the same properties with respect to $X(\sigma, m)$.

Proof. If (i) is true, let $\varphi(t) = (Hx(t) - bx(t))^{\mathrm{T}}(Hx(t) - ax(t))e^{-2\sigma_1 t}$ and let

$$\psi(T) = \int_0^T \varphi(\tau)\, d\tau = \langle Hx - ax, Hx - bx \rangle_{\sigma_1, T}.$$

Then $\psi(T) \le 0$ for all $T > 0$ by (i). Integrating by parts yields

$$\langle Hx - ax, Hx - bx \rangle_{\sigma, T}$$

$$= \int_0^T [d\psi(t)/dt]\, \exp(-2\mathscr{E}t)\, dt$$

$$= \psi(T) \exp(-2\mathscr{E}T) + 2\mathscr{E} \int_0^T \left[\int_0^t \varphi(\tau)\, d\tau \right] \exp(-2\mathscr{E}t)\, dt$$

$$= \psi(T) \exp(-2\mathscr{E}T) + 2\mathscr{E} \int_0^T \psi(t) \exp(-2\mathscr{E}t)\, dt \le 0.$$

Similar arguments can also be employed in the proofs of cases (ii) and (iii). ∎

Since the statements in the following definition can be restated in terms of sector and positivity conditions, these statements will hold for $\sigma \in (\sigma_1, 0)$ whenever they hold for any $\sigma_1 < 0$. In this definition we concern ourselves with multi input–multi output systems of the type considered in Sections 6.2–6.4 (see in particular system $(\mathscr{S})$ in Section 6.4, Part A).

6.6.2. Definition. A multi input–multi output system of the form

$$\begin{aligned} e &= x + w + z, & y &= He \\ f &= u + v + y, & z &= Bf \end{aligned} \qquad (\mathscr{S})$$

viewed as a single-loop system, is said to possess **Property D** with weight $\sigma < 0$ if for some real constants $c, r \ge 0$ and $\delta \in (0, 1)$ and for the space $X(\sigma, m)$, the relation B is interior conic $(c, r(1 - \delta))$ and the relation H satisfies one of the following conditions:

(i) $c^2 > r^2$ and H is exterior conic $(-c(c^2 - r^2)^{-1}, r(c^2 - r^2)^{-1})$, or
(ii) $r^2 > c^2$ and H is interior conic $(c(r^2 - c^2)^{-1}, r(r^2 - c^2)^{-1})$, or
(iii) $r^2 = c^2$ and $2cH + I$ is positive.

We now prove the following preliminary result.

6.6.3. Lemma. If system $(\mathscr{S})$ possesses Property D, then in either of the special cases $x = w = 0$ or $u = v = 0$, the input–output relation connecting u or x to f or e, respectively, has finite gain which cannot exceed $(|c| + |r|)(\delta r)^{-1}$.

Proof. This is just Theorem 6.4.14 and its proof restated. If $u = v = 0$, then by the proof of that theorem the transformed entities

$$e' = e + cf, \qquad y' = y = f = f' = H'e'$$

satisfy

$$\|e'\| \le \delta^{-1}\|x + w\|, \qquad \|y\| \le g(H')\|e'\| \le r^{-1}\|e'\|.$$

Thus,

$$\|e\| \le \|e'\| + |c|\,\|f\| = \|e'\| + |c|\,\|y\|$$
$$\le \delta^{-1}\|x + w\| + |c|\,r^{-1}\delta^{-1}\|x + w\|$$
$$= (|c| + r)(\delta r)^{-1}\|x + w\|.$$

If $x = w = 0$, then by the proof of Theorem 6.4.14 we have

$$\|f\| \le \delta^{-1}(1 + r^{-1}|c|)\|u + v\| = (r + |c|)(\delta r)^{-1}\|u + v\|.$$

This concludes the proof. ∎

We now consider interconnected systems described by the set of equations

$$e_i = u_i + B_i y_i, \qquad y_i = H_i e_i, \qquad u_i = x_i + w_i + \sum_{j=1}^{l} C_{ij} y_j, \quad (6.6.4)$$

$i = 1, ..., l$. In Eq. (6.6.4) we have $e_i(t), u_i(t), y_i(t), x_i(t), w_i(t) \in R^{n_i}$, and $n = \sum_{i=1}^{l} n_i$. This system may be viewed as an interconnection of isolated subsystems described by the set of equations

$$e_i = x_i + w_i + B_i y_i, \qquad y_i = H_i e_i, \qquad (6.6.5)$$

$i = 1, ..., l$. The relations B_i, H_i, and C_{ij} have the same form as corresponding ones in Eq. (6.4.15) while X_i is the space $X(\sigma_i, n_i)$ where $\sigma_i < 0$ for $i = 1, ..., l$. The structure of system (6.6.4) is depicted in Fig. 6.3.

In the subsequent results we require the notion of L_1-memory (see Definition 6.1.22 and Theorems 6.1.23 and 6.1.24).

6.6.6. Theorem. Assume that all isolated subsystems (6.6.5), defined on $X_e(\sigma_i, n_i)$, satisfy Property D for some constants $\sigma_i < 0$, $r_i \ge 0$, and c_i, $i = 1, ..., l$. Let $\sigma = \max_i\{\sigma_i\} < 0$. If each H_i has decaying L_1-memory, if each mapping $(C_{ij} H_j)$: $X(\sigma, n_j) \to X(\sigma, n_i)$ has finite gain and if the successive principal minors of the test matrix $R = [r_{ij}]$ defined by

$$r_{ij} = \begin{cases} \delta_i r_i (|c_i| + |r_i|)^{-1} - g(C_{ii} H_i), & \text{if } i = j \\ -g(C_{ij} H_j), & \text{if } i \neq j \end{cases}$$

are all positive, then the output relations Y_i, $i = 1, ..., l$, associated with composite system (6.6.4) are all bounded with respect to $(L_\infty^n, L_\infty^{n_i})$.

Proof. By Lemma 6.6.1 we can assume that $\sigma = \sigma_1 = \cdots = \sigma_l$. Furthermore, Lemma 6.6.3 and Eq. (6.6.4) imply that for each i,

$$\|e_i\|_{\sigma, T} \leq (|c_i| + r_i)(\delta_i r_i)^{-1} \|u_i\|_{\sigma, T}$$

$$\leq (|c_i| + r_i)(\delta_i r_i)^{-1} \left\{ \|x_i + w_i\|_{\sigma, T} + \sum_{j=1}^{l} g(C_{ij} H_j) \|e_j\|_{\sigma, T} \right\}.$$

Let $E_T = [\|e_1\|_{\sigma, T}, ..., \|e_l\|_{\sigma, T}]^T$ and define $(X + W)_T$ similarly. Then the above inequalities can be written as

$$R E_T \leq (X + W)_T$$

where R is the test matrix. Observing once more that R is an M-matrix, it follows that R^{-1} exists and that $R^{-1} = D = [d_{ij}] \geq 0$, so that

$$E_T \leq D(X + W)_T$$

or

$$\|e_i\|_{\sigma, T} \leq \sum_{j=1}^{l} d_{ij} \|x_j + w_j\|_{\sigma, T} \tag{6.6.7}$$

whenever $x, w \in X(\sigma, n)$.

If $x, w \in L_\infty^n$, then certainly $x, w \in X(\sigma, n)$ for any $\sigma < 0$. Furthermore, we have

$$\|x_j + w_j\|_{\sigma, T} \leq \left(\int_0^T \|x_j + w_j\|_{L_\infty}^2 e^{-2\sigma t} dt \right)^{1/2}$$

$$\leq \|x_j + w_j\|_{L_\infty} [(e^{-2\sigma T} - 1)/(-2\sigma)]^{1/2}$$

$$\leq \|x_j + w_j\|_{L_\infty} e^{-\sigma T}/(-2\sigma)^{1/2},$$

so that

$$e^{\sigma T} \|e_i\|_{\sigma, T} \leq \sum_{j=1}^{l} (d_{ij}/(-2\sigma)^{1/2}) \|x_j + w_j\|_{L_\infty}.$$

By Theorem 6.1.24 it now follows that $\|y\|_{L_\infty} < \infty$. ∎

When C_{ij} and H_j are linear time-invariant operators belonging to $\mathscr{L}$, with $C_{ij} e^\sigma \in \mathscr{L}$ and $H_j e^\sigma \in \mathscr{L}$ for some $\sigma < 0$, then the gains $g(C_{ij} H_j)$ can be obtained as the gains of the relations $e^{-\sigma}C_{ij}H_j e^{\sigma}$ from $L_2(R^+, R^{n_j})$ to $L_2(R^+, R^{n_i})$, that is,

$$g(C_{ik} H_k) = \operatorname*{ess\,sup}_{\omega \geq 0} |C_{ik}^*(\sigma + j\omega) H_k^*(\sigma + j\omega)|.$$

In this case we may replace the gains $g(C_{ik} H_k)$ in R by

$$\operatorname*{ess\,sup}_{\omega \geq 0} |C_{ik}^*(j\omega) H_k^*(j\omega)| \qquad (6.6.8)$$

and check that the successive principal minors of the resulting matrix are all positive. For it then follows by continuity that the test matrix R will have the desired properties for $|\sigma|$ sufficiently small.

Using Theorem 6.6.6 as a model, an incremental counterpart of this theorem can easily be stated and proved.

Analogous discrete time systems can be treated in a similar manner with L_∞-spaces replaced by l_∞-spaces. In this case Definition 6.6.2 and Lemma 6.6.3 remain valid without change, while in the proof of Lemma 6.6.1 we need to replace integration by parts with summation by parts, given by

$$\sum_{i=1}^{N} u_i(v_{i+1} - v_i) = u_{N+1} v_{N+1} - u_1 v_1 - \sum_{i=1}^{N} (u_{i+1} - u_i) v_{i+1}.$$

Letting $I = \{0, 1, 2, \ldots\}$, we use in the present case the notation

$$Z(\sigma, m) = (l_2(I, R^m))_g$$

where for $t \in I$, $g(t) = \exp(\sigma t)$.

For the discrete time version of composite system (6.6.4) with corresponding isolated subsystems of the form (6.6.5) the following result is now easily established.

6.6.9. Theorem. Assume that all isolated subsystems (6.6.5) of composite system (6.6.4) possess Property D for some constants $\sigma_i < 0$, $r_i \geq 0$, and c_i. Let $\sigma = \max\{\sigma_1, \ldots, \sigma_l\}$. If each H_i has decaying l_1-memory, if each mapping $C_{ij} H_j$ maps $Z(\sigma, n_j)$ into $Z(\sigma, n_i)$ with finite gain $g(C_{ij} H_j)$ and if all successive principal minors of the test matrix $R = [r_{ij}]$ determined by

$$r_{ij} = \begin{cases} \delta_i r_i (|c_i| + r_i)^{-1} - g(C_{ij} H_j), & \text{if } i = j \\ -g(C_{ij} H_j), & \text{if } i \neq j \end{cases}$$

are positive, then the input–output relations Y_i associated with composite system (6.6.4) are bounded with respect to $(l_\infty^n, l_\infty^{n_i})$.

We conclude this section with a specific example.

6.6.10. Example. Consider the continuous time interconnected system shown in Fig. 6.13. Here T_1, T_2, T_3, and a are positive constants. The operators

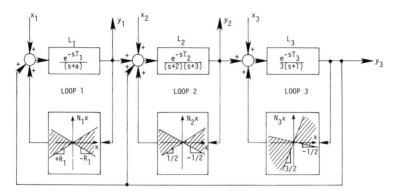

Figure 6.13   Block diagram of Example 6.6.10 where a, T_1, T_2, T_3 are positive constants.

L_1, L_2, L_3 characterized by the indicated Laplace transforms belong to class $\mathscr{L}_e$ and for some constants $\sigma_i < 0$, $e^{\sigma_i} L_i \in \mathscr{L}$, $i = 1, 2, 3$. The operators N_1, N_2, and N_3 are assumed to belong to class $\mathscr{N}$ and are assumed to have graphs in the indicated shaded regions. Figure 6.13 indicates the identification $l = 3$, $C_{21} = C_{13} = C_{32} = C_{23} = 1$, and all other $C_{ij} = 0$.

In the sequel we take full advantage of the remark associated with (6.6.8). Thus, we consider the unshifted Nyquist plots of the linear elements L_i, $i = 1, 2, 3$. When a conicity condition is referred to, no mention will be made of a particular weight. As indicated in the remark for (6.6.8), if the conditions which we will derive are satisfied, then a negative weight σ, with $|\sigma|$ sufficiently small, can be found so that L_∞-boundedness is guaranteed by Theorem 6.6.6 (with all conicity conditions, positivity conditions, etc., being interpreted with respect to that weight). Specifically, in what follows we seek to establish a relationship between the two positive parameters a and R_1 in Fig. 6.13 to guarantee that this system be L_∞-bounded. We first analyze the isolated subsystems determined by the indicated loops.

Loop 1. N_1 is interior conic $(0, R_1)$. Setting $c_1/(r_1{}^2 - c_1{}^2) = 0$ and $r_1/(r_1{}^2 - c_1{}^2) = R_1$, we find $c_1 = 0$ and $r_1 = 1/R_1$. This subsystem will have margin of boundedness δ_1 provided that L_1 is interior conic with center 0 and radius $(1 - \delta_1) r_1$. This will be the case if $|e^{-j\omega T_1}(a + j\omega)^{-1}| \leq (1 - \delta_1) r_1$ for all $\omega \in R^+$. This is true for *all* $T_1 \geq 0$ if $1/a \leq (1 - \delta_1)/R_1$. The best value of δ_1 is therefore given by

$$\delta_1 = 1 - (R_1/a). \tag{6.6.11}$$

Loop 2. N_2 is interior conic $(0, \tfrac{1}{2})$. Setting $c_2/(r_2{}^2 - c_2{}^2) = 0$ and $r_2/(r_2{}^2 - c_2{}^2) = \tfrac{1}{2}$, we find $c_2 = 0$ and $r_2 = 2$. This subsystem will have margin of boundedness δ_2 provided that L_2 is interior conic $(0, 2(1 - \delta_2))$. This will be the case if $|e^{-j\omega T_2}(2 + j\omega)^{-1}(3 + j\omega)^{-1}| \leq 2(1 - \delta_2)$ for all $\omega \in R^+$. This is true for *all* $T_2 \geq 0$ if $\frac{1}{6} \leq 2(1 - \delta_2)$. The best value of δ_2 is therefore given by

$$\delta_2 = 1 - \tfrac{1}{12} = \tfrac{11}{12}.$$

Loop 3. N_3 is interior conic $(\frac{1}{2}, 1)$. Setting $c_3/(r_3{}^2 - c_3{}^2) = \frac{1}{2}$ and $r_3/(r_3{}^2 - c_3{}^2) = 1$, we obtain $c_3 = \frac{2}{3}$ and $r_3 = \frac{4}{3}$. This subsystem will have margin of boundedness δ_3 if L_3 is interior conic $(\frac{2}{3}, (1 - \delta_3)\frac{4}{3})$. This will be the case if $|e^{-j\omega T_3}[3(1 + j\omega)]^{-1} - \frac{2}{3}| \leq (1 - \delta_3)\frac{4}{3}$ for all $\omega \in R^+$. This is true for *all* $T_3 \geq 0$ if $1 \leq (1 - \delta_3)\frac{4}{3}$. The best value of δ_3 is therefore given by

$$\delta_3 = 1 - \tfrac{3}{4} = \tfrac{1}{4}.$$

Using the above calculations together with the gains $g(L_1) = 1/a$, $g(L_2) = \frac{1}{6}$, $g(L_3) = \frac{1}{3}$, we obtain the test matrix

$$R = \begin{bmatrix} \delta_1 & 0 & -1/3 \\ -1/a & 11/12 & -1/3 \\ 0 & -1/6 & 1/6 \end{bmatrix}.$$

The successive principal minors of R are all positive if and only if

$$\delta_1 > 4/(7a). \tag{6.6.12}$$

Combining (6.6.11) and (6.6.12) it follows from Theorem 6.6.6 that the system of Fig. 6.13 is L_∞-bounded if

$$a > R_1 + 4/7,$$

i.e., if the single pole of L_1 is at least a distance $R_1 + 4/7$ into the left-half plane. Note that this condition is independent of $T_1 \geq 0$, $T_2 \geq 0$, $T_3 \geq 0$. As expected, the condition on L_1 becomes more restrictive as the condition on N_1 becomes less restrictive, and conversely.

6.7 Analysis and Design Procedure

The results of Sections 6.2–6.6 can readily be used in analysis and design procedures of large scale systems. For purposes of discussion, we consider Theorem 6.4.17 in the following procedure. Similar statements apply to a great portion of the other results.

Step 1. Impose the constraints that each isolated subsystem (described by Eq. (6.4.16)) have margins of boundedness δ_i, $0 < \delta_i < 1$. Calculate the corresponding gain factors μ_i.

Step 2. Form the test matrix R (of Theorem 6.4.17). Boundedness conditions are obtained by requiring that each of the successive principal minors of matrix R be positive.

Step 3. If the boundedness conditions are not satisfied, modify some or all of the isolated subsystems so as to reduce μ_i. Repeat Step 2.

Although the present results may not represent the "best" stability conditions, they have the desirable property that they single out a class of modifications that can be made to each isolated subsystem separately to enhance overall system stability for the types of systems considered herein. This is possible in part because we frequently view large scale systems (i.e., multi input–multi output systems) as interconnected systems and in part because we often focus attention on large scale systems with subsystems which are single-loop feedback systems. Such a viewpoint is often possible when there is nontrivial feedback in a large scale system, and is a natural one when one attempts to stabilize such systems using *local feedback*. To be more specific, consider a system consisting of l "forward-loop" relations H_i interconnected by a number of feedback relations C_{ij}, as shown in Fig. 6.14. If the conditions of Theorem 6.2.3 do not already yield boundedness, place a compensating feedback relation B_i around each forward loop relation H_i, each B_i being chosen so that the resulting single-loop is bounded with margin of boundedness δ_i. This procedure presupposes only that each H_i satisfies one of the conicity conditions of Theorem 6.4.14 for *some* choices of parameters r, c, and δ, which is not a very restrictive requirement. The *compensated system* then has the structure of Eq. (6.4.15). If the boundedness condition of Theorem 6.4.17 is not satisfied, then alter each of the feedback relations B_i, and thus the corresponding δ_i and μ_i, until it is. This can be accomplished by making the quantity μ_i sufficiently small. (In this connection, note that the test matrix $R = [r_{ij}]$ of Theorem 6.4.17 is required to be an M-matrix. Furthermore, recall that a sufficient condition for R to be an M-matrix is the set of inequalities $r_{ii} - \sum_{j=1, i \neq j}^{l} |r_{ij}| > 0$, $i = 1, \ldots, l$.)

At this point we hasten to emphasize that in some cases, application of the preceding compensation procedure may not be practical nor desirable. For example, certain "forward loop" relations may not be physically accessible to compensation, the cost may be prohibitive, etc.

The similarity between the above procedure and those suggested in earlier

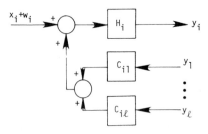

Figure 6.14   Typical interconnected feedback system where $i = 1, \ldots, l$.

chapters is of course not coincidental. After all, we have pursued throughout this book the same basic method of analysis. Only the technical details, frequently not trivial, have changed under vastly different situations.

6.8 Notes and References

Good references on systems theory (dealing with input–output properties of feedback systems) include the books by Willems [1], Holtzman [1], and Desoer and Vidyasagar [1]. A great deal of the fundamental work dealing with this subject is due to Sandberg (see, e.g., Sandberg [1–3, 7–9]), Zames [1–4], and Popov [1].

For additional details concerning the variety of spaces considered in Section 6.1, refer to the books by Corduneanu [2, Chapter I] and Miller [1, Section 5 of Chapter III and Chapter V]. Almost periodic functions are discussed in detail by Besicovitch [1] and Fink [1]. For applications of ultimately periodic or almost periodic spaces, see Beneš [3, 4], Beneš and Sandberg [1], and Miller [2]. Weighted spaces are dicussed in Corduneanu [2], Massera and Schaeffer [2], Miller [1], and Gollwitzer [1], as well as in Willems [1], Desoer and Vidyasagar [1] and the work by Sandberg and Zames for the special case of shifted L_p-spaces. In Definition 6.1.4 the term "compatible space" is used while in Corduneanu [2], Massera and Schaeffer [2], Gollwitzer [1], and Dotseth [1] the term "X is a subspace of X_e with stronger topology than that inherited from X_e" is used to express the same idea. Results on boundedness and continuity of operators $H \in \mathscr{L}_e$ defined on $X = L_2{}^m$ can be found in the work of Sandberg and Zames and in Desoer and Vidyasagar [1], Holtzman [1], and Willems [1]. Boundedness results of operators $H \in \mathscr{L}_e$ defined on various spaces are also given in Corduneanu [2], Massera and Schaeffer [2], Miller [1], Gollwitzer [1], and Dotseth [1]. Shifted L_p-spaces and decaying L_1-memory are discussed in detail in Zames [2]. Finally, for a connection between Lyapunov stability and input–output stability, refer to Willems [2].

The results in Section 6.2 are due to Lasley and Michel [1–6] and Lasley [1]. As presented here, these results are stated in sufficiently general form to include the interesting results of Araki [3], Cook [2], Tokumaru, Adachi, and Amemiya [1], and part of the work of Porter and Michel [3–5] and Porter [1].

The instability results in Section 6.3 are an adaptation of the interesting work of Takeda and Bergen [1]. For related work which generalizes or partially generalizes the results of Takeda and Bergen, refer to Vidyasagar [1] and Miller and Michel [1]. An integral equation example illustrating instability is given in the next chapter.

The results in Section 6.4 are an adaptation of the work of Porter and Michel [3–5], Porter [1], Lasley and Michel [1, 4, 6], and Lasley [1]. For related results see Araki [3], Cook [2], and Sundareshan and Vidyasagar [1].

The Popov-like results of Section 6.5 are due to Lasley and Michel [2, 4, 6] and Lasley [1].

The material presented in Section 6.6 dealing with L_∞- and l_∞-stability is based on results by Lasley and Michel [3, 5, 6] and Lasley [1].

For additional related work refer to the papers by Cook [1], Miller and Michel [2], Rosenbrock [3], and Callier, Chan, and Desoer [1, 2]. Also, refer to Kevorkian [1] and Özgüner and Perkins [1], where structural aspects of interconnected systems are considered.

CHAPTER VII

Integrodifferential Systems

In the present chapter we apply the general stability and instability results of Chapter VI to interconnected systems described by Volterra integrodifferential equations. Throughout this chapter we employ the notation and conventions of the previous chapter.

In the first section some preliminary background material is provided. In the second section L_2-stability and instability results for interconnected systems described by Volterra integrodifferential equations are established. The emphasis in the third section is on linear systems and nearly linear systems. In the fourth section results of this chapter are applied to the point kinetics model of a coupled nuclear reactor with several cores. This chapter is concluded with a discussion of references in the fifth section.

7.1 Preliminary Results

A **linear integrodifferential equation** with **kernel** $H \in \mathscr{L}_{1e}$ is an equation of the form

$$\dot{y}(t) = Hy(t) + f(t) \tag{L}$$

for $t \geq 0$ and $y(0) = y_0$ and input $f \in X_e$. The **resolvent** of (L) is the function $R \in \mathscr{L}_{0e}$ such that

$$R^*(s) = (sI - H^*(s))^{-1}$$

where $R^*(s)$ and $H^*(s)$ denote the Laplace transforms of the operators R and H, respectively. We say that H possesses **Property F** if

$$\det[j\omega I - H^*(j\omega)] \neq 0, \qquad -\infty < \omega < \infty.$$

Also, we say that H possesses **Property Q** if

$$\det(sI - H^*(s)) \neq 0$$

for all complex numbers $s = \sigma + j\omega$ such that $\sigma = \operatorname{Re} s \geq 0$.

If R is the resolvent of H, then a simple argument involving Laplace transforms shows that the solution of (L) is given by

$$y(t) = R(t)y_0 + \int_0^t R(t-\tau)f(\tau)\,d\tau. \qquad (7.1.1)$$

Moreover, the following results are proved in Miller [4] and Grossman and Miller [1].

7.1.2. Theorem. If $H \in \mathscr{L}_1$ then $R \in \mathscr{L}_0$ if and only if H has Property Q. In this case the derivative $\dot{R}$ is also in $\mathscr{L}_0$.

7.1.3. Theorem. If $H \in \mathscr{L}_1$ and H has Property F then there is a finite (possibly empty) set $P = \{s_1, \ldots, s_N\}$ of complex numbers such that $\operatorname{Re} s_j > 0$, there is a set $\{M_{jk}\}$ of matrices, there are integers $m_j \geq 1$, and there is an operator S such that $|S|$ and $|\dot{S}|$ belong to $L_p(R^+)$ for all $p \in [1, \infty]$ such that

$$R^*(s) = S^*(s) + \sum_{j=1}^{N} \sum_{k=1}^{m_j} M_{jk}(s - s_j)^{-k}.$$

The operator S in Theorem 7.1.3 is called the **residual resolvent** of Eq. (L). When $P = \varnothing$ then $R \in \mathscr{L}_0$ and $S = R$, by Theorem 7.1.2.

All **nonlinear integrodifferential equations** considered in this chapter are of the form

$$\dot{x}(t) = Hx(t) + B(x_t, t) + f(t) \qquad (N)$$

for $t \geq 0$ with $x(0) = x_0$ given. Here $H \in \mathscr{L}_{1e}, f \in X_e$, B is a continuous mapping on $X_e \times R^+$ into R^n, and x_t denotes the truncated function. These assumptions ensure that Eq. (N) has solutions which are unique if B satisfies a Lipschitz condition in x (see Miller [1, Chapter II]). If one thinks of $B(x_t, t) + f(t)$ as known, then from Eqs. (7.1.1) and (N) the **variation of constants formula**

$$x(t) = R(t)x_0 + R[f(t) + B(x_t, t)] \qquad (V)$$

follows.

7.2 L_2-Stability and Instability of Interconnected Systems

Now in particular, let $X_e = L_{2e}^n$ and let $X = L_2^n$. In the following, we apply some of the results from Sections 6.2 and 6.3 to equation (N).

Let us assume that Eq. (N) can be written in the form

$$\dot{z}_i(t) = H_i z_i(t) + \sum_{j=1}^{l} B_{ij}(z_{jt}, t) + f_i(t), \qquad (\mathscr{S})$$

$i = 1, \ldots, l$, for $t \geq 0$ with $z_i(0) = z_{i_0}$ given. Here $H_i \in \mathscr{L}_{1e}$ is $n_i \times n_i$ matrix valued, $B_{ij} \colon L_{2e}(R^+, R^{n_j}) \times R^+ \to L_{2e}(R^+, R^{n_i})$, $f_i \in X_{ie} = L_{2e}(R^+, R^{n_i})$, and $n = \sum_{i=1}^{l} n_i$. Following our earlier viewpoint, we consider **composite system** ($\mathscr{S}$) as an interconnection of l **isolated subsystems** ($\mathscr{S}_i$) described by equations of the form

$$\dot{y}_i(t) = H_i y_i(t) + f_i(t), \qquad y_i(0) = z_{i_0}, \qquad (\mathscr{S}_i)$$

with **interconnecting structure** specified by B_{ij}, $i, j = 1, \ldots, l$.

7.2.1. Theorem. Assume that $H_i \in \mathscr{L}_1$ possesses Property Q and that $f_i \in X_i$ for $i = 1, \ldots, l$. Assume that all B_{ij} are bounded with respect to X and assume that the gains $g(B_{ij}) < \infty$ for $i, j = 1, \ldots, l$. If the successive principal minors of the test matrix $A = [a_{ij}]$ defined by

$$a_{ij} = \begin{cases} 1 - g(R_i)\, g(B_{ij}), & \text{if} \quad i = j \\ -g(R_i)\, g(B_{ij}), & \text{if} \quad i \neq j \end{cases}$$

are all positive, then the input–output relations Z_i for composite system ($\mathscr{S}$) are stable with respect to X. (Here the f_i are inputs and the z_i are outputs.)

Proof. By Theorem 7.1.2 the operator $H = \text{diag}[H_1, \ldots, H_l]$ has resolvent $R = \text{diag}[R_1, \ldots, R_l] \in \mathscr{L}_0$. By the results in Section 6.1, Part D, R maps X_e into X_e with finite gain. Thus, Eq. ($\mathscr{S}$) is equivalent to the variation of constants equation (V) and Theorem 6.2.3 can be applied to (V). The conclusion of the theorem follows. ■

7.2.2. Remark. Note that the above theorem remains valid if X_e is any extended space such that R maps X_e into itself and X is the completely compatible subspace. For example, we may have $X_e = L_{pe}^n$ and $X = L_p^n$ for $1 \leq p \leq \infty$, or we may have $X_e = C$ and $X = BC$. Of course the gains must always be computed with respect to the given space X.

The conclusion of Theorem 7.2.1 can be strengthened as follows.

7.2.3. Theorem. Assume that all hypotheses of Theorem 7.2.1 are true. Then the solution $x(t)$ of composite system ($\mathscr{S}$) is in $C_0 \cap L_2^n$.

Proof. By Theorem 7.2.1, $f \in L_2^n = X$ implies that $x \in X$. Hence, $F(t) \triangleq f(t) + B(x_t, t) \in X$. By Theorems 7.1.2 and 7.1.3 we also have $R \in L_p(R^+, R^{n^2})$ for $p = 1, 2$ and $\dot{R} \in L_p(R^+, R^{n^2})$ for $p = 1, 2$. Thus, the convolution

$$(RF)(t) = \int_0^t R(t-\tau) F(\tau) \, d\tau$$

is in BC (by the Schwarz inequality) and $|RF(t)| \le \|R\|_{L_2} \|F\|_{L_2}$ for $t \ge 0$. Furthermore, the derivative

$$\frac{d}{dt}[(RF)(t)] = R(0) F(t) + \int_0^t \frac{d}{dt}[R(t-\tau)] F(\tau) \, d\tau$$

is also in L_2^n. Thus, $(RF)(t) \to 0$ as $t \to \infty$ and

$$x(t) = R(t) x_0 + (RF)(t) \in C_0. \quad \blacksquare$$

It is also possible to establish reasonable conditions for composite system $(\mathscr{S})$ such that instability of one of the isolated subsystems $(\mathscr{S}_i)$ implies instability of $(\mathscr{S})$.

7.2.4. Theorem. Assume that $H_i \in \mathscr{L}_1$ and $f_i \in X$ for $i = 1, \ldots, l$. Assume that all B_{ij} are stable with respect to X with finite gains $g(B_{ij})$ for $i, j = 1, \ldots, l$. Suppose that each H_i possesses either Property F or Property Q and suppose that for at least one value $i = i_0$, H_i satisfies Property F only. Define $M = [m_{ij}] = [g(B_{ij}) g_c(R_j)]$ and let $r(M)$ be its spectral radius, i.e.,

$$r(M) = \max\{|\lambda_i| : \lambda_i \text{ is an eigenvalue of } M\}.$$

If either $r(M) < 1$ or if $r(M) = 1$ and $m_{ij} > 0$ for all i and j, then the input–output relations for composite system $(\mathscr{S})$ are not bounded with respect to X.

Proof. We shall apply Theorem 6.3.1. Pick $z_{i0} = 0$ for $i = 1, \ldots, l$ so that Eq. (V) becomes $z_i = R_i(f_i + \sum_{j=1}^l B_{ij} z_j)$, or

$$e_i = f_i + \sum_{j=1}^l B_{ij} z_j \qquad (7.2.5)$$

$$z_i = R_i e_i$$

for $i = 1, \ldots, l$. Here f_i is in the orthogonal complement of the stable manifold of R_i, i.e., $f_i \in S(R_i)^\perp$. In order to apply Theorem 6.3.1 it is necessary to show that $g_c(R_i) < \infty$ for all i and that $S(R_i)^\perp \ne \varnothing$ when $R_i \notin \mathscr{L}_0$.

Since each R_i satisfies Property F it follows that $R_i^*(j\omega)$ is bounded, say $|R_i^*(j\omega)| \le M_i$ on $0 \le \omega < \infty$. If $y = R_i \varphi$ and if $\varphi \in X_i$, then by the Parseval equation and the bound for $|R_i^*(j\omega)|$ we have

$$\|y\|_{L_2} = \frac{1}{\sqrt{2\pi}} \|R_i^* \varphi^*\|_{L_2} \le \frac{M_i}{\sqrt{2\pi}} \|\varphi^*\|_{L_2} = M_i \|\varphi\|_{L_2}.$$

Thus, $g_c(R_i) \le M_i$ for all i.

For $i = i_0$ we assume that R_i possesses Property F but not Property Q. By Theorem 7.1.3 there is a nonempty set $P = \{s_1, \ldots, s_N\}$ in the right half of the complex plane such that

$$R_i \varphi(t) = S\varphi(t) + \sum_{j=1}^{N} \sum_{i=1}^{m_j} M_{ji} \int_0^t e^{s_j(t-\tau)} \frac{(t-\tau)^{i-1}}{(i-1)!} \varphi(\tau) \, d\tau.$$

Since

$$\frac{d^m}{ds^m} \varphi^*(s_j) = \int_0^\infty (-\tau)^m e^{-s_j\tau} \varphi(\tau) \, d\tau \triangleq d_{jm}(\varphi)$$

then $|R_i \varphi|$ can be in L_2 if and only if $d_{jm}(\varphi) = 0$ for $0 \le m \le m_j - 1$ and $j = 1, \ldots, N$. Thus, when $i = i_0$ we have

$$S(R_i)^\perp = \{\varphi \in X_i : d_{jm}(\varphi) \ne 0 \text{ for at least one } j \text{ and } m\} \ne \varnothing.$$

The conclusion of the theorem now follows from Theorem 6.3.1. ∎

Note that by Lemma 6.2.8 a sufficient condition for $r(M) < 1$ is that all successive principal minors of the test matrix $A = [a_{ij}]$ defined by

$$a_{ij} = \begin{cases} 1 - g(B_{ii}) g_c(R_i), & \text{if } i = j \\ -g(B_{ij}) g_c(R_j), & \text{if } i \ne j \end{cases}$$

are all positive.

7.3 Linear Equations and Linearized Equations

We now consider some properties of linear equations (L) and nearly linear equations (N). In the present section we define boundedness of system (N) as follows.

7.3.1. Definition. Given a space X we say that **system (N) is bounded** with respect to X if for each $f \in X$ and each initial condition $x_0 \in R^n$ the solution of (N) is in X.

Note that system (L) is included as a special case of (N). We will require the following result.

7.3.2. Theorem. Consider system (L) where $H \in \mathscr{L}_1$ is defined as

$$H\varphi(t) = h_0 \varphi(t) + \int_0^t h(t-\tau) \varphi(\tau) \, d\tau$$

and $t^r |h(t)| \in L_2(R^+)$ for some $r > 1$. Then the following conditions are equivalent.

   (i)   The resolvent $R \in \mathscr{L}_0$;
   (ii) $\det(sI - h_0 - h^*(s)) \neq 0$ for all s such that $\operatorname{Re} s \geq 0$;
   (iii) system (L) is bounded with respect to $X = L_2{}^n$.

Proof. The equivalence of (i) and (ii) is just a restatement of Theorem 7.1.2. To show that (ii) is equivalent to (iii), assume that (ii) is true. Then by Theorems 7.1.2 and 7.1.3 it follows that $R = S$ and $|S| \in L_p$ for $p = 1, 2$. Since $|R| \in L_2$ we have $R(t) y_0 \in X$. Since $|R| \in L_1$ then by Theorem 6.1.13, $Rf \in X = L_2{}^n$ if $f \in X$. Applying these facts to Eq. (7.1.1) we see that the solution $y \in X$.

Conversely, to prove that (iii) implies (ii), assume that (ii) is not true. Then there is a nonzero vector y_0 and a complex number s_0 with $\operatorname{Re} s_0 \geq 0$ such that

$$[s_0 I - h_0 - h^*(s_0)] y_0 = 0.$$

Define $y(t) = \exp(s_0 t) y_0$ and

$$f(t) = \int_{-\infty}^{0} h(t - \tau) y(\tau) \, d\tau = \int_{t}^{\infty} h(\tau) y(t - \tau) \, d\tau.$$

Then

$$\dot{y}(t) - h_0 y(t) - \int_{0}^{t} h(t - \tau) y(\tau) \, d\tau - f(t)$$

$$= \dot{y}(t) - h_0 y(t) - \int_{-\infty}^{t} h(t - \tau) y(\tau) \, d\tau$$

$$= [s_0 y_0 - h_0 y_0 - h^*(s_0) y_0] \exp(s_0 t) = 0.$$

So y solves Eq. (L) and y is not in X. But the Schwarz inequality shows that

$$\int_{1}^{\infty} |f(t)|^2 \, dt \leq |y_0|^2 \int_{1}^{\infty} \left(\int_{t}^{\infty} |h(\tau)| \, d\tau \right)^2 dt$$

$$= |y_0|^2 \int_{1}^{\infty} \left(\int_{t}^{\infty} |h(\tau)| \tau^r \cdot \tau^{-r} \, d\tau \right)^2 dt$$

$$\leq |y_0|^2 \int_{1}^{\infty} \left(\int_{t}^{\infty} |\tau^r h(\tau)|^2 \, d\tau \right) (t^{-2r+1}/(2r-1)) \, dt$$

$$\leq |y_0|^2 \int_{0}^{\infty} |\tau^r h(\tau)|^2 \, d\tau \, (2r-1)^{-1} \int_{1}^{\infty} t^{-2r+1} \, dt < \infty.$$

Thus $f \in X$ but the solution $y \notin X$. ∎

Combining Theorems 7.2.1 and 7.3.2 we obtain the following result for composite system ($\mathscr{S}$).

7.3.3. Theorem. For composite system $(\mathscr{S})$ (described in Section 7.2) assume that $X = L_2{}^n$ and that all H_i and B_{ij} are linear and are specified by the formulas

$$(H_i \varphi)(t) \triangleq h_{i0}\, \varphi(t) + \int_0^t h_i(t-\tau)\, \varphi(\tau)\, d\tau$$

$$(B_{ij} \varphi)(t) \triangleq B_{ij0}\, \varphi(t) + \int_0^t b_{ij}(t-\tau)\, \varphi(\tau)\, d\tau$$

(7.3.4)

where $h_i \in L_1(R^+, R^{n^2})$, $b_{ij} \in L_1(R^+, R^{n^2})$, and $t^r |h_i| \in L_2$, $t^r |b_{ij}| \in L_2$ for some $r > 1$ and for $i,j = 1, ..., l$. If the successive principal minors of the test matrix $A = [a_{ij}]$ defined by

$$a_{ij} = \begin{cases} 1 - g(R_i)\, g(B_{ij}), & \text{if } i = j \\ -g(R_i)\, g(B_{ij}), & \text{if } i \neq j \end{cases}$$

are all positive, then the resolvent of the linear composite system $(\mathscr{S})$ is in $\mathscr{L}_0$.

Of course the gains of R_i and B_{ij} in the above theorem can be computed as

$$g(R_k) = \max_{\omega \geq 0} |R_k{}^*(j\omega)|$$

and

$$g(B_{ik}) = \max_{\omega \geq 0} |B_{ik}^*(j\omega)|.$$

Note that in the scalar case these gains can effectively be estimated by graphical means.

Theorem 7.3.3 can be combined with Theorems 6.1.13 and 6.1.14 to prove the following result.

7.3.5. Theorem. If the hypotheses of Theorem 7.3.3 are satisfied and if X is any of the spaces BC, C_1, C_0, A_ω, AP, or $L_p{}^n$ for some p in the interval $1 \leq p \leq \infty$, then (linear) composite system $(\mathscr{S})$ is bounded with respect to X.

Theorems 7.2.4 and 7.3.2 can also be combined, to yield the following result.

7.3.6. Theorem. Suppose that $X = L_2{}^n$, H_i and B_{ij} satisfy Eq. (7.3.4) with $|h_i| \in L_1$, $|b_{ij}| \in L_1$, and $t^r |h_i| \in L_2$, $t^r |b_{ij}| \in L_2$ for some $r > 1$ and for $i,j = 1, ..., l$. Suppose that each H_i satisfies either Property Q or Property F and for at least one $i = i_0$ it satisfies Property F only. Let $M = [m_{ij}] = [g(B_{ij})g_c(R_j)]$ and let $r(M)$ be the spectral radius of M. If either $r(M) < 1$ or $r(M) = 1$ and $m_{ij} > 0$ for all i and j, then the resolvent of linear composite system $(\mathscr{S})$ is not in $\mathscr{L}_0$. Thus, $(\mathscr{S})$ is also not bounded with respect to $L_2{}^n$ or with respect to BC.

An alternate form of the above instability theorem can be established.

7.3.7. Theorem. Consider composite system $(\mathscr{S})$ with $X = L_2{}^n$ and with Eq.

(7.3.4) true and $|h_i| \in L_1$ and $|b_{ij}| \in L_1$ for all $i,j = 1, \ldots, l$. Suppose that for each i, H_i satisfies either Property Q or Property F and that for at least one $i = i_0$ there is a complex number s_0 with $\text{Re } s_0 > 0$ such that $(s_0 I - H_{i_0}^*(s)) = 0$. Let $M = [g(B_{ij})g_c(R_j)]$ have spectral radius $r(M) < 1$. Then the resolvent of linear composite system $(\mathscr{S})$ is not in $\mathscr{L}_0$. Indeed, there is a complex number s_1 with $\text{Re } s_1 > 0$ such that

$$\det[sI - H_i^*(s)\delta_{ij} - B_{ij}^*(s)] = 0 \qquad \text{when} \quad s = s_1 \qquad (7.3.8)$$

where δ_{ij} denotes the Kronecker delta.

Proof. Pick $\gamma > 0$ and in Eq. $(\mathscr{S})$ replace $z_i(t)$ by $z_i(t)e^{-\gamma t}$, $h_i(t)$ by $h_i(t)e^{-\gamma t}$, etc. The stability equations for the modified subsystems are now of the form

$$\det[(s+\gamma)I - H_i^*(s+\gamma)] = 0, \qquad \text{Re } s \geq 0.$$

If γ is sufficiently small, then Property Q or F for the original subsystems will be preserved for the modified subsystems and $r(M') < 1$ for the modified matrix of gains M'. By Theorem 7.3.6 the modified system $(\mathscr{S})$ is unbounded with respect to $X = L_2{}^n$ and thus also with respect to BC. By Theorem 7.1.2 it follows that

$$\det[(s+\gamma)I - H_i^*(s+\gamma)\delta_{ij} - B_{ij}^*(s+\gamma)] = 0$$

has a root s_2 with $\text{Re } s_2 \geq 0$. Thus, Eq. (7.3.8) is true with $s_1 = s_2 + \gamma$.   ∎

A major reason for the importance of determining stability of a linear system, i.e., for determining whether or not its resolvent is in $\mathscr{L}_0$, is that local stability or instability results for nonlinear systems can be determined by linearization. In order to illustrate this, let us consider systems described by equations of the form

$$\dot{y}(t) = n(y(t)) + \int_0^t b(t-\tau)m(y(\tau))\, d\tau + f(t) \qquad (7.3.9)$$

with $y(0) = y_0 \in R^n$ given. Suppose that n and m are continuously differentiable. For y_0 and f "small" we hope that $y(t)$ will be "small" so that we may replace Eq. (7.3.9) by

$$\dot{y}(t) = n'(0)y(t) + \int_0^t b(t-\tau)m'(0)y(\tau)\, d\tau + f(t), \qquad (7.3.10)$$

where $n^{\mathrm{T}} = (n_1{}^{\mathrm{T}}, \ldots, n_l{}^{\mathrm{T}})$, $m^{\mathrm{T}} = (m_1{}^{\mathrm{T}}, \ldots, m_l{}^{\mathrm{T}})$, and $n'(0)$ and $m'(0)$ are the matrices determined by

$$n'(0) = \left[\frac{\partial n_i}{\partial y_j}(0)\right], \qquad m'(0) = \left[\frac{\partial m_i}{\partial y_j}(0)\right].$$

Once the stability or instability of Eq. (7.3.10) is analyzed, we can return to the stability problem for the nonlinear equation (7.3.9) in the following fashion.

7.3.11. Theorem. If the resolvent of the operator H defined by

$$H\varphi(t) = n'(0)\,\varphi(t) + \int_0^t b(t-\tau)\,m'(0)\,\varphi(\tau)\,d\tau \qquad (7.3.12)$$

is in $\mathscr{L}_0$, then Eq. (7.3.9) is **locally stable** in the sense that given $\mathscr{E} > 0$ there is a $\delta > 0$ such that whenever $|y_0| < \delta$, $f \in BC$, and $\|f\|_{L_\infty} < \delta$, then $y \in BC$ and $|y(t)| < \mathscr{E}$ for all $t \geq 0$.

For the proof of Theorem 7.3.11 refer to Grossman and Miller [1].

Instability of Eq. (7.3.10) corresponds to having a solution s_0 to the equation

$$\det[sI - n'(0) - b^*(s)\,m'(0)] = 0 \qquad (7.3.13)$$

with $\operatorname{Re} s \geq 0$. In the critical case where all roots satisfy $\operatorname{Re} s_0 = 0$ it is not possible to conclude anything about stability or instability of Eq. (7.3.9) from the stability properties of Eq. (7.3.10). However, in certain noncritical cases $\operatorname{Re} s_0 > 0$, the following is known.

7.3.14. Theorem. If H is given by Eq. (7.3.12), if Eq. (7.3.13) has at least one solution s_0 with $\operatorname{Re} s_0 > 0$ and no critical solutions s with $\operatorname{Re} s = 0$, then there is a constant $\delta > 0$ and points y_0 inside the ball $B(\delta)$ with y_0 arbitrarily close to the origin such that the solution of Eq. (7.3.9) with $f(t) \equiv 0$ and $y(0) = y_0$ must leave the ball $B(\delta)$ in finite time.

For a proof of Theorem 7.3.14 see Miller and Nohel [1].

7.4   An Example

In order to demonstrate the applicability and usefulness of the ideas and methods discussed above, we reconsider once more the nuclear reactor problem treated in Example 5.7.18. Throughout this section we use the same notation as was used in that example.

Specifically, consider the **point kinetics model of a coupled nuclear reactor** with l cores (see Akcasu, Lillouche, and Shotkin [1] and Plaza and Kohler [1]) given by

$$\dot{p}_j(t) = \frac{\rho_j - \mathscr{E}_j - \beta_j}{\Lambda_j}\,p_j(t) + \frac{\rho_j}{\Lambda_j} + \sum_{i=1}^{6} \frac{\beta_{ij}}{\Lambda_j}\,c_{ij}(t)$$

$$+ \frac{1}{\Lambda_j} \sum_{k=1}^{l} \mathscr{E}_{kj}\,\frac{P_{k0}}{P_{j0}} \int_0^t h_{kj}(t-\tau)\,p_k(\tau)\,dt$$

$$\dot{c}_{ij}(t) = \lambda_{ij}[p_j(t) - c_{ij}(t)]$$

for $i = 1, ..., 6$ and $j = 1, ..., l$. Assume that the expression for reactivity ρ_j, given by

$$\rho_j(t) = \int_0^t W_j(t-\tau) p_j(\tau) \, d\tau$$

is correct at least to linear terms and that the feedback function $W_j \in L_1$. Solving for the $c_{ij}(t)$ in terms of p_j we see that

$$c_{ij}(t) = c_{ij}(0) e^{-\lambda_{ij} t} + \lambda_{ij} (e^{\lambda_{ij} t} * p_j)$$

where $*$ denotes the convolution integral. Substituting this expression into the equation for $\dot{p}_j(t)$ and linearizing, we obtain

$$\dot{p}_j(t) = f_j(t) - \left(\frac{\mathscr{E}_j + \beta_j}{\Lambda_j}\right) p_j + \left(\frac{W_j}{\Lambda_j} + \sum_{i=1}^{6} \frac{\beta_{ij} \lambda_{ij} e^{-\lambda_{ij} t}}{\Lambda_j}\right) * p_j$$

$$+ \frac{1}{\Lambda_j} \sum_{k=1}^{l} \frac{P_{k0}}{P_{j0}} h_{kj} * p_k, \qquad (7.4.1)$$

$j = 1, ..., l$. This system has the form of $(\mathscr{S})$ (see Section 7.2) with $z_j = p_j$, all $n_j = 1$,

$$(H_j \varphi)(t) \triangleq -\left(\frac{\mathscr{E}_j + \beta_j}{\Lambda_j}\right) \varphi(t) + \frac{1}{\Lambda_j}\left(W_j + \sum_{i=1}^{6} \beta_{ij} \lambda_{ij} e^{-\lambda_{ij} t}\right) * \varphi(t)$$

$$(7.4.2)$$

$$B_{jk}(\varphi_t, t) \triangleq \frac{1}{\Lambda_j} \frac{P_{k0}}{P_{j0}} (h_{kj} * \varphi)(t) \qquad (7.4.3)$$

and

$$f_j(t) \triangleq \sum_{i=1}^{6} c_{ij}(0) e^{-\lambda_{ij} t}.$$

The resolvent R_k for an isolated subsystem (one core) has Laplace transform $R_k*(s) = 1/D_k(s)$ and is in $\mathscr{L}_0$ if

$$D_k(s) \triangleq s + \left(\frac{\mathscr{E}_k + \beta_k}{\Lambda_k}\right) - \frac{1}{\Lambda_k}\left(W_k*(s) + \sum_{i=1}^{6} \frac{\beta_{ik} \lambda_{ik}}{s + \lambda_{ik}}\right) \neq 0 \qquad (7.4.4)$$

when $\operatorname{Re} s \geq 0$. This condition can be checked graphically from a plot of $D_k(j\omega)$ on $0 \leq \omega < \infty$ in the complex plane. The gain $g(R_k)$ can also be determined since $1/g(R_k)$ is the minimum distance from the graph of $D_k(j\omega)$ on $0 \leq \omega < \infty$ to the origin. Moreover,

$$g(B_{ik}) = \max_{\omega \geq 0} \left\{ \frac{P_{k0}}{\Lambda_i P_{i0}} |h_{ki}^*(j\omega)| \right\}.$$

If the successive principal minors of the test matrix A,

$$A = I - [g(R_i) g(B_{ij})]$$

are all positive, then by Theorem 7.2.3 the solutions of Eq. (7.4.1) are in $C_0 \cap L_2$. Moreover, by Theorem 7.3.11, the original nonlinear point kinetics model is locally stable in this case.

An alternate method of analysis would be to put $B_{jk} = 0$ in Eq. (7.4.3) when $j = k$ and add a term $\Lambda_j^{-1}(h_{jj} * z_j)(t)$ to Eq. (7.4.2). The left-hand side of Eq. (7.4.4) will have the added term $\Lambda_j^{-1} h_{jj}^*(s)$ while $g(B_{jj}) = 0$ for all j. The rest of the stability analysis proceeds as before.

If $D_k(j\omega) \neq 0$ for $-\infty < \omega < \infty$ and $k = 1, ..., l$, and $D_k(s) = 0$ for some $k = k_0$ and $s = s_0$ with $\mathrm{Re}\, s_0 > 0$, then the instability results can be applied. As above, $1/g_c(R_k)$ is the minimum distance from the graph of $D_k(j\omega)$ on $-\infty < \omega < \infty$ to the origin. If the spectral radius of the matrix $M = [g(B_{ij})\, g_c(R_j)]$ is less than one, then system (7.4.1) is L_2-unstable (see Theorem 7.3.7). Finally, from Theorem 7.3.14 we see that the original nonlinear point kinetics model is in this case also unstable in the sense of that theorem.

7.5 Notes and References

Basic background material on Volterra integral and integrodifferential equations can be found in the books by Bellman and Cooke [1, Chapter 7], Volterra [1], Miller [1], and Corduneanu [2]. The stability aspects of Property Q and local stability via linearization of integrodifferential equations are studied in Grossman and Miller [1] and Corduneanu [1]. Unstable linear systems satisfying Property F are investigated in Miller [4] while local instability via linearization is studied in Miller and Nohel [1]. In addition, related results for integral equations can be found in several other papers. In particular, the significant work of Sandberg [1–9] develops the systems theoretic viewpoint used in studying both L_2 and L_∞ stability. The work of Beneš [1–4] and Beneš and Sandberg [1] exploits fixed point theorems while Wu and Desoer [1] use operator techniques. Hale [3] develops C_0-semigroup theory for a class of linear problems while Levin [1] and Driver [1] use Lyapunov functions to study Volterra integrodifferential equations. For additional references, see also the bibliography in Miller [1].

The results in Sections 7.2 and 7.4 are based on Miller and Michel [1]. Related results for integral equations can also be found in Miller and Michel [2]. Theorem 7.3.2 is an adaptation and expansion of a result given in Grossman and Miller [1]. Theorems 7.3.3 and 7.3.5–7.3.7 are new. An introduction to reactor dynamics and to the related engineering literature can be found in Akcasu, Lillouche, and Shotkin [1]. For a more mathematical treatment of the stability of point kinetics models of nuclear reactors refer to Bronikovski, Hall, and Nohel [1], Bronikovski [1], Levin and Nohel [1], Ergen, Lipkin, and Nohel [1], Hsu [1], and the bibliographies in these references.

References

Aizerman, M. A. and Gantmacher, F. R.
   [1] *Absolute Stability of Regulator Systems*, Holden-Day, Inc. San Francisco, California, 1964.

Akcasu, Z., Lillouche, G. S., and Shotkin, L. M.
   [1] *Mathematical Methods in Nuclear Reactor Dynamics*, Academic Press, New York, 1971.

Apostol, T. M.
   [1] *Mathematical Analysis*, Addison-Wesley, Reading, Massachusetts, 1957.

Araki, M.
   [1] Stability of composite control systems, Ph.D. Dissertation, Dept. of Electronics, Kyoto Univ., Kyoto, Japan (1971).
   [2] M-matrices, Dept. of Computing and Control, Imperial College of Science and Technol., London, England, Rep. No. 74/19 (March 1974).
   [3] Input-output stability of composite feedback systems, Dept. of Computing and Control, Imperial College of Science and Technol., London, England, Rep. No. 75/1 (January 1975).
   [4] Application of M-matrices to the stability problems of composite dynamical systems, *J. Math. Anal. Appl.* **52** (1975), 309–321.

Araki, M. and Kondo, B.
   [1] Stability and transient behavior of composite nonlinear systems, *IEEE Trans. Automatic Control* **17** (1972), 537–541.

Araki, M., Ando, K., and Kondo, B.
[1] Stability of sampled-data composite systems with many nonlinearities, *IEEE Trans. Automatic Control* **16** (1971), 22–27.

Arnold, L.
[1] *Stochastic Differential Equations: Theory and Applications*, Wiley, New York, 1974.

Athans, M., Sandell, N., and Varaiya, P.
[1] Stability of interconnected systems, *Proceedings of the 1975 IEEE Conference on Decision and Control, Houston, Texas*, IEEE, December 1975, pp. 456–462.

Bailey, F. N.
[1] Stability of interconnected systems, Ph.D. Dissertation, Univ. of Michigan, Ann Arbor, Michigan (1964).
[2] The application of Lyapunov's second method to interconnected systems, *SIAM J. Control* **3** (1966), 443–462.

Barbu, V. and Grossman, S. I.
[1] Asymptotic behavior of linear integrodifferential systems, *Trans. Amer. Math. Soc.* **171** (1972), 277–288.

Bellman, R.
[1] The boundedness of solutions of infinite systems of linear differential equations, *Duke Math. J.* **14** (1947), 695–706.
[2] Vector Lyapunov functions, *SIAM J. Control* **1** (1962), 32–34.
[3] *Introduction to Matrix Analysis*, McGraw-Hill, New York, 1970.
[4] Large systems, *IEEE Trans. Automatic Control* **19** (1974), 465.

Bellman, R. and Cooke, K. L.
[1] *Differential Difference Equations*, Academic Press, New York, 1963.

Beneš, V.
[1] A fixed point method for studying the stability of a class of integrodifferential equations, *J. Math. and Physics* **40** (1961), 55–67.
[2] A nonlinear integral equation from the theory of servomechanisms, *Bell System Tech. J.* **40** (1961), 1309–1321.
[3] Ultimately periodic solutions to a nonlinear integrodifferential equation, *Bell System Tech. J.* **41** (1962), 257–268.
[4] Ultimately periodic behavior in a class of nonlinear servomechanisms, *Nonlinear Differential Equations and Nonlinear Mechanics*, Academic Press, New York, 1963.

Beneš, V. and Sandberg, I. W.
[1] On the response of a time variable nonlinear system to almost periodic signals, *J. Math. Anal. Appl.* **10** (1965), 245–268.

Bers, L., John, F., and Schechter, M.
[1] *Partial Differential Equations*, Amer. Math. Soc., Providence, Rhode Island, 1964.

Bertram, J. E. and Sarachik, P. E.
[1] Stability of circuits with randomly time-varying parameters, *IRE Trans. Circuit Theory* **6** (1959), 260–270.

Besicovitch, A. S.
[1] *Almost Periodic Functions*, Dover, New York, 1954.

Bhatia, N. P. and Szegö, G. P.
[1] *Stability Theory of Dynamical Systems*, Springer-Verlag, Berlin and New York, 1970.

Blight, J. D. and McClamroch, N. H.
[1] Graphical stability criteria for large-scale nonlinear multiloop systems, *Preprints, Sixth IFAC World Congress, Boston, Massachusetts, August 1975*, IFAC, 1975 (Paper No. 44.5).

Bose, A.
   [1] Stability and compensation of systems with multiple nonlinearities, Ph.D. Dissertation, Dept. of Electrical Engineering, Iowa State Univ., Ames, Iowa (1974).
Bose, A. and Michel, A. N.
   [1] Qualitative analysis of large-scale systems, *Proceedings of the 17th Midwest Symposium on Circuit Theory, Univ. of Kansas, Lawrence, Kansas, September 1974*, Western Periodicals, North Hollywood, California, 1974, pp. 1–8.
   [2] Qualitative analysis of large-scale systems: stability, instability and boundedness, *Z. Angew. Math. Mech.* **56** (1976), 13–20.
Brezis, H.
   [1] Monotonicity methods in Hilbert spaces and some applications to nonlinear partial differential equations, in *Contributions to Nonlinear Functional Analysis*, E. H. Zarantonello, Ed., Academic Press, New York, 1971, pp. 101–156.
   [2] *Operateurs Maximaux Monotones*, North-Holland Publ., Amsterdam, 1973.
Bronikovski, T. A.
   [1] An integrodifferential system which occurs in reactor dynamics, *Arch. Rational Mech. Anal.* **37** (1970), 363–380.
Bronikovski, T. A., Hall, J. E., and Nohel, J. A.
   [1] Quantitative estimates for a nonlinear system of integrodifferential equations arising in reactor dynamics, *SIAM J. Math. Anal.* **3** (1972), 567–588.
Buck, R. C.
   [1] *Advanced Calculus*, McGraw-Hill, New York, 1956.
Burns, J. A. and Herdman, T. L.
   [1] An application of adjoint semigroup theory for a functional differential equation with infinite delays to a Volterra integrodifferential system, *J. Math. Anal. Appl.* (to appear).
Callier, F. M., Chan, W. S., and Desoer, C. A.
   [1] Stability theory of interconnected systems—Part I: Arbitrary interconnections, Electronics Research Laboratory, Univ. of California, Berkeley, California, Rep. No. ERL-M565 (August 1975).
   [2] Stability theory of interconnected systems—Part II: Strongly connected subsystems, Electronics Research Laboratory, Univ. of California, Berkeley, California, Rep. No. ERL-M565 (November 1975).
Chaffee, N.
   [1] A stability analysis for a semilinear parabolic partial differential equation, *J. Differential Equations* **15** (1974), 522–540.
Chaffee, N. and Infante, E. F.
   [1] A bifurcation problem for a nonlinear partial differential equation of parabolic type, *Applicable Anal.* **4** (1974), 17–37.
Coddington, E. A. and Levinson, N.
   [1] *Theory of Ordinary Differential Equations*, McGraw-Hill, New York, 1955.
Coleman, B. D. and Mizel, V. J.
   [1] Norms and semigroups in the theory of fading memory, *Arch. Rational Mech. Anal.* **23** (1966), 87–123.
   [2] On the stability of solutions of functional differential equations, *Arch. Rational Mech. Anal.* **30** (1968), 173–196.
Cook, P. A.
   [1] Modified multivariable circle theorems, in *Recent Mathematical Developments in Control*, D. J. Bell, Ed., Academic Press, London, 1973, pp. 367–372.
   [2] On the stability of interconnected systems, *Internat. J. Control* **20** (1974), 407–415.

Corduneanu, C.
  [1] Some differential equations with delay, *Proceedings of EQUADIFF 3, Brno, Czechoslovakia, 1972* (Czechoslovak Conference on Differential Equations and Their Applications), pp. 105–114.
  [2] *Integral Equations and Stability of Feedback Systems*, Academic Press, New York, 1973.
Crandall, M. G.
  [1] Semigroups of nonlinear transformations in Banach spaces, in *Contributions to Nonlinear Functional Analysis*, E. H. Zarantonello, Ed., Academic Press, New York, 1971.
Crandall, M. G. and Liggett, T. M.
  [1] Generation of semigroups of nonlinear transformations on general Banach spaces, *Amer. J. Math.* **93** (1971), 265–298.
Dantzig, G. B.
  [1] *Linear Programming and Extensions*, Princeton Univ. Press, Princeton, New Jersey, 1963.
Defermos, C. M. and Slemrod, M.
  [1] Asymptotic behavior of nonlinear contraction semigroups, *J. Functional Analysis* **13** (1973), 97–106.
Desoer, C. A. and Vidyasagar, M.
  [1] *Feedback Systems: Input–Output Properties*, Academic Press, New York, 1975.
Dieudonné, J.
  [1] *Foundations of Modern Analysis*, Academic Press, New York, 1960.
Doetsch, G.
  [1] *Theorie und Anwendung der Laplace-Transformation*, Springer-Verlag, Berlin and New York, 1937.
Doob, J. L.
  [1] *Stochastic Processes*, Wiley, New York, 1953.
Dotseth, G. M.
  [1] Admissibility results on subspaces of $C(R,R^n)$ and $LL^p(R,R^n)$, *Math. Systems Theory* **9** (1975), 9–17.
Driver, R. D.
  [1] Existence and stability of solutions of a delay-differential system, *Arch. Rational Mech. Anal.* **10** (1962), 401–426.
Dunford, N. and Schwarz, J. T.
  [1] *Linear Operators*, Part I, Wiley (Interscience), New York, 1958.
Ergen, W. K., Lipkin, H. J., and Nohel, J. A.
  [1] Application of Lyapunov's second method in reactor dynamics, *J. Math. and Phys.* **36** (1957), 36–48.
Fiedler, M. and Ptak, V.
  [1] On matrices with non-positive off-diagonal elements and positive principal minors, *Czechoslovak Math. J.* **12** (1962), 382–400.
Fink, A. M.
  [1] *Almost Periodic Differential Equations* (Lecture Notes in Mathematics), Springer-Verlag, Berlin and New York, 1974.
Gantmacher, F. R.
  [1] *The Theory of Matrices*, Vol. I, Chelsea, Bronx, New York, 1959.
  [2] *The Theory of Matrices*, Vol. II, Chelsea, Bronx, New York, 1959.
Godunov, A. N.
  [1] Peano's theorem in Banach spaces, *Functional Anal. Appl.* **9** (1975), 53–56.

Gollwitzer, H. E.
[1] Admissibility theory and the global behavior of solutions of functional equations, in *Delay and Functional Differential Equations*, K. Schmitt, Ed., Academic Press, New York, 1972.

Grimmer, R.
[1] Stability of a scalar differential equation, *Proc. Amer. Math. Soc.* **32** (1972), 452–456.

Grossman, S. I. and Miller, R. K.
[1] Nonlinear Volterra integrodifferential systems with L^1-kernels, *J. Differential Equations* **13** (1973), 551–566.

Grujić, Lj. T. and Siljak, D. D.
[1] On stability of discrete composite systems, *IEEE Trans. Automatic Control* **18** (1973), 522–524.
[2] Asymptotic stability and instability of large-scale systems, *IEEE Trans. Automatic Control* **18** (1973), 636–645.
[3] Exponential stability of large-scale discrete systems, *Internat. J. Control* **19** (1974), 481–491.

Hahn, W.
[1] *Theorie und Anwendung der Direkten Methode von Ljapunov*, Springer-Verlag, Berlin and New York, 1959.
[2] *Stability of Motion*, Springer-Verlag, Berlin and New York, 1967.

Halanay, A.
[1] *Differential Equations: Stability, Oscillations, Time Lags*, Academic Press, New York, 1966.

Hale, J. K.
[1] Dynamical systems and stability, *J. Math. Anal. Appl.* **26** (1969), 39–59.
[2] *Functional Differential Equations*, Springer-Verlag, Berlin and New York, 1971.
[3] Functional differential equations with infinite delays, *J. Math. Anal. Appl.* **48** (1974), 276–283.
[4] The solution operator with infinite delays, *International Conference on Differential Equations, Brown Univ., Providence, Rhode Island, August 1974*, Academic Press, New York, 1975, pp. 330–336.

Hille, E. and Phillips, R. S.
[1] *Functional Analysis and Semi-Groups* (Amer. Math. Soc. Colloquium Publ., Vol. 33), Amer. Math. Soc., Providence, Rhode Island, 1957.

Holtzman, J. M.
[1] *Nonlinear System Theory: A Functional Analysis Approach*, Prentice-Hall, Englewood Cliffs, New Jersey, 1970.

Hsu, C.
[1] Stability of reactor systems via Lyapunov's second method, *Trans. ASME J. Basic Eng.* **89** (1967), 307–310.

Infante, E. F. and Walker, J. A.
[1] On the stability properties of an equation arising in reactor dynamics, Lefschetz Center for Dynamical Systems, Lecture Notes 74–10, Brown Univ., Providence, Rhode Island (1974).

Kalman, R. E.
[1] Lyapunov functions for the problem of Luré, *Proc. Nat. Acad. Sci. U.S.A.* **49** (1963), 201–205.

Kalman, R. E. and Bertram, J. E.
[1] Control system analysis and design via the "second method" of Lyapunov—

Part I: Continuous-time systems, *Trans. ASME J. Basic Eng.* **82** (1960), 371–393.

[2] Control system analysis and design via the "second method" of Lyapunov—Part II: Discrete-time systems, *Trans. ASME J. Basic Eng.* **82** (1960), 394–400.

Kalman, R. E., Ho, Y. C., and Narendra, K. S.

[1] Controllability of linear dynamical systems, *Contributions to Differential Equations* **1** (1963), 189–213.

Kamke, E.

[1] Zur Theorie der Systeme gewöhnlicher Differentialgleichungen, II, *Acta Math.* **58** (1932), 57–85.

Kats, I. Ia. and Krasovskii, N. N.

[1] On the stability of systems with random parameters, *PMM* **24** (1960), 809–823.

Kevorkian, A. K.

[1] Structural aspects of large dynamic systems, *Preprints, Sixth IFAC World Congress, Boston, Massachusetts, August 1975*, IFAC, 1975 (Paper No. 19.3).

Kozin, F.

[1] A survey of stability of stochastic systems, *Automatika* **5** (1969), 95–112.

Krasovskii, N. N.

[1] *Stability of Motion*, Stanford Univ. Press, Stanford, California, 1963.

Krein, S. G.

[1] *Linear Differential Equations in Banach Space* (Translations of Mathematical Monographs, Vol. 29), Amer. Math. Soc., Providence, Rhode Island, 1970.

Kron, G.

[1] *Diakoptics*, Macdonald, London, 1963.

Kuo, B. C.

[1] *Discrete-Data Control Systems*, Prentice-Hall, Englewood Cliffs, New Jersey, 1970.

Kurtz, T.

[1] Convergence of sequences of semigroups of nonlinear equations with applications to gas kinetics, *Trans. Amer. Math. Soc.* **186** (1973), 259–272.

Kushner, H.

[1] *Stochastic Stability and Control*, Academic Press, New York, 1967.

[2] Converse theorems for stochastic Lyapunov functions, *SIAM J. Control* **5** (1967), 228–233.

[3] *Introduction to Stochastic Control*, Holt, Rinehart and Winston, New York, 1971.

Ladas, G. E. and Lakshmikantham, V.

[1] *Differential Equations in Abstract Spaces*, Academic Press, New York, 1972.

Lakshmikantham, V.

[1] Stability and asymptotic behavior of solutions of differential equations in a Banach space, Technical Rep., Dept. of Mathematics, Univ. of Texas, Arlington, Texas.

Lakshmikantham, V. and Leela, S.

[1] *Differential and Integral Inequalities*, Vol. I, Academic Press, New York, 1969.

[2] *Differential and Integral Inequalities*, Vol. II, Academic Press, New York, 1969.

LaSalle, J. P.

[1] Vector Lyapunov functions, *Bulletin of the Institute of Mathematics, Academia Sinica* **3** (1975), 139–150.

LaSalle, J. P. and Lefschetz, S.

[1] *Stability by Lyapunov's Direct Method with Applications*, Academic Press, New York, 1961.

Lasdon, L. S.

[1] *Optimization Theory of Large Systems* Macmillan, New York, 1970.

Lasley, E. L.
[1]   The qualitative analysis of composite systems, Ph.D. Dissertation, Dept. of
      Electrical Engineering, Iowa State Univ., Ames, Iowa, (1975).
Lasley, E. L. and Michel, A. N.
[1]   Input-output stability of composite systems, *Proceedings of the 8th Asilomar
      Conference on Circuits, Systems, and Computers, Pacific Grove, California,
      December 1974*, Western Periodicals, North Hollywood, California, 1974, pp.
      472–482.
[2]   Input-output stability of interconnected systems, *Proceedings of the IEEE
      International Symposium on Circuits and Systems, Boston, Massachusetts, April
      1975*, IEEE, 1975, pp. 131–134.
[3] L_∞- and l_∞-stability of interconnected systems, *Proceedings of the IEEE Conference
      on Decision and Control, Houston, Texas, December 1975*, IEEE, 1975, pp. 375–382.
[4]   Input-output stability of interconnected systems, *IEEE Trans. Automatic Control*
      21 (1976), 84–89.
[5] L_∞- and l_∞- stability of interconnected systems, *IEEE Trans. Circuits and Systems*
      23 (1976), 261–270.
[6]   Input-output stability of large-scale systems, in *Large-Scale Dynamical Systems*,
      R. Saeks, Ed., Point Lobos Press, Los Angeles, California, 1976, pp. 195–220.
Lasota, A. and Yorke, J. A.
[1]   The generic property of existence of solutions of differential equations in Banach
      space, *J. Differential Equations* **13** (1973), 1–12.
Lefschetz, S.
[1]   *Stability of Nonlinear Control Systems*, Academic Press, New York, 1965.
Levin, J. J.
[1]   The asymptotic behavior of the solution of a Volterra equation, *Proc. Amer.
      Math. Soc.* **14** (1963), 534–547.
Levin, J. J. and Nohel, J. A.
[1]   On a system of integro-differential equations occurring in reactor dynamics,
      J. Math. Mech. **9** (1960), 347–368.
Massera, J. L.
[1]   Contributions to stability theory, *Annals of Math.* **64** (1956), 182–206.
Massera, J. L. and Schaeffer, J. J.
[1]   Linear differential equations and functional analysis—Part I, *Annals of Math.*
      67 (1958), 517–573.
[2]   *Linear Differential Equations and Function Spaces*, Academic Press, New York,
      1966.
Matrosov, V. M.
[1]   Method of Lyapunov-vector functions in feedback systems, *Automat. Remote
      Control* **33** (1972), 1458–1469.
[2]   The method of vector Lyapunov functions in analysis of composite systems with
      distributed parameters, *Automat. Remote Control* **34** (1973), 1–16.
Matzer, E.
[1]   Zur Stabilität gekoppelter Systeme, D. Sc. Dissertation, Technical Univ. of
      Graz, Graz, Austria (1973).
McClamroch, N. H.
[1]   A representation for multivariable feedback systems and its use in stability
      analysis, Dept. of Engineering, Univ. of Cambridge, England, Report No.
      CUED/B-Control TR/98 (1975).
McClamroch, N. H. and Ianculescu, G. D.
[1]   Global stability of two linearly interconnected nonlinear systems, *IEEE Trans.
      Automatic Control* **20** (1975), 678–682.

McKenzie, L. W.
[1] The matrix with dominant diagonal and economic theory, *Proceedings of a Symposium on Mathematical Methods in the Social Sciences*, Stanford Univ. Press, Palo Alto, California, 1960, pp. 277–292.

Mesarović, M. D. and Takahara, Y.
[1] *General Systems Theory: Mathematical Foundations*, Academic Press, New York, 1975.

Mesarović, M. D., Macko, D., and Takahara, Y.
[1] *Theory of Hierarchical Multilevel Systems*, Academic Press, New York, 1970.

Metzler, L.
[1] Stability of multiple markets: The Hicks Conditions, *Econometrica* **13** (1945), 277–292.

Michel, A. N.
[1] On the bounds of the trajectories of differential systems, *Internat. J. Control* **10** (1969), 593–600.
[2] Quantitative analysis of systems: stability, boundedness and trajectory behavior, *Arch. Rational Mech. Anal.* **38** (1970), 107–122.
[3] Stability, transient behavior and trajectory bounds of interconnected systems, *Internat. J. Control* **11** (1970), 703–715.
[4] Quantitative analysis of simple and interconnected systems: stability, boundedness, and trajectory behavior, *IEEE Trans. Circuit Theory* **17** (1970), 292–301.
[5] Stability analysis of interconnected systems, Berichte der Mathematisch-Statistischen Sektion im Forschungszentrum Graz, Bericht Nr. 4, Technical Univ. of Graz, Graz, Austria (1973).
[6] Stability analysis and trajectory behavior of composite systems, *Proceedings of the 1974 IEEE International Symposium on Circuits and Systems, San Francisco, California*, IEEE, April 1974, pp. 240–244.
[7] Stability analysis of interconnected systems, *SIAM J. Control* **12** (1974), 554–579.
[8] Stability analysis of stochastic large-scale systems, *Proceedings of the 8th Annual Princeton Conference on Information Sciences and Systems*, Princeton Univ., Princeton, New Jersey, March 1974, pp. 285–288.
[9] Stability analysis and trajectory behavior of composite systems, *IEEE Trans. Circuits and Systems* **22** (1975), 305–312.
[10] Stability analysis of stochastic large-scale systems, *Z. Angew. Math. Mech.* **55** (1975), 93–105.
[11] Stability analysis of stochastic composite systems, *IEEE Trans. Automatic Control* **20** (1975), 246–250.

Michel, A. N. and Heinen, J. A.
[1] Quantitative stability of dynamical systems, *Internat. J. Systems Sci.* **1** (1971), 303–306.
[2] Quantitative Mengenstabilität von Systemen, *Regelungstech. Prozeß-Datenverarbeit.* **20** (1972), 113–121.
[3] Comparison theorems for set stability of differential equations, *Internat. J. Systems Sci.* **12** (1972), 317–324.
[4] Quantitative and practical stability of systems, *Automatic Control Theory Appl.* **1** (1972), 9–15.

Michel, A. N. and Porter, D. W.
[1] Analysis of discontinuous large-scale systems: Stability, transient behavior, and trajectory bounds, *Proceedings of the 1970 IEEE International Symposium on Circuit Theory, Atlanta, Georgia*, IEEE, December 1970, pp. 92–94.

[2] Analysis of discontinuous large-scale systems: stability, transient behavior, and trajectory bounds, *Int. J. Systems Sci.* **2** (1971), 77–95.

[3] Stability analysis of composite systems, *IEEE Trans. Automatic Control* **17** (1972), 222–226.

[4] Practical stability and finite-time stability of discontinuous systems, *IEEE Trans. Circuit Theory* **19** (1972), 123–129.

Michel, A. N. and Rasmussen, R. D.

[1] Stability analysis of stochastic interconnected systems, *Proceedings of the 12th Annual Allerton Conference on Circuit and System Theory*, Univ. of Illinois, Urbana, Illinois, October 1974, pp. 77–86.

[2] Stability of stochastic composite systems, *IEEE Trans. Automatic Control* **21** (1976), 89–94.

[3] Stability anlysis of stochastic interconnected systems, in *Advances in Control and Dynamical Systems: Theory and Applications*, Vol. 13, C. T. Leondes, Ed., Academic Press, New York, 1976.

Miller, R. K.

[1] *Nonlinear Volterra Integral Equations*, Benjamin, Menlo Park, California, 1971.

[2] Almost-periodic behavior of solutions of a nonlinear Volterra system, *Quart. Appl. Math.* **28** (1971), 553–570.

[3] Linear Volterra integrodifferential equations as semigroups, *Funkcial. Ekvac.* **17** (1974), 39–55.

[4] Structure of solutions of unstable linear Volterra integrodifferential equations, *J. Diff. Equations* **15** (1974), 129–157.

Miller, R. K. and Michel, A. N.

[1] L_2-stability and instability of large-scale systems described by integrodifferential equations, *SIAM J. Math. Anal.* (to appear).

[2] Stability of multivariable feedback systems containing elements which are open-loop unstable, *Proceedings of the 13th Annual Allerton Conference on Circuit and System Theory*, Univ. of Illinois, Urbana, Illinois, October 1975, pp. 580–589.

Miller, R. K. and Nohel, J. A.

[1] A stable manifold theorem for a system of Volterra integrodifferential equations, *SIAM J. Math. Anal.* **6** (1975), 506–522.

Mitra, D. and So, H. C.

[1] Existence conditions for L_1 Lyapunov functions for a class of nonautonomous systems, *IEEE Trans. Circuit Theory* **19** (1972), 594–598.

Müller, M.

[1] Über das Fundamentaltheorem in der Theorie der gewöhnlichen Differentialgleichungen, *Math. Z.* **26** (1926), 619–645.

Narendra, K. S. and Neuman, C. P.

[1] Stability of continuous time systems with n-feedback nonlinearities, *AIAA J.* **11** (1967), 2021–2027.

Narendra, K. S. and Taylor, H. J.

[1] *Frequency Domain Criteria for Absolute Stability*, Academic Press, New York, 1973.

Ostrowski, A.

[1] Determinanten mit überwiegender Hauptdiagonale und die absolute Konvergenz von linearen Iterationsprozessen, *Comment. Math. Helv.* **30** (1956), 175–210.

Özgüner, Ü. and Perkins, W. R.

[1] On the multilevel structure of large scale composite systems, *IEEE Trans. Circuits and Systems* **22** (1975), 618–622.

Pai, M. A. and Narayan, C. L.
[1] Stability of large scale power systems, *Preprints, Sixth IFAC World Congress, Boston, Massachusetts, August 1975*, IFAC, 1975 (Paper No. 31.6).

Pao, C. V.
[1] Semigroups and asymptotic stability of nonlinear differential equations, *SIAM J. Control* **3** (1972), 371–379.

Pazy, A.
[1] Semigroups of linear operators and applications to partial differential equations, Univ. of Maryland, Dept. of Mathematics, Lecture Notes No. 10 (1974).

Piontkovskii, A. A. and Rutkovskaya, L. D.
[1] Investigation of stability theory problems by the vector Lyapunov function method, *Automat. Remote Control* **10** (1967), 1422–1429.

Plant, R. H. and Infante, E. F.
[1] Bounds on motions of some lumped and continuous dynamical systems, *Trans. ASME Ser. E. J. Appl. Mech.* **38** (1972), 251–256.

Plaza, H. and Kohler, W. H.
[1] Coupled-reactor kinetics equations, *Nuclear Science and Engineering*, **22** (1966), 419–422.

Popov, V. M.
[1] Absolute stability of nonlinear systems of automatic control, *Automat. Remote Control* **22** (1961), 857–875.

Porter, D. W.
[1] Stability of multiple-loop nonlinear time-varying systems, Ph.D. Dissertation, Iowa State Univ., Ames, Iowa (1972).

Porter, D. W. and Michel, A. N.
[1] Stability of composite systems, *Proceedings of the 4th Asilomar Conference on Circuits and Systems, Pacific Grove, California, November 1970*, Western Periodicals, North Hollywood, California, 1970, pp. 634–638.
[2] Stability analysis of composite systems with nonlinear interconnections, *Proceedings of the 14th Midwest Symposium on Circuit Theory, Univ. of Denver, Denver, Colorado, May 1971*, Western Periodicals, North Hollywood, California, 1971, pp. 6.6.1–6.6.10.
[3] Stability of multiple-loop nonlinear time-varying systems, Rept. No. ISU-ERI-AMES-73167, Iowa State Univ., Ames, Iowa (1973).
[4] Input-output stability of multiloop feedback systems, *Proceedings of the 1974 Joint Automatic Control Conference, Austin, Texas*, IEEE, June 1974, pp. 75–82.
[5] Input-output stability of time varying nonlinear multiloop feedback systems, *IEEE Trans. Automatic Control* **19** (1974), 422–427.

Quirk, J. and Saposnik, R.
[1] *Introduction to General Equilibrium Theory and Welfare Economics*, McGraw-Hill, New York, 1968.

Rasmussen, R. D.
[1] Lyapunov stability of large-scale dynamical systems, Ph.D. Dissertation, Dept. of Electrical Engineering, Iowa State Univ., Ames, Iowa (1976).

Rasmussen, R. D. and Michel, A. N.
[1] On vector Lyapunov functions for stochastic dynamical systems, *Proceedings of the 3rd Milwaukee Symposium on Automatic Computation and Control, Milwaukee, Wisconsin, April 1975*, Western Periodicals, North Hollywood, California, 1975, pp. 161–166.
[2] Stability analysis of large scale dynamical systems, *Proceedings of the 18th Midwest*

Symposium on Circuit Theory, Montreal, Canada, August 1975, Western Periodicals, North Hollywood, California, 1975, pp. 144–148.

[3] On vector Lyapunov functions for stochastic dynamical systems, *IEEE Trans. Automatic Control* **21** (1976), 250–254.

[4] Stability of interconnected dynamical systems described on Banach spaces, *IEEE Trans. Automatic Control* **21** (1976), 464–471.

Rosenbrock, H. H.

[1] A Lyapunov function with applications to some nonlinear physical systems, *Automatika* **1** (1963), 31–53.

[2] A Lyapunov function for some naturally occurring linear homogenous time dependent equations, *Automatika* **1** (1963), 97–109.

[3] Multivariable circle theorems, in *Recent Mathematical Developments in Control*, D, J. Bell, Ed., Academic Press, London, 1973, pp. 345–364.

Sandberg, I. W.

[1] On the response of nonlinear control systems to periodic input signals, *Bell System Tech. J.* **43** (1964), 911–926.

[2] On the L_2-boundedness of solutions of nonlinear functional equations, *Bell System Tech. J.* **43** (1964), 1581–1599.

[3] A frequency domain condition for the stability of feedback systems containing a single time-varying nonlinear element, *Bell System Tech. J.* **43** (1964), 1601–1608.

[4] A condition for the L_∞-stability of feedback systems containing a single time-varying nonlinear element, *Bell System Tech. J.* **43** (1964), 1815–1817.

[5] On truncation techniques in approximate analysis of periodically time varying nonlinear networks, *IEEE Trans. Circuit Theory* **11** (1964), 195–201.

[6] A stability criterion for linear networks containing time varying capacitors, *IEEE Trans. Circuit Theory* **12** (1965), 2–11.

[7] On the boundedness of solutions of nonlinear integral equations, *Bell System Tech. J.* **44** (1965), 439–453.

[8] Some results on the theory of physical systems governed by nonlinear functional equations, *Bell System Tech. J.* **44** (1965), 871–898.

[9] Some stability results related to those of V. M. Popov, *Bell System Tech. J.* **44** (1965), 2133–2148.

[10] Some theorems on dynamic response of nonlinear transistor networks, *Bell System Tech. J.* **48** (1969), 35–54.

Shaw, L.

[1] Existence and approximation of solutions to an infinite set of linear time-invariant differential equations, *SIAM J. Appl. Math.* **22** (1972), 266–279.

Sirazetdinov, T. K.

[1] On the theory of stability of processes with distributed parameters, *PMM* **31** (1967), 37–48.

Slemrod, M.

[1] Asymptotic behavior of a class of abstract dynamical systems, *J. Differential Equations* **7** (1970), 584–600.

[2] Asymptotic behavior of C_0-semigroups as determined by the spectrum of the generator, *Indiana Univ. J. Math.* (to appear).

Suhadloc, A.

[1] On a system of integrodifferential equations, *SIAM J. Appl. Math.* **21** (1971), 195–206.

Sundareshan, M. K. and Vidyasagar, M.

[1] L_2-stability of large-scale dynamical systems: criteria via positive operator theory (to appear).

Szarski, J.
[1] *Differential Inequalities*, Polish Scientific Publ., Warzawa, 1965.
Takeda, S. and Bergen, A. R.
[1] Instability of feedback systems by orthogonal decomposition of L_2, *IEEE Trans. Automatic Control* **18** (1973), 631–636.
Taussky, O.
[1] A recurring theorem on determinants, *Amer. Math. Monthly* **56** (1949), 672–676.
Tewarson, R. P.
[1] *Sparse Matrices*, Academic Press, New York, 1973.
Thompson, W. E.
[1] Stability of interconnected systems, Ph.D. Dissertation, Michigan State Univ., East Lansing, Michigan (1968).
[2] Exponential stability of interconnected systems, *IEEE Trans. Automatic Control* **15** (1970), 504–506.
Thompson, W. E. and Koenig, H. E.
[1] Stability of a class of interconnected systems, *Internat. J. Control* **15** (1972), 751–763.
Tokumaru, H., Adachi, N., and Amemiya, T.
[1] On the input-output stability of interconnected systems, *Systems and Control* (J. Japan Association of Automatic Control Engineers) **17** (1973), 121–125.
[2] Macroscopic stability of interconnected systems, *Preprints, Sixth IFAC World Congress, Boston, Massachusetts, August 1975*, IFAC, 1975 (Paper No. 44.4).
Vidyasagar, M.
[1] L_2-instability criteria for interconnected feedback systems, *Proceedings of the 1976 IEEE International Symposium on Circuits and Systems, Munich, Germany, April 1976*, IEEE, 1976, pp. 386–389.
Volterra, V.
[1] *Theory of Functionals and of Integral and Integrodifferential Equations*, Dover, New York, 1959.
Walker, J. A.
[1] On the application of Lyapunov's direct method to linear dynamical systems, Lefschetz Center for Dynamical Systems, Lecture Notes TR 74-7, Brown Univ., Providence, Rhode Island (1974).
Walker, J. A. and Infante, E. F.
[1] Some results on the precompactness of orbits of dynamical systems, Lefschetz Center for Dynamical Systems, Technical Rep. 74-2, Brown Univ., Providence, Rhode Island (1974).
Walter, W.
[1] *Differential and Integral Inequalities*, Springer-Verlag, Berlin and New York, 1970.
Wang, P. K. C.
[1] Asymptotic stability of distributed parameter systems with feedback controls, *IEEE Trans. Automatic Control* **11** (1966), 46–54.
[2] Stability analysis of elastic and aeroelastic systems via Lyapunov's direct method, *J. Franklin Inst*, **281** (1966), 51–72.
[3] On the stability of equilibrium of a mixed distributed and lumped parameter control system, *Internat. J. Control* **3** (1966), 130–147.

[4]   Stability for a mixed distributed and lumped parameter control system, *Internat. J. Control* **6** (1967), 65–73.
[5]   Theory of stability and control for distributed parameter systems (a bibliography), *Internat. J. Control* **7** (1968), 101–116.

Wazewski, T.
[1]   Systémes des equations et des inégalités différentielles ordinaires aux deuxiémes membres monotones et leurs applications, *Ann. Soc. Poln. Mat.* **23** (1950), 112–166.

Webb, G. F.
[1]   Autonomous nonlinear functional differential equations and nonlinear semigroups, *J. Math. Anal. Appl.* **46** (1974), 1–12.

Weiss, L. and Infante, E. F.
[1]   Finite time stability under perturbing forces and on product spaces, *IEEE Trans. Automatic Control* **12** (1967), 54–59.

Weissenberger, S.
[1]   Stability regions of large-scale systems, *Automatika* **9** (1973), 653–663.

Willems, J. C.
[1]   *The Analysis of Feedback Systems*, M. I. T. Press, Cambridge, Massachusetts, 1971.
[2]   The generation of Lyapunov functions for input-output stable systems, *SIAM J. Control* **9** (1971), 105–134.

Wong, E.
[1]   *Stochastic Processes in Information and Dynamical Systems*, McGraw-Hill, New York, 1971.

Wu, M. Y. and Desoer, C. A.
[1] L^p-stability ($1 \le p \le \infty$) of nonlinear time-varying feedback systems, *SIAM J. Control* **7** (1969), 356–364.

Yacubovich, V. A.
[1]   Solution of certain matrix inequalities occurring in the theory of automatic control, *Dokl. Acad. Nauk SSSR* **143** (1962), 1304–1307.

Yoshizawa, T.
[1]   *Stability Theory by Lyapunov's Second Method*, Math. Soc. Japan, Tokyo, Japan, 1966.

Zames, G.
[1]   Functional analysis applied to nonlinear feedback systems, *IEEE Trans. Circuit Theory* **10** (1963), 393–404.
[2]   Nonlinear time-varying feedback systems—conditions for L_∞-boundedness derived using conic operators on exponentially weighted spaces, *Proceedings of the 3rd Annual Allerton Conference on Circuit and System Theory*, Univ. of Illinois, Urbana, Illinois, October *1965*, pp. 460–471.
[3]   On the input-output stability of time-varying nonlinear feedback systems— Part I: Conditions derived using the concept of loop, gain, conicity and positivity, *IEEE Trans. Automatic Control* **11** (1966), 228–238.
[4]   On the input-output stability of time-varying nonlinear feedback systems— Part II: Conditions involving circles in the frequency plane and sector non-linearities, *IEEE Trans. Automatic Control* **11** (1966), 465–476.

Zemanian, A. H.
[1]   *Distribution Theory and Transform Analysis*, McGraw-Hill, New York, 1965.

Zubov, V. I.
[1]   *Methods of A. M. Lyapunov and Their Application*, P. Noordhoff Ltd., The Netherlands, 1964.

Index

A
B 7
C 8
D 9
E 0
F 1
G 2
H 3
I 4
J 5